T0269175

VISIBILITY
THE SEEING OF NEAR AND DISTANT LANDSCAPE FEATURES

VISIBILITY

THE SEEING OF
NEAR AND DISTANT
LANDSCAPE FEATURES

WILLIAM C. MALM

Cooperative Institute for Research in the Atmosphere
Colorado State University, Fort Collins, CO, USA

ELSEVIER

AMSTERDAM • BOSTON • HEIDELBERG • LONDON
NEW YORK • OXFORD • PARIS • SAN DIEGO
SAN FRANCISCO • SINGAPORE • SYDNEY • TOKYO

Elsevier
Radarweg 29, PO Box 211, 1000 AE Amsterdam, Netherlands
The Boulevard, Langford Lane, Kidlington, Oxford OX5 1GB, UK
50 Hampshire Street, 5th Floor, Cambridge, MA 02139, USA

Notices
Knowledge and best practice in this field are constantly changing. As new research and experience broaden our understanding, changes in research methods, professional practices, or medical treatment may become necessary.

Practitioners and researchers must always rely on their own experience and knowledge in evaluating and using any information, methods, compounds, or experiments described herein. In using such information or methods they should be mindful of their own safety and the safety of others, including parties for whom they have a professional responsibility.

To the fullest extent of the law, neither the Publisher nor the authors, contributors, or editors assumes any liability for any injury and/or damage to persons or property as a matter of products liability, negligence or otherwise, or from any use or operation of any methods, products, instructions, or ideas contained in the material herein.

British Library Cataloguing-in-Publication Data
A catalogue record for this book is available from the British Library.

Library of Congress Cataloging-in-Publication Data
A catalog record for this book is available from the Library of Congress.

ISBN: 978-0-12-804450-6

For information on all Elsevier publications
visit our website at https://www.elsevier.com/

  Working together
to grow libraries in
developing countries

www.elsevier.com • www.bookaid.org

Publisher: Candice Janco
Acquisition Editor: Laura Kelleher
Editorial Project Manager: Emily Thompson
Production Project Manager: Paul Prasad Chandramohan
Designer: Victoria Pearson

Typeset by Thomson Digital

Contents

4. Visibility Metrics

5. Human Perception of Haze and Landscape Features

6. Image Processing Techniques for Displaying Haze Effects on Landscape Features

7. Monitoring Visibility

8. History of Visibility as an Esthetic Concern in the United States

Preface

Vision through the Atmosphere by Middleton (1963) was the last book to be written on the topic of visibility, with the last printing in 1963. It is no longer in print. In the preface, Middleton states, "What originally concerned mainly the meteorologists has become of great importance in military tactics as well as in peacetime transportation, and it is not too much to say that the subject has entirely changed its aspect in the last ten years." He further writes in the introductory chapter "…and at the present time a worker in the field of vision through the atmosphere is as likely to be a physiologist or a psychologist as a meteorologist or physicist." This last statement is especially true today, though it should also include chemists. It has been over 50 years since the writing of his book, but the use of the term "visibility" to represent only visual range, the farthest distance at which a dark feature on the horizon can be seen, has profoundly changed to represent something akin to atmospheric clarity, the ability to see and appreciate vistas, whether they be scenic landscape features found in national parks and pristine remote areas of the world or the urban areas in which most of us live. To that end, and to his credit, Bennet (1930), who actually introduced the term "visual range," did say that he used the term to express atmospheric clarity and the clearness with which objects stand out from their surroundings under conditions of good seeing.

Just as military tactics and peacetime transportation issues motivated a better understanding of visibility as documented by Middleton, concerns about the reduction of our ability to clearly see the scenic wonders of Grand Canyon and other national parks, associated with energy development and urban growth in the southwestern United States, have motivated a renewed interest in the visibility sciences. From the mid-1970s to the early 2000s, this interest in existing and potential future visibility impairment resulted in many millions of dollars being spent by government and industry on visibility research. This was coincident with and certainly related to the development of federal laws and regulations for the protection of visibility in U.S. national parks and wilderness areas. The research addressed the questions needed to write and implement these laws and regulations. These included knowledge of current visibility conditions; causes of current visibility impairment; sensitivity of visual changes to changes in pollutant levels; methods to monitor, model, and communicate visibility conditions; and the value of good visual conditions. While building on the previously developed understanding of visibility, these were fundamentally different issues than the earlier work addressed, so they required additional thought and investigation. This book attempts to present information that builds upon the base as documented by Middleton and features the more recently developed insights needed to understand visibility impairment of landscape features. The author of this book can only hope that topics in visibility that have evolved over the last 50 years are covered with the same "clarity" as those presented by Middleton so many years ago.

William C. Malm, PhD, started his career as an assistant professor of environmental science at Northern Arizona University, Flagstaff, Arizona, where he, other faculty, and his students made some of the first visibility and air quality measurements in the National Park Service system at the Grand Canyon in 1972. He continued to work on the subject of visibility while employed by the United States Environmental Protection Agency (EPA) and later the National Park Service (NPS), where he led the visibility research program for 30 years. He retired from the NPS and works part time at Colorado State University, Fort Collins, Colorado. The team of scientists assembled by Dr. Malm at the NPS continues under the leadership of Dr. Bret Schichtel to conduct visibility research and maintain the monitoring and assessment activities that supply the insights needed to protect what is now recognized as one of the most-valued features of national parks, the ability of visitors to clearly see the wonders of those treasured places.

Over his decades-long career, Dr. Malm designed and built instrumentation to measure the effects of atmospheric aerosols on the scenic qualities of landscape features, as well as their optical and chemical properties. He formulated radiation transfer algorithms that allow pictorial visualization of aerosol scattering and absorption effects on scenic landscape features. He pioneered studies of visibility perception that elicit human responses, in terms of both psychophysical and value assessment, to changes in scenic quality as a function of aerosol optical properties. He designed and led large field campaigns to better characterize aerosol physical and optical properties, especially as they relate to aerosol hygroscopic properties, and to assess the relative contributions of various source types to visibility impacts in a number of national parks and wilderness areas. He also helped pioneer a number of back-trajectory receptor modeling methodologies that allow estimates of the relative contributions of source areas to aerosol concentrations or visibility effects at selected receptor sites. Many of the results from this work have been incorporated into the Interagency Monitoring of Protected Visual Environments (IMPROVE) program and the EPA's Regional Haze Rule (RHR). Several of these topics are covered in this book.

The approach employed in this book is to start each chapter with a simplified presentation of topics to be covered in that chapter that can be read and understood by the layperson and in many cases are exemplified by photographs. This is especially true in Chapter 1, where the many and varied types of visibility impairment are shown as a function of illumination and haze spatial distributions. Then, readers schooled in mathematics can explore the topics in more detail and in many cases are referred to the literature that goes into the various mathematical formulations in even greater depth.

This book is not meant to be a general literature review of the many topics covered. Where certain subjects, such as radiation transfer, are covered in some detail in other definitive publications, only the basic principles are presented and discussed here, while other issues germane to the subject of visibility that are not easily found in the literature are covered in more detail. Also, several topics are not covered that are epithetical to visibility but not related to the seeing of landscape features, such as the use of visual range to estimate atmospheric mass concentrations and the whole subject of aerosol optical depth, whether it be measured by sun photometers or estimated from satellite measurements.

Chapter 1 is an introduction to visibility concepts and explains how the intervening atmosphere between an observer and landscape feature attenuates the appearance of that feature, while at the same time adding light to the sight path and ultimately modulating the original scene. The concept of uniform and layered hazes is introduced and discussed within the context of how haze might affect the contextual detail of scenic landscape features. These concepts are illustrated with photographs showing various types of visibility impairment such as uniform and layered hazes within both urban and pristine settings and under different lighting conditions illustrating the dramatic effect of sun-angle–observer geometry. The goal of Chapter 1 is to sensitize the reader to visibility as it relates to the seeing and appreciating of landscape features, whether in a national park, wilderness area, or rural or urban area.

Chapter 2 covers the fundamental physical properties of light and introduces the idea of light being both wavelike and particle-like depending on the conditions of its environment. The idea of wavelength and color is introduced, as is the concept of index of refraction as it relates to the description of how the direction of light is altered as it passes from one medium to another. These ideas are then extended to the interaction of light with atmospheric particles, and the concepts of light scattering and absorption are discussed. Light is removed from the sight path by both scattering and absorption by atmospheric particles, the sum of which is referred to as extinction. The importance of particle size and composition to their efficiency for light extinction and the direction of light scattering are described. Various approaches and issues related to apportionment of light extinction among the various aerosol components are identified and discussed, and algorithms to estimate extinction from measured particle species are also presented.

Chapter 3 builds on the concepts discussed in Chapter 2. The reader is introduced to how the wavelength-dependent scattering properties, observer–illumination geometry as a function of particle physiochemical characteristics, and the wavelength of light manifest themselves in sky color and altered image visual characteristics. The radiation transfer equation is introduced and solved under some very limiting assumptions to show that the radiance difference between two landscape features is attenuated as it passes through the atmosphere according to transmittance or Beer's law and how that relationship leads to the idea of contrast and contrast reduction as a function of atmospheric transmittance. A simple equation for estimating air light is introduced, as is the general equation of image transmittance in the form of the atmospheric modulation transfer function. The concepts of mean square radiance fluctuation and equivalent contrast, which led to the ability to express image textural characteristics in the form of a contrast that is sensitive to the spatial resolution or characteristics of the landscape, are discussed. Photographic images are presented that show the effects of air light and atmospheric transmittance independently on scenic quality as a function of observer–sun geometry.

Given the extinction coefficient and spatial distribution of radiation representing an observed scene, a number of visibility metrics can be defined. These are covered in Chapter 4. There are two types of metrics: the so-called universal metrics and scene-dependent metrics. The universal metrics, such as the extinction coefficient, various forms of visual range, and deciview, do not say anything about the scenic quality as a function of haze but are a representation of the amount of haze at one point in space and time. None of the universal metrics inform as to how any specific scene will look as a function of haze levels and viewing conditions. Scene-dependent visibility metrics are also introduced and discussed. All scene-dependent metrics are some form of scene contrast or landscape radiance difference. These metrics can be calculated at one or more wavelengths or be combined in such a way as to be proportional to the response of the human visual system. The idea of a just noticeable difference or change of the appearance of a landscape feature as a function of an incremental change in haze is discussed, as is the idea of how much haze, in the form of layered or uniform haze, is required to be just noticed. These ideas are expressed in the context of threshold and suprathreshold problems.

One of the more interesting visibility-related problems is understanding the human observer dimension of seeing and appreciating a scenic or landscape resource as a function of varying haze levels, scene, and illumination characteristics. Chapter 5 addresses these topics. It starts out by defining the problem of varying illumination and haze spatial distribution in the context of visual air quality and scenic beauty judgments and visibility acceptability levels and how varying visibility affects the human experience in general. Threshold and suprathreshold topics are also discussed in some detail, especially as they relate to the sensitivity of the human visual system to varying spatial frequencies in landscape feature content and on size and shape of haze distributions.

Chapter 6 covers the development of image processing techniques to photographically display the visual appearance of modeled visibility impairment either in the form of uniform or nonuniform distributions of haze such as suspended plumes of ground-based layered haze. While Chapter 3 presented a simplified solution to the radiative transfer equation, the solution to this equation accounting for earth curvature and nonuniform particle distributions required the development of sophisticated radiation transfer models. The models are discussed, and a number of interesting examples of visibility impairment resulting from changes in illumination and particle concentration distributions are presented.

Chapter 7 discusses the topic of visibility monitoring. It starts out by presenting a simplified discussion of optical and particle monitoring and presents the triad idea of visibility monitoring that includes optic and particle measurements along with a photographic program to capture the appearance of a scene under the many and varied illumination conditions. Particle monitoring techniques are not covered; however, the reader is referred to a number of books and reviews on the subject. Various types of optical measurements are discussed but not the specific instrumentation that is currently commercially available. Chapter 7 discusses the types and limitations of information that can be extracted from teleradiometer measurements and photographic images. A wide array of visibility metrics, discussed in Chapter 4, can be extracted from webcam or photographic images.

Chapter 8 covers, in some detail, the evolution of visibility from being a concern of how far you can just see a landscape feature, or in the case of military applications, a target, to the esthetic issue of seeing, enjoying, and appreciating the many scenic landscape features available to us as we visit wild and scenic natural areas or, for that matter, as we drive to work on a daily basis. A brief history of the energy development in the southwestern United States and its impacts on visibility at national parks in the region is presented as the prelude to legislation that was passed in 1977 protecting visibility as an esthetic concern across all of the United States. The subsequent monitoring and difficult task of trying to figure out how much each source of visibility-reducing gases and particles could be attributed to visibility impairment in places such as Grand Canyon National Park are discussed. The legal haggling over who is responsible for visibility impairing haze in protected areas is presented. Finally, the much-improved visibility levels, resulting from a number of emission reduction programs, are presented in "before" and "after" photographs of some of the national parks.

This book summarizes the scientific underpinnings of visibility protection efforts of the last 40 years in the United States. The intent in writing it was to provide an overview for those U.S. scientists and policy makers who have the responsibility to continue the process, as well as those who will be initiating similar activities elsewhere. In the United States, the primary motivation was to prevent the loss of extremely good visibility in remote areas of the country that were beginning to experience industrial and urban pollution. These same approaches can be applied to the extreme haze conditions caused by poorly controlled industrial and urban emissions in China, India, and other developing countries that have been the topics of much concern in recent years. While the levels of haze are much higher in these circumstances, the science needed to understand the causes and effects and to assess the visual response to proposed emission changes are the same as those summarized in this book.

References

Bennet, M.G., 1930. The physical conditions controlling visibility through the atmosphere. Quar. J. Roy. Meteorol. Soc. 56, 1–29.

Middleton, W.E.K., 1963. Vision through the Atmosphere, corrected edition University of Toronto Press, Toronto, ON, 250 pp.

Acknowledgments

There are so many people who have contributed, directly and indirectly, to this book and the development of a visibility protection program for the United States, that there isn't enough room to list them all. But I would be remiss not to mention a few. First and foremost, I would like to thank Marc Pitchford for reviewing, critiquing, and making many helpful suggestions that hopefully have added "clarity" to the various discussions. In the mid-1970s as an Environmental Protection Agency (EPA) staff scientist for several years, then for many years as a National Oceanic and Atmospheric Administration research meteorologist, he was responsible for initiating and guiding many of the early visibility investigative studies in the western United States, and contributed to the development of the U.S. Regional Haze Rule, initiated and chaired the IMPROVE program steering committee for 25 years, and authored many journal articles on varied visibility topics. Special thanks to Bruce Polkowsky, who, when he worked for the EPA, was the principal author of the Regional Haze Rule. He later joined the National Park Service (NPS) Air Resources Division (ARD), where he was instrumental in executing the rule through his leadership with the five Regional Air Partnerships. Bruce reviewed and contributed to Chapter 8, History of Visibility as an Esthetic Concern in the United States, which covers the history of the development and evolution of the US visibility regulatory programs. Finally, Bret Schichtel reviewed and made many helpful clarifying suggestions.

I would like to thank the National Park Service (NPS) for giving me freedom to pursue many and varied visibility research topics that ultimately contributed to the understanding of how haze affects the seeing and appreciating of near and distant landscape features. Special mention goes to Barbara Brown and Dick Briceland, who were instrumental in forming the Washington-DC-based Air Quality Division (ARD) within the NPS administrative structure, and Dave Shaver, who was directly responsible for hiring me into the NPS air quality program. Also thanks to the following past and present members of the ARD: Phil Wondra, Marc Scruggs, John Vimont, John Christiano, Chris Shaver, and Carol McCoy, who all, in their division management roles, supported the research and monitoring efforts that led to the formal protection of visibility in the United States. Without the cooperation and encouragement of national park and other federal land manager field staff, it would not have been possible to operate the hundreds of monitoring sites deployed throughout the United States. Within that context, special mention of Jim Renfro (Great Smoky Mountains National Park) and Carl Bowman (Grand Canyon National Park) is appropriate.

Thanks to Ken Odell and Eric Walther for "turning me on" to the concerns over the potential degradation of the ability to see and appreciate the many iconic vistas on the Colorado Plateau during my years at Northern Arizona University. Interactions and discussions with friends and co-workers were especially helpful in designing and implementing field programs to address specific issues pertaining to understanding visibility as an esthetic concern. I acknowledge many students, especially John Molenar and Scott Archer, who spent countless hours hiking into and out of the Grand Canyon, servicing monitoring equipment within and on the canyon floor. Mr. Molenar went on to form an air quality/visibility consulting company whose services were instrumental in operating many field programs carried out over the last 30–40 years, and Mr. Archer was the Bureau of Land Management visibility protection program leader for many years.

I would like to specifically mention my friends and scientists in the electric power generation industry, Peter Mueller, Rob Farber, and Prem Bhardwaja, who contributed greatly to making sure concepts and measurement programs were well thought out and critically reviewed. Special thanks to scientists at the University of California, Davis (UCD), the Desert Research Institute (DRI) in Reno, Nevada, and Optec, Inc., all of whom were critical to designing instrumentation and implementing field programs. Thomas Cahill of UCD was instrumental in designing the first particle monitor that was used across the western United States; Warren White of UCD always added critical insight into the interpretation of data; John Watson and Judy Chow of DRI were instrumental in the design of critical analytic techniques; and Jerry Pershaw of Optec designed and built many of the optical monitoring gadgets used over the years. Ron Henry, University of Southern California, was instrumental in identifying new and more useful visibility metrics and must be credited for introducing the idea of using linear systems theory in the field of visibility. Scientists and students at Colorado State University, especially Hari Iyer of the statistics department, contributed greatly to the success of the many data analysis and field monitoring programs.

Last, but not least, I would like to thank my immediate and extended family for their support and encouragement over the years and for tolerating my being away from home for many weeks and months and Helene Bennett for assisting in organizing and proofing the manuscript.

1

Introduction

WHAT IS VISIBILITY?

Visibility has varied and diverse meanings. Usually, it is interpreted as the degree of being visible, as in the social or business context of an advertisement or individual being visible. Other uses of the word "visibility" include, but are not limited to, the design of windows, architectural design, and even programming languages (Yasunori and Kenji, 1998). A dictionary definition of visibility usually includes a statement about the degree of atmospheric clarity, specifically the greatest distance through the atmosphere at which a prominent object can be identified with the naked eye, typically referred to as the atmospheric or meteorological visual range. Visibility as a contemporary issue of seeing and appreciating landscape features is discussed in a review article by Hyslop (2009).

In the early 1900s, the aviator highlighted the issue of visual range when he asked the meteorologist, "How far can I see?" This is also the title of a paper by Middleton (1935) addressing this question. Up until about 1970, visibility research was mostly concerned with aircraft operations and the military's need to identify and recognize targets from the air, the ocean, underwater, and to some degree, space. Answers were sought to questions such as the following: Can some particular object of interest be seen? How far can it be seen? How rapidly can it move and yet be visible? What is the effect of object color? Will filters help? How much magnification is necessary to make the object visible at a given distance? What is the optimum procedure for a visual search? What is the probability of success for sighting the object being sought? Under what circumstances can it be recognized? Is visual identification possible? How is visual performance affected by fatigue, discomfort, distraction, apprehension, motivation, and kindred factors (Duntley et al., 1964)?

During this time period, "visibility" was almost exclusively used to denote the human capability to detect, recognize, and identify objects by means of the human visual mechanism without the aid of intervening

Visibility: The Seeing of Near and Distant Landscape Features. http://dx.doi.org/10.1016/B978-0-12-804450-6.00001-2

sensor systems. Light reflected or generated by an object forms a body of image-forming flux that, after transmission and alteration by the intervening media, forms a retinal image that in turn is transmitted to the brain and perceived by the observer. In like fashion, the background against which the object is seen generates flux from a different part of object space, and this signal follows a corresponding path to the perceptual level of the observer. Discrimination of the object from its background depends upon the thresholds of the human visual system.

As a consequence of the need for energy and water to support urban and agricultural growth in the western and especially the southwestern United States, visibility took on a more diverse and esthetically related meaning. Rivers such as the Colorado and Columbia were dammed to create reservoirs from which water was pumped to the urban and industrial centers and that allowed the concurrent development of large, coal-fired electric-generating facilities to support expanding populations. This development raised concerns of a general degradation of the environment, but especially on the Colorado Plateau, a geographic region that includes many national parks, such as Grand Canyon, Canyonlands, and Bryce Canyon national parks, with striking geological features (see Fig. 8.1, Chapter 8). This concern escalated in the late 1960s and the early 1970s with the building of the Glen Canyon Dam, forming Lake Powell, which in turn allowed for the development of the Navajo Generating Station (NGS), a 2250-MW, coal-fired electric-generating facility adjacent to Grand Canyon National Park. During the same time period, other large, coal-fired electric-generating facilities were also constructed in the region (see Chapter 8 for a more detailed discussion).

As a consequence, visibility emerged as an esthetic concern in that emissions from industrial sources, such as the coal-fired power plants, and associated urbanization were perceived to be having a deleterious effect on the ability to see and appreciate the scenic wonders found in many of the Colorado Plateau's national parks. The use of the term "visibility" became synonymous with atmospheric clarity and the ability to clearly see and appreciate near and more-distant landscape features, as opposed to the earlier question of "How far do I have to back away from the mountain or how much haze must be added to the atmosphere before the vista just disappears?" In this new paradigm, an observer might remark on the color of the mountain, on the textural content of geological features, or on the amount of snow cover resulting from a recent storm system. Approaching landscape features such as those shown in Fig. 1.1a–d, the observer may comment on the contrast detail of nearby geological structures, on shadows cast by overhead clouds, or even on the shape, color, and beauty of the clouds themselves.

There is a need for the management of visual resources in terms of their inherent scenic qualities or beauty, instead of only the distance at which the resource disappears or can just be barely detected or seen. Urban area managers have also identified good visibility in the urban setting as an important value worthy of being protected. To this end, research and visibility

FIGURE 1.1 (a) The farthest scenic feature is the 130-km-distant Navajo Mountain, as seen from Bryce Canyon National Park. (b) The La Sal Mountains, as seen from the Colorado River, are a dominant form on the distant horizon. (c) View in Canyonlands National Park showing the highly textured foreground canyon walls against the backdrop of the La Sal Mountains. The La Sals are 50 km away from the observation point. (d) Bryce Canyon as seen from Sunset Point. Notice the highly textured and brightly colored foreground features.

concerns generally shifted from studying the atmospheric effects of being able to see and detect objects or targets to studying the seeing and detecting of the haze itself and its effects on landscape feature scenic quality.

WHAT IS VISIBILITY IMPAIRMENT?

Important factors involved in seeing an object are outlined in Fig. 1.2 and summarized as follows:

- Illumination of the overall scene by the sun. This includes illumination resulting from sunlight scattered by clouds and the atmosphere, as well as reflections by the ground and vegetation.
- Landscape characteristics that include color, texture, form, and brightness.
- Optical characteristics of the intervening atmosphere.
  - Image-forming information (radiation) originating from landscape features is scattered and absorbed (attenuated) as it passes through the atmosphere toward the observer.

- Sunlight, ground-reflected light, and light reflected by other objects are scattered by the intervening atmosphere into the sight path.
- Psychophysical responses of the eye–brain system to incoming radiation.

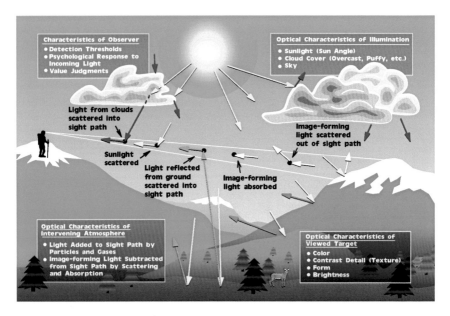

FIGURE 1.2 Important factors involved in seeing a scenic vista. Image-forming information from an object is reduced (scattered and absorbed) as it passes through the atmosphere to the human observer. Air light (technically referred to as path radiance) is also added to the sight path by scattering processes. Sunlight, light from clouds, and ground-reflected light all impinge on and scatter from particulates located in the sight path. Some of this scattered light remains in the sight path, and at times it can become so bright that the image essentially disappears. A final important factor in seeing and appreciating a scenic vista is the characteristics of the human observer.

Consider a sight path between some observer and distant landscape feature. Light reflected by landscape features forms an image whose characteristics are altered as it is transmitted through the atmosphere. The image-forming information is scattered out of and absorbed within the sight path, while at the same time light from the sun and surrounding features is scattered into the sight path. The light scattered by air molecules or particles in the sight path toward the observer is referred to as air light, or, more technically, path radiance. When the air light that reaches the eye of the observer is sufficient to outcompete the contribution of image-forming information, the appearance of the image is degraded, and with enough intervening atmospheric particles eventually disappears, much like driving a car in a snowstorm.

It is of interest to highlight the significance of air light, that is, the light that is scattered in the sight path toward the observer. The amount of light scattered by the atmosphere and the particles between the object and the observer can be so bright and dominant that the light reflected by the landscape features becomes insignificant. This is somewhat analogous to viewing a candle in a brightly lit room and then in a room that would otherwise be in total darkness. In the first case, the candle can hardly be seen, while in the other case it becomes the dominant feature in the room.

A real-world example of the role air light or path radiance plays in the seeing of scenic landscape features is highlighted in Fig. 1.3, photographs of Eagle Mesa in Monument Valley in northern Arizona, taken on a day when the atmosphere was nearly free of atmospheric particulate matter. Fig. 1.3a is a picture of the mesa viewed at a distance of about 1 km, while Fig. 1.3b shows the same mesa but viewed at a distance of about 15 km. The pictures were taken within a few minutes of each other. The contrast between the sky and mesa is somewhat reduced, but the most significant effect of increased haze (optical depth) between the observer

FIGURE 1.3 Photographs of a butte in Monument Valley in northern Arizona. Photograph (a) was taken at a distance of about 1 km, while photograph (b) was taken at a distance of about 15 km.

and landscape feature is the dramatic color shift toward the blue end of the spectrum. Color change or shift from red to purple is caused primarily by added air light that is the color of the sky, or bluish in nature.

In general, the appearance or visibility of haze is determined by the concentration and distribution of light scattering and absorbing particles. A common denominator in the perception of all types of visibility impairment resulting from haze spatial distributions is contrast, defined to be the fractional difference in light coming from adjacent features. This could be the contrast between a haze layer and its background or the contrast between two different elements within a landscape feature, which is generally referred to as contiguous contrast. Contrast can be expressed in terms of a single color—green, for instance—or as a color contrast. The reduction of contrast between two features to the point where one of the features can just be seen is referred to as threshold contrast, whereas contrast change that is just noticeable when a landscape feature is clearly visible is referred to as a suprathreshold value.

Fig. 1.4a shows a graphical representation of a clean-air scene; Fig. 1.4b–d shows three types of typical visibility impairment. Fig. 1.4b shows a plume from some sort of industrial source. A plume is typically characterized by lower and upper contrast edges. It constitutes visibility impairment because it can be seen in and of itself, but it also obscures scenic landscape features behind the plume. Fig. 1.4c is an example of uniform

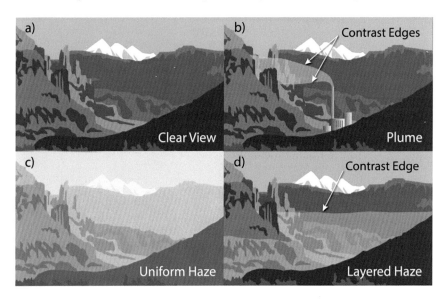

FIGURE 1.4   Three ways that air pollution can visually degrade a scenic vista. (a) is a graphical representation of a clean-air scene. When there is sufficient sunlight to cause the atmosphere to become turbulent, pollutants emitted into the atmosphere become well mixed and appear as a uniform haze (c). On the other hand, during cold winter months the atmosphere becomes stagnant. Pollutants emitted during these periods will appear either as a coherent plume (b) or as a layered haze (d).

haze, which generally obscures all landscape features in the scene and may be harder to differentiate from naturally occurring background haze. Finally, Fig. 1.4d shows a haze layer trapped near the ground. It is generally characterized by a contrast edge when viewed against a background scene or sky. It not only is recognized as a visible haze layer but also obscures landscape features behind the haze layer. Detection of plume and layered hazes is straightforward because of the presence of a contrast edge between the haze and its background. Detection of a uniform haze is more difficult, because it is sometimes difficult to differentiate a particle haze from naturally occurring background haze due to air molecules. Uniform haze becomes easier to detect and assess if the observer has the luxury of seeing the vista or landscape features under clear conditions.

It should be pointed out that the notion of uniform haze should not be confused with being synonymous with regional haze, which refers more to a large spatial scale of visible pollution. Whether a haze appears uniform or layered is at least in part a result of the geometry between the observer and the haze. If an observer is situated above the mixed layer, as often occurs when hiking in mountainous regions, a haze can be regional in extent but can appear to be layered in that the observer is above the mixed layer. Likewise, an observer who finds himself in a smoke plume would most likely judge the haze to be uniform in that he would not observe a haze with a defined edge.

Some distributions and concentrations of atmospheric particles and related visibility impairment may be perceived to be more objectionable than others. Are visitors or local residents more concerned with landscape features disappearing from a view or with changes in color and general appearance of nearby landscape features? Are they more interested in protecting the frequency of clear days or would they prefer a general protection of all days? Is a uniform haze more or less intrusive than a ground-based or suspended, layered haze? These issues and others will be discussed in more detail in Chapter 5 on visibility perception. A number of examples of visibility impairment that highlight these questions are presented in the following section.

Examples of Visibility Impairment

The camera can be an effective tool for capturing the visual impact that aerosols have on a visual resource. The following pictures show how the visual appearance of haze depends on the mixing characteristics of the atmosphere; the chemical, physical, and optical properties of the aerosol; and, just as important, the geometry of the landscape features and haze relative to the observer. In the following sections of this book, mathematical formalisms will be developed that will quantitatively build the relationships between these many variables.

The following pictures are of haze that appears to be uniform, primarily due to scattering of light by concentrations of particulate matter, combined with scattering by gases in the naturally occurring atmosphere. Fig. 1.5 shows the effects of various levels of uniform haze on the appearance of the La Sal Mountains in Utah. These photographs were taken from Island in the Sky, Canyonlands National Park, which is about 50 km due east of the La Sals. Sky–La Sal mountain contrasts, the fractional difference in light coming from the sky versus that arriving from the mountain, are -0.46, -0.21, -0.07, and -0.04, for parts a–d, respectively.

FIGURE 1.5   Effects of regional or uniform haze on a Canyonlands National Park vista. The distant landscape feature is the La Sal Mountains.

Figs. 1.6 and 1.7 show similar hazes of vistas at Mesa Verde and Bryce Canyon national parks. The Chuska Mountains in Fig. 1.6 are 95 km away, with the contrast at -0.26. Navajo Mountain is 130 km distant (Fig. 1.7), and in this photograph the sky–mountain contrast is -0.08. This photograph should be compared with Fig. 1.1a, a photograph of Navajo Mountain taken on a day on which the particulate concentration in the atmosphere was near zero.

FIGURE 1.6   Effects of uniform haze on the 95-km-distant Chuska Mountains as seen from Mesa Verde National Park.

FIGURE 1.7   Uniform haze degrades visual air quality at Bryce Canyon National Park. The just barely visible, 130-km-distant landscape feature is Navajo Mountain.

Of course, haze is not limited to scenic areas within the United States. In fact, haze enshrouds many scenic areas of the world. Fig. 1.8 shows Huashan Mountain outside of Xi'an in Shaanxi province, China. The farthest distance one could see when this photograph was taken was about 10 km, which corresponds to pollution levels considered to be unhealthy by North American standards.

Haze also obscures urban areas. In fact, haze tends to be worse in and surrounding populated areas because of localized emissions of industrial and mobile sources. Fig. 1.9 shows a uniform haze surrounding the

FIGURE 1.8   Uniform haze degrades visual air quality at Huashan Mountain, Xi'an, China. The farthest distance one could see at the time this photograph was taken was about 10 km, which corresponds to about 100 $\mu g/m^3$ particle loading.

FIGURE 1.9   Photograph of downtown Denver, Colorado, with mountains behind the city obscured.

Denver metropolitan area. The Denver skyline is about 25 km distant, and the particle loading on the day this photograph was taken was approximately 10–15 $\mu g/m^3$. Hazes such as this are typical of urban areas around the globe and in many cases are many times denser than shown in this figure.

Figs. 1.10 and 1.11 show a typical uniform haze in Xi'an, China, where buildings only a few kilometers distant are substantially obscured.

FIGURE 1.10 Haze obscuring buildings in downtown Xi'an, China. The visual range is only a few kilometers.

FIGURE 1.11 Haze obscuring buildings in downtown Xi'an, China. The visual range is only a few kilometers.

Uniform hazes in many parts of China are so dense that visual obscuration is evident over distances on the order of meters, as shown in Fig. 1.12.

One more example of urban haze is shown in Fig. 1.13, where the 5-km-distant mountains just outside of Monterrey, Mexico, are marginally visible.

FIGURE 1.12   Haze is sufficiently dense in this photograph to obscure features that are only a few meters distant.

FIGURE 1.13   Photograph taken just on the outskirts of Monterrey, Mexico, where mountains that are 5 km distant are obscured.

Haze does not always appear to be uniform. Under stagnant air mass conditions, aerosols can be "trapped" and can produce a visibility condition that is referred to as layered haze. Fig. 1.14 shows Navajo Mountain viewed from Bryce Canyon National Park, with a bright layer of haze that extends from the ground to about halfway up the mountain. Fig. 1.15 shows a similar example of layered haze but with the top portion of the mountain obscured.

FIGURE 1.14 Navajo Mountain as seen from Bryce Canyon National Park. Navajo Mountain, which is 130 km distant, has a bright white haze obscuring the bottom portion of the mountain.

FIGURE 1.15 The 130-km-distant Navajo Mountain obscured by an elevated white haze layer.

Fig. 1.16 is a classic example of plume blight. In plume blight instances, specific sources such as those shown in Fig. 1.17 emit pollutants into a stable atmosphere. The pollutants are then transported in some direction with little or no vertical mixing.

FIGURE 1.16   Dark plume over Navajo Mountain.

FIGURE 1.17   (a) Plume from an uncontrolled coal-fired power plant, near Mazatlán, Mexico, viewed a few kilometers downwind from the plant. (b) The same plume emitting into a stable atmosphere that exhibits little to no mixing. (c) The same plume is still visible tens of kilometers away from the plant.

Figs. 1.18–1.20 show other layered haze conditions that frequently occur at the Grand Canyon. Fig. 1.18 shows smoke from a fire that occurred below the rim of the canyon, while Fig. 1.19 is a more uniform haze primarily made up of emissions from nearby coal-fired power plants. Naturally occurring layered haze also effectively obscures landscape features. Occasionally during winter months, cloud layers build up below the rim of the canyon as shown in Fig. 1.20.

FIGURE 1.18   Smoke from a fire that occurred within the Grand Canyon causing a nonuniform layered haze that obscures many of the landscape features.

FIGURE 1.19   Example of a sulfate haze associated with coal-fired power plant emissions that filled the Grand Canyon, obscuring nearly all scenic landscape features.

FIGURE 1.20 A naturally occurring haze layer also can obscure scenic features. In this photograph, a cloud layer has developed within the Grand Canyon.

As with uniform haze, layers of haze are observed in and over urban areas as well. Fig. 1.21 shows a layered haze in Denver, Colorado, where buildings extend up through the pollutants trapped by a well-developed inversion layer. Hazes such as these are typical in all urban areas during the colder winter months when vertical dispersion is at times minimal.

To some degree, the observance of haze layers is geometry dependent in that the observer must be above or at some distance from the haze layer

FIGURE 1.21 A deep inversion has trapped emissions, causing a shallow haze layer in Denver, Colorado. The depth of the haze layer is only a few meters, with buildings extending well above the layer.

for it to appear to be layered. An observer who found himself in the haze layer would judge the obscuration to be uniform.

Fig. 1.22 shows the appearance of a black or light absorbing carbon plume, in this case the result of burning tires. Black carbon absorbs all wavelengths of light and always appears dark or "black," independent of lighting and illumination–observer geometry.

Fig. 1.23, on the other hand, is a plume from a forest fire that contains both light absorbing carbon and carbon that scatters light. Notice that the portion of the plume that is illuminated appears a bright gray, as opposed to the background sky, which appears blue, and clouds that are white. Also notice that the part of the plume that is not directly illuminated by sunlight appears dark, showing the dramatic effect of illumination on the "visibility"

FIGURE 1.22   Light absorbing dark plume resulting from the burning of tires.

FIGURE 1.23   Photograph of a wildfire plume, showing the effect of light absorbing carbon. Areas of the plume that are directly illuminated are dark, indicating a predominance of black carbon, while other areas of the plume are made up of particles that both scatter and absorb light, resulting in a gray plume.

of the plume. Viewing landscape features through a light absorbing plume will usually result in an overall "graying," independent of illumination geometries, as opposed to a shift in perceived landscape color.

Fig. 1.24 is another example of the effect of illumination on the appearance of plumes, in this case, power plant emissions. The two plumes on the left are particulate plumes, whereas the two plumes on the right consist mainly of water droplets. The plume on the far right, which is illuminated by direct sunlight, appears to be white. The second, an identical water-droplet plume that is shaded, appears dark. The amount of illumination can have a significant effect on the appearance of particulate concentrations. Second, the optical and physical characteristics of particle and water droplet plumes are quite different and thus have a substantial effect on how the plumes scatter light.

FIGURE 1.24   Two types of coal-fired power plant plumes. The two plumes on the left are primarily fly ash plumes, while the plumes on the right are primarily water vapor plumes associated with scrubber technology designed to remove fly ash.

Although black carbon absorbs all wavelengths of light, nitrogen dioxide gas absorbs more in the blue than in the red portion of the spectrum. Fig. 1.25 demonstrates how nitrogen dioxide gas (NO_2) can combine with varied background illumination to yield a very brown atmospheric discoloration. If a volume of atmosphere containing NO_2 is shaded and if light passes through this shaded portion of the atmosphere, the light reaching the eye will be deficient in photons in the blue part of the spectrum. As a consequence, the light will appear brown or reddish in color. However, if light is allowed to shine on, but not through, that same portion of the atmosphere, scattered light reaches the observer's eye and the light can appear to be gray in nature. Both these conditions are shown in Fig. 1.25. On the right side of the figure, the mixture of NO_2 and particulates is shaded by clouds. The

FIGURE 1.25   Brown discoloration resulting from an atmosphere containing nitrogen dioxide (NO$_2$) being shaded by clouds but viewed against a clear blue sky. Light scattered by particulate matter in that atmosphere can dominate light absorbed by NO$_2$, causing a gray- or blue-appearing haze (left side of photograph).

same atmosphere, illuminated because the cloud cover has disappeared, appears almost gray in the middle portion of the photograph.

Fig. 1.26 is another example of an urban layered haze over Denver, but this haze is made up of primarily NO$_2$, as indicated by its brownish appearance.

FIGURE 1.26   Inversion layer over Denver, Colorado, that has trapped NO$_2$ emissions. NO$_2$ gas absorbs more in the blue portion of the spectrum, resulting in a brown haze layer.

Effects of the angle between the source of illumination, the haze, and the observer are illustrated in Fig. 1.27. Fig. 1.27 is an easterly view of the La Sal Mountains in southeastern Utah as seen from an elevated point that is some 100 km distant. The haze in the photographs shown in Fig. 1.27 is nearly the same, the only difference being that Fig. 1.27a was taken at

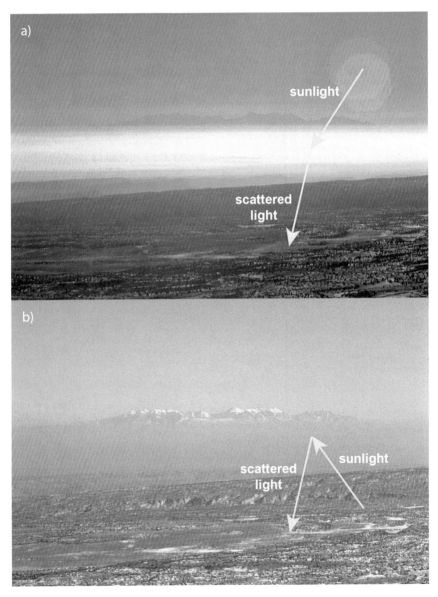

FIGURE 1.27   Photographs showing how the vistas appeared on a day when pollutants were trapped under an inversion layer. In (a), the haze appears white; in (b), the identical haze is dark or gray.

9:00 am, whereas Fig. 1.27b was taken on the same day but at about 3:00 pm local time. As indicated by the arrows, in Fig. 1.27a in the morning, the sun–haze–observer geometry corresponds to light being scattered from the particles to the observer in the forward direction, while in Fig. 1.27b the geometry corresponds to light being scattered backward from the haze to the observer. In Fig. 1.27a, the haze appears white and brighter than the sky, whereas in Fig. 1.27b it is dark gray. Although the background sky is blue, the haze in both cases is achromatic or lacking in color.

Light-scattering properties associated with particle size and physical/chemical characteristics will be discussed in more detail in Chapter 2. However, it is convenient to point out here that haze particles typically occur in a size range between 0.1×10^{-6} and 1.0×10^{-6} m (1×10^{-6} m = 1 μm = 1 micron), while air molecules are about 10,000 times smaller or about 1×10^{-10} m. The larger haze particles tend to scatter all wavelengths of light about equally and mostly in the forward direction. Very little light is scattered in the backward direction. Air molecules, on the other hand, scatter light about equally in the forward and backward directions and scatter more blue light than red, hence the blue sky. The photographs in Fig. 1.27 are excellent examples of this effect.

Fig. 1.28 shows another example of plume blight from a coal-fired power plant, with a number of interesting lighting and atmospheric effects. Notice how the constrained plume follows the terrain features, indicating a very stable atmosphere. Also notice how the brightness of the plume

FIGURE 1.28 Photograph of a constrained coherent plume, showing the effects of forward and backward scattering. With forward-scattering geometry, the plume appears white, while backward-scattering geometry results in a dark plume.

changes as a function of illumination, plume, and observer geometry. The plume appears white with forward-scattering geometry and dark when the geometry is conducive to backward scattering.

Furthermore, the angle at which the sun illuminates a vista or landscape feature (sun angle) plays another important role. Fig. 1.29 exemplifies this effect. The view again is from Island in the Sky, Canyonlands National Park, looking out over Canyonlands with its many colorful features toward the 50-km-distant La Sal Mountains. Fig. 1.29a shows how the canyon appears when it is in total shadow (6:00 am local time). Fig. 1.29a–c shows a progressively higher sun angle, until in Fig. 1.29d the scene is entirely illuminated. In each case, the air quality is the same. The only change is in the angle at which the sun illuminated the vista. There are primarily two reasons for the apparent change in visual air quality. First, at higher sun angles, there is less scattering of light by the intervening atmosphere in the direction of the observer. Second, the vista reflects more light; consequently, more image-forming information (reflected photons from the vista) reaches the eye. The contrast detail and scene are enhanced.

FIGURE 1.29   Four photographs showing the effects of a progressively shifting sun angle on the appearance of a vista as seen from Island in the Sky, Canyonlands National Park. In each photograph, the air quality is the same. In (a) (6:00 am), the sun angle-observer-vista geometry results in a large amount of scattered air light (forward scattering) added to the sight path but a minimal amount of imaging light reflected from the vista; (d) (12:00 noon) shows just the opposite. Scattered light is minimized, and reflected imaging light is at a maximum.

Are Haze and Impaired Visibility Only Contemporary Issues?

While visibility as it relates to seeing and appreciating landscape features is a contemporary topic, it seems clear that visibility impairment resulting from human activity must have been around for as long as man has had the ability to alter his environment. Analysis of flints at an archaeological site near the river Jordan suggests that humans had the ability to create fire nearly 790,000 years ago (Alperson-Afil, 2008). Where there is fire there is smoke, and where there is smoke there is visibility impairment.

Célia Sapart of Utrecht University in the Netherlands charted the chemical signature of methane in ice samples spanning 2100 years. She notes that human activity such as metallurgy and large-scale agricultural practices dating back to 100 BC contributed to elevated methane gas worldwide: "The ice core data show that as far back as the time of the Roman Empire, human [activities] emitted enough methane gas to have had an impact on the methane signature of the entire atmosphere" (Sapart et al., 2012). Emissions associated with metallurgy, which was utilized worldwide, undoubtedly contributed at some level to ambient levels of particulate concentrations and in turn visibility impairment.

As civilization progressed and urbanization grew, more metal smelters, kilns, and fossil-fuel burning had an ever-increasing impact on daily life. Stern (1973) quotes Seneca, a Roman philosopher in 61 AD: "As soon as I had gotten out of the heavy air of Rome, and from the stink of the chimneys thereof ... I felt an alteration to my disposition." Even as far back as early Roman times, it was recognized that "heavy air," which is likely associated with poor visibility, may contribute to a person's "disposition." It has been demonstrated that depression and the absence of a feeling of well-being, both of which are contemporary concerns, are associated with poor visibility and high levels of air pollution (Evans and Cohen, 1987; Evans et al., 1987; Zeidner and Schechter, 1988).

During the Renaissance era, from the 14th to 17th centuries, scholars and artists attempted to move toward realism in various art forms. There was an attempt to capture the sky and distant mountains disappearing, as well as changes in sky color, haze, and mist. Giotto di Bondone (1267–1337) is credited with first treating a painting as a "window into space" (White, 1967). Filippo Brunelleschi (1377–1446) helped formalize perspective as an artistic technique (Edgerton, 2009). In 1891, Monet said, "For me, a landscape does not exist in its own right, since its appearance changes at every moment; but its surroundings bring it to life—the air and the light, which vary continually. ... For me, it is only the surrounding atmosphere which gives objects their true value" (Gordon and Forge, 1989).

Because artists during this time period attempted to capture a realistic impression of natural landscapes, it is interesting to examine their paintings in light of their attempt to capture haze and atmospheric color under the various atmospheric conditions of the time. In the many works of the

Renaissance time period, almost all artists painted some pictures that included landscape elements. Most depict landscape elements that are distant, have cool tones, and become hazed out and disappear near the center focal point of the painting. Some pictures show rather intense haze, while others seem to depict a high level of clarity, even for features that appear to be quite distant. For instance, the painting "The Tuileries" by Claude Monet shows rather extreme haze with nearby as well as more-distant features nearly disappearing. Many of Monet's works show pictures with some level of haze present. Even the famous "Mona Lisa" by Leonardo da Vinci depicts substantial haze in the background.

On the other hand, paintings that depict more-distant landscape features with rather clear conditions seem to be in the majority. One fine example is "The Last Supper" by Domenico Ghirlandaio, in which the background sky is blue, and high, wispy clouds are clearly visible, something rarely seen in the modern urban setting. Other examples of background landscapes that are depicted as being quite clear and distant can be found in "The Testament and Death of Moses," a painting in the Sistine Chapel by Luca Signorelli. It is fascinating to explore the paintings made during the Renaissance years with an eye toward observing the atmospheric effects and specifically the visibility conditions that are depicted. There are far too many to mention here, but images are easily accessible on the Web. Note that even though literature does not highlight or focus on "haze" during this time period, it was obviously present, regionally in the form of naturally occurring aerosols and locally from agricultural practices, metallurgy, and fire. In fact, during the building of Brunelleschi's dome for the Cathedral of Florence, Santa Maria del Fiore, it was pointed out that black and gray smoke belched from the forges surrounding the building site (King, 2001).

Even before the industrial revolution, King Edward I banned the burning of "sea-coal," coal that was washed up on shore by ocean currents (Hutchinson, 1794; Campbell, 1871). However, the coming of the industrial revolution marked the beginning of serious concerns over air quality. The industrial revolution began in Britain but spread quickly to the rest of Europe and the United States. The main culprit in degradation of air quality was the burning of coal, not only for industrial purposes but also for cooking and household heating.

Barry Rutherford (http://barry-rutherford.newsvine.com/_news/2007/07/13/835368-the-london-smog-of-1952) describes a pollution case in December 1813 in which a mail coach from London to Birmingham took seven hours. Accounts tell of fog so thick that visibility was limited to across the street. In London during one week in 1873, over 700 people died from poor air quality (Schneider, 1992). However, the worst recorded extended air pollution episode in the area occurred in December 1952, killing over 4000 people. It was said the visibility was so degraded that people could not see their own feet. Visibility was below 500 m continuously for 114 h and below 50 m for two days. At Heathrow Airport, visibility was below 10 m for 48 h (see the Rutherford note above). Fig. 1.30 shows visibility conditions during this time period (photographs from http://www. metoffice.gov.uk/education/teens/case-studies/great-smog and http:// www.todayifoundout.com/index.php/2013/01/the-deadly-london-smog-of-1952/).

FIGURE 1.30 In December 1952, the city of London experienced a 5-day bout of "fog" that killed at least 4000 people and made an estimated 100,000 sick. Visibility was degraded to the point that seeing only a few feet was difficult.

The Environmental Institute of Houston has published a table of worldwide air quality incidents with some legislation that was consequently passed (http://prtl.uhcl.edu/portal/page/portal/EIH/outreach/tfors/history). Some notable air pollution and haze episodes are the 1930 Meuse Valley, Belgium, episode in which 60 people died. There were four coke ovens, three steel mills, four glass factories, and three zinc smelters along the 15 miles of the narrow valley of the Meuse River from Huy to Liege. Smog during a 1939 episode in St. Louis was so thick that lanterns were needed during daylight hours. In 1943, an episode of smog in Los Angeles reduced visibility to three blocks and was blamed on a nearby industrial plant that when shut down had no effect on air pollution. In October 1948, Donora, a small industrial town near Pittsburgh, Pennsylvania, had a haze episode that killed 20 people (Davis, 2002). The picture below shows the visibility at noon in the town during the episode (Fig. 1.31).

In *A History of the Chemical Industry in Widnes* (http://commons. wikimedia.org/wiki/File:Widnes_Smoke.jpg), D.W.F. Hardie presents a picture of factories, as shown in Fig. 1.32, in the industrial town of Widnes

FIGURE 1.31   Visibility in Donora, PA, during an extreme haze episode in 1948.

FIGURE 1.32   Smoke stacks in Widnes, England, that formed the center of many industrial and associated urban areas during the industrial revolution and times thereafter.

in the borough of Halton, Cheshire, England, around the late 19th century. Sherard (1897) writes of every tree and every blade of grass being dead for miles around.

Even though prior to about 1900 visibility was not addressed as a singular topic of concern, there is documented evidence that populations were aware of haze and expressed concern about haze episodes over the course of the last 2000 years. Furthermore, wildfires have occurred for millennia, and with wildfire there is smoke, and with smoke we have a natural form of haze. One can only surmise that early humans recognized the degrading effects that smoke would have had on the seeing of landscape features, what we now refer to as visibility impairment.

References

Alperson-Afil, N., 2008. Continual fire-making by hominins at Gesher Benot Ya'aqov, Israel. Quaternary Sci. Rev. 27 (17–18), 1733–1739.

Campbell, G.D., 1871. (Duke of Argyll). George Douglas Campbell Argyll Report of the Commissioners Appointed to Inquire into the Several Matters Relating to Coal, Volume 3, printed by G.E. Eyre and W. Spottiswoode for H. M. Stationery, Great Britain.

Davis, D., 2002. When Smoke Ran Like Water: Tales of Environmental Deception and the Battle against Pollution. Basic Books, New York.

Duntley, S.Q., Gordon, J.I., Taylor, J.H., White, C.T., Boileau, A.R., Tyler, J.E., et al., 1964. Visibility. Appl. Opt. 3 (5), 549–598.

Edgerton, S.Y., 2009. The Mirror, the Window & the Telescope: How Renaissance Linear Perspective Changed Our Vision of the Universe. Cornell University Press, Ithaca, NY.

Evans, G.W., Cohen, S., 1987. Environmental stress. In: Stokols, D., Altman, I. (Eds.), Handbook of Environmental Psychology. Wiley, New York.

Evans, G.W., Jacobs, S.V., Dooley, D., Catalano, R., 1987. The interaction of stressful life events and chronic strains on community mental-health. Am. J. Commun. Psychol. 15, 23–24.

Gordon, R., Forge, A., 1989. Monet. Harry N Abrams Press, New York.

Hutchinson, W., 1794. The History and Antiquities of the County Palatine of Durham. Hodgson & Robinsons, London, vol. 3 S.

Hyslop, N.P., 2009. Impaired visibility: the air pollution people see. Atmos. Environ. 43, 189–202.

King, R., 2001. Brunelleschi's Dome: How a Renaissance Genius Reinvented Architecture. Penguin Books, New York.

Middleton, W.E.K., 1935. How far can I see? Sci. Mon. 41, 343–346.

Sapart, C.J., Monteil, G., Prokopiou, M., Van de Wal, R.S.W., Kaplan, J.O., Sperlich, P., et al., 2012. Natural and anthropogenic variations in methane sources during the past two millennia. Nature 490, 85–88.

Schneider, T. (Ed.), 1992. Air Pollution in the 21st Century: Priority Issues and Policy. Elsevier, Amsterdam.

Sherard, R.H., 1897. The White Slaves of England Being True Pictures of Certain Social Conditions in the Kingdom of England in the Year 1897. James Bowden, London, p. 47.

Stern, A.C., 1973. Fundamentals of Air Pollution, third ed. Academic Press, London.

White, J., 1967. The Birth and Rebirth of Pictorial Space. Faber and Faber, London.

Yasunori, H., Kenji, M., 1998. Programming languages based on visibility. Trans. Inform. Process. Soc. Japan 39, 70–76.

Zeidner, M., Schechter, M., 1988. Psychological responses towards air pollution: some personality and demographic correlates. J. Environ. Psych. 8, 191–208.

On the Nature of Light and Its Interaction with Atmospheric Particles

All the visible effects of haze shown in Chapter 1 are dependent on the nature of light and its interaction with naturally occurring air molecules and small particles found in the atmosphere. Not only are we personally dependent on light to carry visual information, but also much of what we know about the stars and the solar system is derived from light waves registering on our eyes and on optical instruments.

Light can be thought of as waves, and to a certain extent they are analogous to water and sound waves. Fig. 2.1 shows water waves with the distance

One Wavelength (λ)

FIGURE 2.1   Water waves can be used to show the concept of wavelengths. A wavelength is defined as the distance from one crest to the next.

Visibility: The Seeing of Near and Distant Landscape Features. http://dx.doi.org/10.1016/B978-0-12-804450-6.00002-4

from trough to trough denoted as one wavelength λ. Fig. 2.1 is very interesting because it shows two waves traveling perpendicular to each other, resulting in an interference pattern. When both waves combine in phase (they move up together), the amplitude increases, and conversely, if they combine out of phase, they cancel each other out. Light waves do the same thing.

A wave passes a fixed point in space with some velocity V, and the relationship between the frequency f with which the crest or trough passes by that point in space, V, and λ is given by

$$V = f\lambda. \tag{2.1}$$

Similar oscillations of electrical and magnetic fields are called electromagnetic radiation. Ordinary light is a form of electromagnetic radiation, as are X-rays, ultraviolet and infrared light, radar, and radio waves. All of these travel at approximately 300,000 km/s (186,000 mi/s) and only differ from one another in wavelength. Fig. 2.2 is a schematic representation of the electromagnetic spectrum, with the visible portion shown in color to emphasize the portion of the spectrum to which the human eye is sensitive. The visible spectrum is white light separated into its component wavelengths or colors. The wavelength of light, typically measured in terms of millionths of a meter (microns, μm), extends from about 0.4 to 0.7 μm.

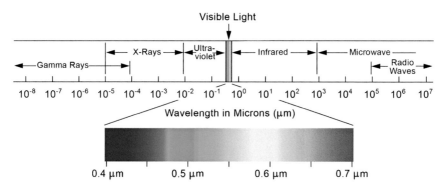

FIGURE 2.2   Vibrations of electric and magnetic fields are referred to as electromagnetic radiation. This diagram shows the wavelengths of various types of electromagnetic radiation, including visible light. The wavelength of the visible spectrum varies from 0.4 (blue) to 0.7 μm (red). One μm equals one-millionth of a meter.

The human eye has a characteristic response to different wavelengths of radiant energy as shown in Fig. 2.3. Maximum response to a unit of radiant energy is at the wavelength of 0.550 μm. When radiant energy is discussed and measured in terms of the response of the human eye, photometric concepts and units are used, and the radiation will be referred to as light. The corresponding radiometric concepts and units are used in the discussion and measurement of light energy in an absolute

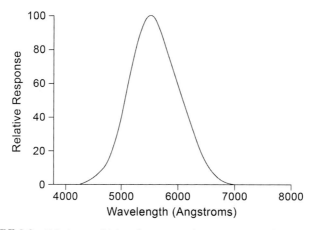

FIGURE 2.3   Relative sensitivity of an average human eye to various wavelengths of light.

TABLE 2.1   Radiometric and Photometric Concepts and Units

Radiometric	Symbol	Units	Photometric	Symbol	Units
Radiant energy	U	Joule	Luminous energy	Q	Talbot
Radiant flux	P	Watt	Luminous flux	F	Lumen
Radiant intensity	J	Watt/ steradian	Luminous intensity	I	Lumen/ steradian
Radiance	N	Watt/m^2/ steradian	Luminance	B	Lumen/m^2/ steradian
Irradiance	H	Watt/m^2	Illuminance	E	Lumen/m^2

sense. Radiometric and photometric concepts of importance are listed in Table 2.1 and in more detail in Appendix 1.

Waves of all kinds, including light waves, carry energy. Electromagnetic energy is unique in that energy is carried in small, discrete parcels called photons. As shown in Fig. 2.2, the color of visible light depends on its wavelength. Schematic representations of blue, green, and red photons are shown in Fig. 2.4. Blue, green, and red photons have wavelengths of around 0.45, 0.55, and 0.65 μm, respectively. The color properties of light depend on its behavior both as waves and as particles.

When light passes from a low-density to a higher-density medium, its velocity decreases, and as such, its wavelength shortens. The index of refraction, expressed as $n = \lambda_1/\lambda_2$, is a measure of the change in wavelength or velocity as light passes from one medium (1) to another (2). Strictly speaking, the index of refraction is defined relative to light passing through a vacuum or a medium devoid of molecules or particles. If light

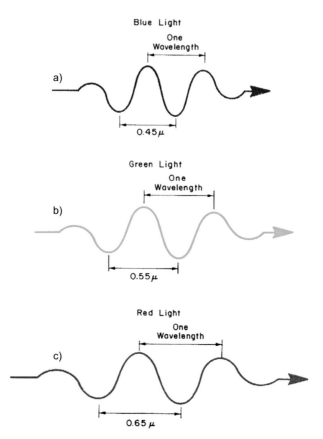

FIGURE 2.4   Light exhibits wave and particle-like characteristics. Light can be thought of as consisting of bundles of vibrating electrical and magnetic waves. These bundles of energy are called photons, and the wavelengths of radiant energy making up the photon determine its "color." The figure schematically shows blue, green, and red photons.

strikes a surface at some angle θ, it will bend toward the normal of the surface, according to the relationship (Snell's law) (Born and Wolf, 1999)

$$n_1 \sin(\theta_1) = n_2 \sin(\theta_2) \tag{2.2}$$

where n_1 and n_2 are the indexes of refraction of mediums one and two, and θ_1 and θ_2 are the angles between the incident and immerging beam and a normal or perpendicular line to the surface.

When light is shone on a prism at some angle with respect to the surface normal as indicated in Fig. 2.5, differing wavelengths are dispersed into the component colors or wavelengths of the incident light. Because each color is bent at a different angle, we can conclude that the index of refraction of the prism medium is wavelength dependent, with the index of refraction being greater on shorter wavelengths. Colors in a rainbow are the result of water droplets acting like small prisms dispersed through the atmosphere. Each water droplet refracts light into the component colors of the visible spectrum.

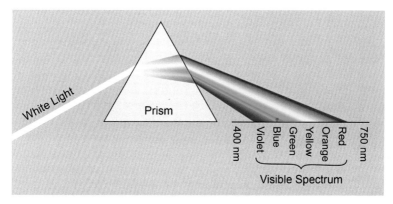

FIGURE 2.5   Schematic of white light being split into its component wavelengths by a prism.

The colors of light are more commonly separated in other ways. When light strikes an object, certain color photons are captured by molecules in that object. Different types of molecules capture photons of different colors. The only colors we see are those photons that the surface reflects. For instance, chlorophyll in leaves captures photons of red and blue light and allows green photons to reflect back, thus providing the green appearance of leaves. Nitrogen dioxide, a gas emitted into the atmosphere by combustion sources, captures blue photons. Consequently, nitrogen dioxide gas tends to look reddish brown.

The next section explores how light interacts with gases and particles that are typically found in the atmosphere.

INTERACTION OF LIGHT AND PARTICLES

A photon of light is said to be scattered when it is received by a particle and reradiated at the same wavelength in any direction. Most visibility degradation results from light scattering and absorption by gases and atmospheric particles that are nearly the same diameter as the wavelength of the light.

Whether light is scattered by gases or particles, the associated visibility impairment is not only dependent on the total amount of light scattered but also the directional dependence of the scattering process.

Image-forming light is lost by scattering and absorption in each elementary segment of the sight path, and path radiance, or air light, is generated by the scattering of ambient light that reaches the path segment from all directions. These processes result in a reduction of the contrast of an object against its background, impairing our ability to see it.

The loss of light due to scattering and absorption within any path segment is proportional to the amount of light present; the coefficient of proportionality is b_{ext}, the attenuation or extinction coefficient at position r. As a fractional loss of light per unit of distance along the light path, b_{ext} has units of 1 over length, typically either km^{-1} or Mm^{-1} (i.e., inverse kilometers or inverse megameters). b_{ext} is a function of position within the sight path; it does not depend upon the transmission direction. It is independent of the manner in which the path segment is lighted by the sun or sky; it is a physical property of the atmosphere alone. Absorption refers to light being absorbed by a molecule or particle, resulting in an increase in its internal energy. Attenuation by scattering results from any change of direction of radiant energy sufficient to cause the radiation to fall outside an area of detection either by the eye or electro-optical detection device.

Therefore,

$$b_{ext} = b_{scat} + b_{abs} = b_{sg} + b_{ag} + b_{sp} + b_{ap} \qquad (2.3)$$

where b_{scat} and b_{abs} are the atmospheric scattering and absorption coefficients, respectively. The scattering coefficient is further broken down into scattering by gases b_g and particles b_{sp}. Likewise, b_{ag} and b_{ap} refer to absorption by gases and particles, respectively.

Light does not scatter from gases or particles equally in all directions. Therefore,

$$b_{scat} = \int_{4\pi} \sigma(r, \beta) d\Omega \qquad (2.4)$$

where r and β are the position and scattering angles, respectively, $d\Omega$ is an increment of solid angle, and $\sigma(r, \beta)$ is defined as the volume scattering function. The volume scattering function is a measure of the atmosphere's ability to scatter light in a given direction.

The scattering function $\sigma'(r, \beta)$ is the normalized volume scattering function defined by the following equation:

$$\sigma'(r, \beta) = \sigma(r, \beta)/b_{scat}(x) \qquad (2.5)$$

and as such,

$$\int_{4\pi} \sigma'(r, \beta) d\Omega = 1. \qquad (2.6)$$

The scattering of light by air molecules, primarily nitrogen and oxygen, is typically referred to as Rayleigh scattering, named after the British physicist John William Strutt, more commonly known as Lord Rayleigh (Born and Wolf, 1999). Rayleigh scattering is applicable to spheres that are less than about 10 times the wavelength of the incident radiation, as is the case for sunlight incident on air molecules. The scattering is elastic in that the amount or energy of incident radiation is equal to that being scattered.

The total volume scattering $\sigma(r, \beta)$ is the sum of scattering by air molecules (Rayleigh scattering) and particles (Mie scattering):

$$\sigma(r, \beta) = \sigma(r, \beta)_R + \sigma(r, \beta)_M. \qquad (2.7)$$

It can be shown that $\sigma(r, \beta)_R$ is given by

$$\sigma(r, \beta)_R = \left[\frac{8\pi^4 d^6 N \left(\dfrac{n^2 - 1}{n^2 + 2} \right)^2}{\lambda^4} \right] \left(1 + \cos^2(\beta) \right) \qquad (2.8)$$

where N is the number of molecules, d is the molecule diameter, β is the angle between the scattered and incident radiant energy, and n is the index of refraction. Fig. 2.6 is meant to schematically show direction and intensity dependence, represented by Eq. 2.8, as a function of distance from the scattering molecule. First, notice that scattering in the forward and backward directions is identical and that scattering at 90° to the incident radiation is half that scattered either forward or backward. Although not shown in Fig. 2.6, the scattering at 90° is polarized, while scattering in the forward and backward directions is completely unpolarized.

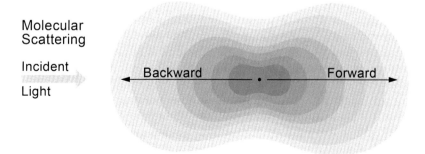

FIGURE 2.6 Relative directional dependence of light scattered by air molecules, usually referred to as Rayleigh scattering.

Eq. 2.8 also shows that scattering by air molecules has a strong wavelength dependence of $1/\lambda^4$. This wavelength dependence is graphically shown in Fig. 2.7 and is primarily responsible for the blue color of a particle-free sky. This also explains the colors of sunrises and sunsets in which sunlight, transmitted through a longer sight path in the atmosphere, has lost more of the blue photons, leaving a greater ratio of the red to yellow photons.

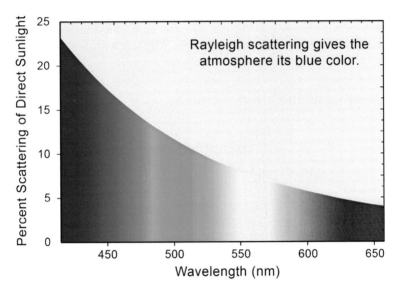

FIGURE 2.7   Wavelength dependence of light scattered by air molecules.

The only gas that is normally found in the atmosphere and absorbs light is nitrogen dioxide (NO_2). Absorption by NO_2 at 0.55 μm is $b_{ag} = 330[NO_2]$, where the units of b_{ag} are Mm^{-1} and the units of $[NO_2]$ are ppm. Furthermore, NO_2 absorbs more in the blue portion of the spectrum than in the red portion. Therefore, NO_2 appears brown or yellowish if viewed against a background sky.

Scattering from particles with sizes that are approximately equal to the wavelength of light is described by what has been commonly referred to as Mie scattering (van de Hulst, 1981; Born and Wolf, 1999). Mie scattering equations are a result of solving Maxwell's equations for describing how light interacts with particles as it passes by and through them. Mie theory really only applies to spherical particles; however, it is easy to find discussions in the literature (Fuller et al., 1999) of scattering from particles of other shapes that are still referred to as Mie scattering. The Mie solution of Maxwell's equations results in the sum of an infinite set of equations that are typically approximated using recursive computer algorithms. The solutions that yield $\sigma(x, \beta)_M$ depend on particle size, wavelength, and index of refraction. Even though Mie scattering is generally used to refer to scattering by particles with sizes on the order of the wavelength of light, the formalism is general and applies equally well to molecular scattering or to scattering from particles many times the size of light wavelengths. It should be pointed out that, in general, the index of refraction is a complex number in which the imaginary component of the index of refraction describes the absorption characteristics of the particle.

Mie scattering is a result of the combination of the phenomena shown schematically in Fig. 2.8a–c. Fig. 2.8a shows diffraction, a phenomenon whereby radiation is bent to "fill in the shadow" behind the particle. Fig. 2.8b depicts light being bent (refracted) as it passes through the particle. A third effect (Fig. 2.8c), resulting from slowing a photon, is a little difficult to understand. Consider two photons approaching a particle, vibrating "in phase" with one another. One passes by the particle, retaining its original speed, while the other, passing through the particle, has its speed altered. When this photon emerges from the particle, it will be vibrating "out of phase" with its neighbor photon; when it vibrates up, its neighbor will vibrate down. As a consequence, they interfere with each other's ability to propagate in certain directions. Fig. 2.8d indicates how a

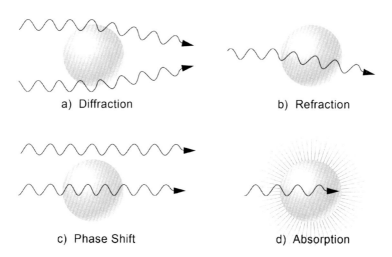

a) Diffraction b) Refraction

c) Phase Shift d) Absorption

FIGURE 2.8 Particle light scattering and absorption. Diffraction (a) and refraction (b) combine to "bend" light to "fill in the shadow" behind the particle. Diffraction, an edge effect, causes photons passing very close to a particle to bend into the shadow area; refraction is a result of the light wavefront slowing down as it enters the particle. In phase shift (c), while the photon is within the particle, its wavelength is also shortened. Thus, when it emerges from the particle, it may vibrate out of phase with adjacent photons and interfere with their ability to propagate in a given direction. As a fourth possibility, the photon may be absorbed by the particle (d). In this case, the internal energy of the particle is increased. The particle may rotate faster, or its molecules may vibrate with greater amplitude.

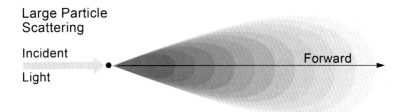

FIGURE 2.9   For particles that are about the same size or larger than the incident radiation, most of the light is scattered in the forward direction.

photon can be absorbed by the particle. The photon energy is transferred to internal molecular energy or heat energy. In the absorption process, the photon is not redistributed into space; the photon ceases to exist.

The efficiency with which a particle can scatter light and the direction in which the incident light is redistributed are dependent on all four of these effects. For particles whose size is on the order of the wavelength of incident light, photons are scattered preferentially in a forward direction as shown in Fig. 2.9. Furthermore, Mie scattering is less wavelength dependent than Rayleigh scattering, typically, $\sigma(\lambda)_M \alpha \, 1/\lambda^n$, where n is about 1 or less.

It is convenient to express the size dependency of particle scattering and absorption in terms of an extinction efficiency factor. The extinction efficiency factor is expressed as a ratio of a particle's effective cross section to its actual cross section. Fig. 2.10 shows how this efficiency varies as a

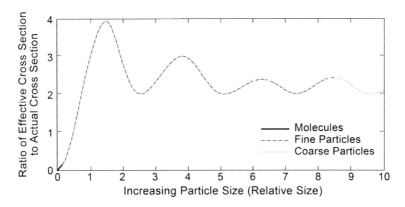

FIGURE 2.10   The relative efficiency with which particles of various sizes scatter light. The solid line corresponds to the scattering efficiency of molecules. The dashed and dotted lines show the efficiency with which fine and coarse particles scatter light. Note that fine particles (0.1–1.0 μm) can be more efficient at scattering light than either molecules or coarse particles.

function of particle size. Very small particles and molecules are very inefficient at scattering light. As a particle increases in size, it becomes a more efficient light scatterer until, at a size that is close to the wavelength of the incident light, its effective cross section can appear to be many times its actual size. Even particles that are very large scatter light as though they were twice as big as they actually measure. Each of these particles removes twice the amount of light intercepted by its geometric cross-sectional area.

The extinction coefficient for a single particle is then

$$b_{ext} = (\pi D^2/4)Q_e(n_i, \chi, \lambda) \tag{2.9}$$

where D and Q_e are the particle diameter and extinction efficiency, respectively. n_i and χ are the complex index of refraction and size parameter, respectively, defined as $\pi D/\lambda$, where D is the diameter of the particle.

Because many atmospheric monitoring programs measure particle mass concentrations, it is convenient to express Eq. 2.9 in terms of mass extinction efficiency or extinction per unit mass $E(n, \chi, \lambda)$. Then,

$$b_{ext} = E(n, \chi, \lambda)m \tag{2.10}$$

where m refers to particle mass concentration and $E(n, \chi, \lambda) = 3Q_e(n_i, \chi, \lambda)/2\rho D$, in which ρ refers to particle density.

Some physio-chemical properties of particles typically found in the atmosphere are summarized in Table 2.2.

Fig. 2.11 shows mass scattering efficiencies as a function of single particle size for elemental carbon, iron, silica, and water. In all cases, $\lambda = 0.55\ \mu m$. For $D \ll \lambda$, E is proportional to D^3, and for $D \gg \lambda$, E is proportional to D^{-1}. On a per unit mass basis, it is clear that both large and small particles are very inefficient scatterers compared to aerosols with

TABLE 2.2   Values of Chemical Component Density and Refractive Index

Compound	Refractive Index	Density (g/cm³)
Ammonium sulfate	1.531	1.76
Ammonium nitrate	1.479	1.78
Sulfuric acid	1.408	1.8
Potassium nitrate	1.531	2.11
Organic carbon	1.55	1.4
Elemental carbon	1.96–0.66i	2.0
Silica	1.55	2.66
Iron	3.53–3.95i	7.86
Water	1.33	1.00

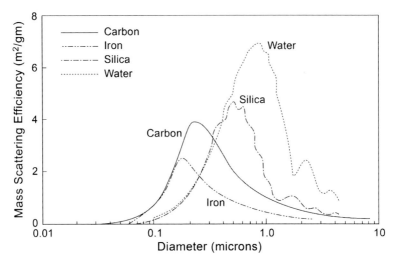

FIGURE 2.11 Calculated scattering cross section per unit mass at a wavelength of 0.55 μm for absorbing and nonabsorbing materials as a function of diameter for single-sized particles. The following refractive indexes and densities (g/cm³) were used: carbon: $n = 1.96-0.66i$, $\rho = 2$; iron: $n = 3.53-3.95i$, $\rho = 7.86$; silica: $n = 1.55$, $\rho = 2.66$; water: $n = 1.33$, $\rho = 1.0$.

sizes approximately 0.1–1.0 μm. The effects of absorption are to decrease the mass scattering efficiency and to shift it to smaller sizes compared to nonabsorbing particles. The density of a particle has an important effect on its mass scattering efficiency in that a high-density particle is smaller than a particle of the same mass that is less dense.

Fig. 2.12 is a plot of the scattering functions for air molecules, water, silica, ammonium sulfate, elemental carbon, and iron. The calculations for silica and iron were done using a diameter of 5.0 μm, since these types of particles are usually associated with windblown dust, while the diameter of the remaining particles was 0.3 μm. Elemental carbon and iron have indexes of refraction that are complex and were set equal to $n = 1.96-0.66i$ and $n = 3.53-3.95i$, respectively. These curves quantitatively show the scattering functions schematically presented in Figs. 2.6 and 2.9. Notice that the scattering function axes are logarithmic, covering five orders of magnitude. Rayleigh scattering, or scattering by air molecules, shows equal scattering in the forward and backward directions, while particles with a diameter of 5.0 μm scatter about 10,000 times more light in the forward than the backward direction, implying that there is almost no backscatter, and a 0.3 μm particle scatters about 100 times more light in the forward direction. Also, notice that for the 0.3 μm particles, the scattering at about $\beta = 40°$ is about the same, independent of particle type. However, for the 5.0 μm particles, scattering at $\beta = 40°$ is about 10 times less. In general, the scattering function is more sensitive to particle size than particle physiochemical characteristics.

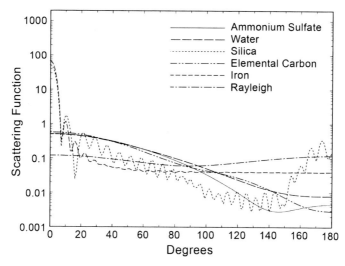

FIGURE 2.12 Modeled scattering functions for particles typically found in the atmosphere. Sulfates, water, and elemental carbon were assumed to have a 0.3 μm diameter each, while silica and iron scattering functions correspond to 5.0 μm.

The effects of forward- and backward-scattering characteristics of particles in this size range on visibility are exemplified in photographs shown in Fig. 2.13a and b. The extinction coefficient associated with the haze in both pictures is the same; however, in Fig. 2.13a, which corresponds to forward scattering, the haze is bright and white, while in Fig. 2.13b, which corresponds to back scattering, that same haze is dark and still devoid of color. The haze being achromatic, or lacking in color, implies that scattering at all wavelengths is approximately the same, which corresponds to larger particles.

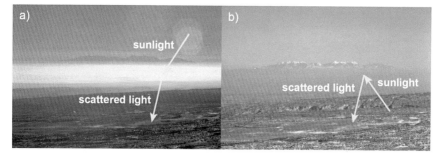

FIGURE 2.13 Photographs of a haze in front of the La Sal Mountains in southern Utah. The haze as represented by the extinction coefficient in both photos is the same, but the scattering angle in (a) corresponds to forward scattering, while in (b) the scattering is in the backward direction.

Typical Characteristics of Atmospheric Particulate Matter

Eqs. 2.4 and 2.5 and Figs. 2.11 and 2.12 refer to the scattering and absorption characteristics of single particles or particles of one size. However, in the ambient atmosphere, particles are distributed over many sizes and have a variety of physical and chemical properties. Fig. 2.14 shows images of typical particles found in the atmosphere, as well as examples of their size distributions expressed as the number, volume, and scattering cross section per increment of size (Whitby et al., 1972; Seinfeld and Pandis, 1998; Brasseur et al., 2003). The size range extends over four orders of magnitude starting at less than 0.01 μm, up to about 10.0 μm. Typically, combustion particulate emissions from manmade and natural sources fall in the 0.1–1.0 μm diameter size range, while mechanically produced particles such as dust, whether natural windblown or from mining and agriculture, tend to fall in the 1.0–10.0 μm diameter size range. Another particle classification system is primary versus secondary. Primary particles are those that are emitted into the atmosphere as particles, such

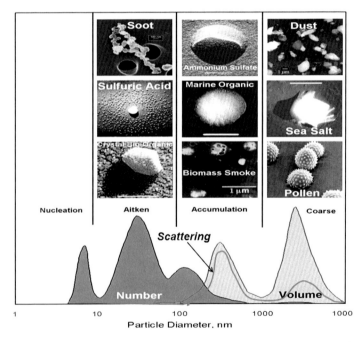

FIGURE 2.14   Scanning electron microscope pictures of typical particles in the nucleation, accumulation, and coarse modes. Also shown are representative number, scattering, and volume distributions as a function of size.

as organic and soot particles in smoke plumes or dust particles. Secondary particles are those that are formed from gas-to-particle conversion in the atmosphere, such as sulfates (from SO_2), nitrates (from NOx), and secondary organics, all of which are smaller than 1.0 μm in diameter (Seinfeld and Pandis, 1998).

With respect to particle size, particles tend to occur in three modes: the nuclei mode (0.005–0.1 μm diameter), accumulation mode (0.1 to 1–3 μm), and coarse mode (1–3 to 50–100 μm).

The nuclei mode dominates with respect to numbers of particles but contributes very little to particle mass or scattering because the particles are so small. This mode forms from the condensation and coagulation of hot, supersaturated vapors during combustion and from homogeneous nucleation of secondary aerosols (nucleation that forms new particles). The accumulation mode predominates with respect to the scattering (visibility) cross section and contributes substantially to aerosol mass. This mode forms from two processes: coagulation of smaller particles and heterogeneous nucleation of secondary particles (condensation of one material on another, e.g., on existing particles). Because of the decrease in particle numbers and surface area at the upper size range of the accumulation mode, particles in this mode do not tend to grow significantly into the coarse mode.

Notice that in spite of the fact that the coarse mode has very few particles relative to the condensation mode, it has more volume/mass than any other mode. However, even though more mass is found in the coarse mode, its scattering cross section is less than the fine mode. It is for this reason that most visibility impairment tends to be associated with fine particles or those particles found in the 0.1–1.0 μm range. Visibility impairment by coarse-mode particles can be significant only if the concentration of coarse-mode particles dominates the fine mode, such as during major dust storm events or in remote desert areas during periods with nearly zero fine-mode particle concentrations.

Another important aspect of aerosol air quality is chemical composition, which, as one would expect from the above discussions, bears a strong interrelationship to particle size. Although it is possible to measure particle mass concentration in multiple size ranges, typically, samples are collected in the size ranges of 0.0–2.5 μm ($PM_{2.5}$, fine mode) and 2.5–10.0 μm (PM_{10}, coarse mode). In almost all cases, dry fine particle mass is dominated by just five types of chemical species: sulfates (typically with ammonium and/or hydrogen cations), organics, ammonium nitrate, soil dust (from the lower tail of the coarse mode), and elemental carbon. In some coastal areas, sea salt is a sixth aerosol component contributing to fine mass in a significant way. Most routine monitoring programs collect 24-h samples on various filter substrates, consistent with the type of analytic procedure used to extract the chemical species of interest. The two largest contributors are usually sulfates and organics, which together typically account for about 60–80% of average dry fine particle mass (Malm et al., 2004).

Dry fine mass, or $PM_{2.5}$, is typically computed using Eqs. 2.11 and 2.12:

$$PM_{2.5} = NH_4 + SO_4 + NO_3 + POM + LAC + Soil + Sea\,salt \qquad (2.11)$$

$$Soil = 2.2\,Al + 2.49\,Si + 1.94\,Ti + 1.63\,Ca + 2.42\,Fe \qquad (2.12)$$

where NH_4, SO_4, and NO_3 refer to the mass concentrations of the ammonium, sulfate, and nitrate ions, respectively. Organic carbon is included as particulate organic material (POM), computed by multiplying organic carbon (OC) concentrations by a molecular weight per carbon weight ratio (POM = R_{oc}·OC). Light absorbing carbon is referred to as LAC. The term LAC is used here because it is more representative of the optical properties of light absorbing carbon rather than elemental or black carbon, although these terms are often used interchangeably in the literature. Fine soil concentrations include the contributions from assumed forms of soil-dust-related elements (Eq. 2.12) (Malm et al., 1994). Sea salt is estimated as 1.8 × Cl. Mass concentrations are given in units of $\mu g/m^3$.

Under ambient conditions, aerosol water constitutes a very important sixth component of fine particle mass. At high relative humidities (e.g., above 80%), water often contributes the majority of ambient fine aerosol mass. Most of this water is drawn into the aerosol phase by the hygroscopic properties of sulfates, nitrates, and some organics. Reducing the atmospheric concentrations of these hygroscopic species would reduce the concentrations of ambient aerosol water.

Coarse mass tends to be dominated by soil dust (e.g., oxides and other salts of Si, Al, Fe, Ca, etc.). Other, usually lesser, contributions to coarse mass come from sea spray (Na, Cl, etc.), plant particles (organics), reactions of gaseous nitric acid with soil dust or sea salt particles (nitrates), and the upper tail of the accumulation mode particles (e.g., sulfates and organics).

Given a number size distribution such as shown in Fig. 2.14, the extinction coefficient associated with species i can be estimated by

$$b_{ext,i} = \frac{\pi}{4} \int_0^\infty D^2 Q_e(n_i, \chi) N_i(D) d(D) \qquad (2.13)$$

where Q_e (typically determined by Mie calculations) is the particle extinction efficiency; $N_i(D)$ is the aerosol number size distribution of the ith species; and D, n_i, and χ are the particle diameter, complex index of refraction, and size parameter ($\pi D/\lambda$), respectively. Since many particle measurement programs are designed to measure mass size distribution, it is convenient to rewrite Eq. 2.5 as

$$b_{ext,i} = \int_{-\infty}^{\infty} E_e(n_i, x, \lambda) f_i(x) dx \qquad (2.14)$$

where E_e is mass extinction efficiency, $f_i(x)$ is the aerosol mass distribution dm/dx of the ith species, $x = \ln[D/Do]$, and λ is the wavelength.

Eq. 2.6 can be rewritten as

$$b_{ext,i} = \alpha_i m_i \qquad (2.15)$$

where

$$\propto_i = \int_0^{\infty} E_e(n_i, x, \lambda) \overline{f_i(x)} dx. \qquad (2.16)$$

$\propto_i$ is the integrated mass extinction efficiency and the function $\overline{f_i(x)}$ is the normalized mass distribution given by $\overline{f_i} = f_i/m_i$, where m_i is the total mass concentration of the ith species. Thus, the total particle extinction associated with all particle types is simply

$$b_{ext} = \sum_i \alpha_i m_i. \qquad (2.17)$$

Any one of the size distribution modes shown in Fig. 2.14 is typically represented by a lognormal size distribution such as

$$\overline{f_i}(\ln D) = \frac{1}{\sqrt{2\pi}\ln(\sigma_g)} \exp\left[-\frac{\left(\ln D - \ln D_g\right)^2}{2 \ln^2 \sigma_g} \right] \qquad (2.18)$$

where D_g is the mass mean diameter and σ_g is the geometric standard deviation.

Fig. 2.15 shows an example of scattering, absorption, and extinction efficiencies as a function of mass mean diameter for a hypothetical, but typical, absorbing aerosol with a complex index of refraction $n = 2.0 + 0.5i$, particle density $\rho = 1.5$ g/cm^3, and wavelength $\lambda = 0.55$ μm. Also shown is the scattering albedo $\omega = b_s/b_{ext}$. The aerosols were assumed to be lognormally distributed with a geometric standard deviation of $\sigma_g = 2.0$. Notice that at mass mean diameters $D_g > 0.2$ μm, scattering and absorption efficiencies are nearly equal, while for $D_g < 0.1$ μm, aerosol absorption dominates and approaches a constant value for $D_g \ll \lambda$. Therefore, for small absorbing aerosols, extinction efficiencies become independent of size. The scattering efficiency reaches a maximum of approximately 3–4 m^2/gm for a mass mean diameter of approximately 0.2 μm. The extinction efficiency at 0.2 μm is approximately 10 m^2/gm. This is typical for most naturally occurring aerosols with commonly found size distributions.

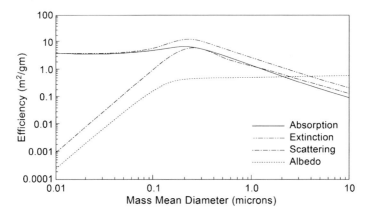

FIGURE 2.15   Calculated scattering, absorption, albedo, and extinction efficiencies per unit mass at a wavelength of 0.55 μm for absorbing spheres as a function of mass mean diameter. A lognormal particle size distribution was assumed with geometric standard deviation $\sigma_g = 2.0$, refractive index $n = 2.0 + 0.5i$, and density $\rho = 1.5$ g/cm³.

As indicated above and illustrated in Fig. 2.13, the scattering function is also critical to the appearance of landscape features as the sun–scene–observer geometry varies. Fig. 2.16 shows typical size-weighted volume scattering functions for mass size distributions of $D_g = 5.0$ μm and $\sigma_g = 2.0$ for dust, $D_g = 0.3$ μm and $\sigma_g = 2.0$ for ammonium sulfate, and two mass

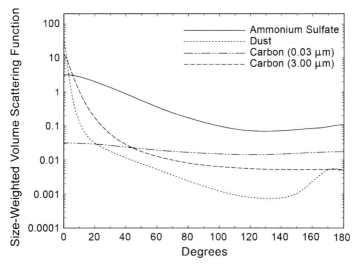

FIGURE 2.16   Calculated size-weighted volume scattering functions at a wavelength of 0.55 μm for sulfate, dust, and two carbon particle lognormal mass size distributions. The following refractive indexes, diameters, geometric standard deviations, and mean mass diameters were used: sulfate: $n = 1.53$, $\rho = 1.76$ g/cm³, $\sigma_g = 2.0$, $D_g = 0.3$ μm; dust: $n = 1.55–0.01i$, $\rho = 1.66$ g/cm³, $\sigma_g = 2.0$, $D_g = 5.0$ μm; and carbon: $n = 1.96–0.66i$, $\rho = 2.0$ g/cm³, $\sigma_g = 2.0$, $D_g = 0.03$ μm, $D_g = 3.0$ μm.

size distributions for elemental carbon of $D_g = 0.03$ μm and $D_g = 3.0$ μm with $\sigma_g = 2.0$. As indicated in Fig. 2.16, large dust particles scatter radiant energy in the forward direction many times more efficiently than particles that are less than 1.0 μm in size; however, only in the extreme forward direction do they scatter more light than a typical particle, such as sulfate, in the accumulation mode. This can be seen as a substantial brightening of the sky very near the sun when viewed during a dust-dominated pollution event. For scattering angles greater than about 40°, fine particle exceeds coarse particle scattering by about a factor of 100. Notice that small absorbing particles scatter in all directions nearly equally. The minimum light scattering for all particles is somewhere between 90° and 140°. Therefore visibility impairment will always be the least noticeable during afternoon hours and be most prevalent during early morning and late afternoon hours.

REAL-WORLD ESTIMATES OF MASS SCATTERING EFFICIENCIES

In most instances, particle scattering and absorption are primarily responsible for visibility reduction. As discussed above, particle scattering and absorption properties can, with a number of limiting assumptions, be calculated using the Mie theory. However, before such calculations are carried out, the distribution of the materials that make up the aerosol must be known or assumed. Typically, particle models assume

- external mixtures—particles exist in the atmosphere as pure chemical species that are mixed without interaction;
- multicomponent aerosols—single particles are made up of two or more species.

If the chemical species are combined in fixed proportions independent of particle size, the aerosol is referred to as internally mixed. Another typical multicomponent aerosol model assumes a solid core encased by a deposited shell of varied thickness and composition.

Aerosol models usually assume that aerosol species are mixed externally; the simplest example is a particle composed of a single chemical species such as ammonium sulfate, and for this case, extinction efficiency is referred to as "mass extinction efficiency." However, realistically, particles in the atmosphere comprise a variety of inorganic and organic species. These types of particles are referred to as internal mixtures, and their extinction efficiencies are termed "specific mass extinction efficiencies." Internally mixed particles can also be externally mixed from other particle populations; the most obvious case would be internally mixed, fine-mode aerosols externally mixed from coarse-mode aerosols.

Eq. 2.17, derived for externally mixed aerosols, also holds for an internally mixed aerosol in which the chemical species are mixed in fixed proportions to each other, the index of refraction is not a function of composition or size, and the aerosol density is independent of volume. When computing total extinction using the Mie theory, the microscopic structure of the aerosol (i.e., the extent of internal or external mixing) is found to be relatively unimportant so that the assumption of internally versus externally mixed particles does not have much impact on the predicted results. The mass extinction (or scattering) efficiency is a stronger function of density and size than optical properties such as refractive index (Ouimette and Flagan, 1982; Sloane, 1986; Malm and Kreidenweis, 1997).

AEROSOL WATER

The issue of water uptake by hygroscopic species must be addressed. Implicit in the use of Eq. 2.17 is an assumed linear relationship between aerosol mass and extinction. It is well known that sulfates and other hygroscopic species form solution droplets that increase in size as a function of relative humidity (*RH*) (Seinfeld and Pandis, 1998). Therefore, if scattering is measured at various *RH*s, the relationship between measured scattering and hygroscopic species mass can be quite nonlinear. A number of authors have attempted to linearize the model, in an empirical way, by multiplying the hygroscopic species by something like $1/(1 - RH)$ to account for the presence of water mass (White and Roberts, 1977; Malm et al., 1986). However, Malm et al. (1989) and Gebhart and Malm (1990) proposed a different approach. The scattering coefficient at some *RH* is

$$b_{sp}(RH) = b_{sp}(RH=0)\frac{b_{sp}(RH)}{b_{sp}(RH=0)} \tag{2.19}$$

where $b_{sp}(RH = 0) = \alpha_{i,d}M_i$, and $f(RH) = \dfrac{b_{sp}(RH)}{b_{sp}(RH=0)}$.

Therefore

$$b_{sp,i}(RH) = \alpha_{i,d}M_i f_i(RH) \tag{2.20}$$

where $\alpha_{i,d}$ and $_dM_i$ are the mass scattering coefficient and dry mass, respectively, of the *i*th hygroscopic species. $f_i(RH)$ is the *RH* enhancement factor (Tang et al., 1981). $f(RH)$ is calculated on a sampling-period-by-sampling-period basis using Mie theory and an assumed size distribution and laboratory or modeled aerosol growth curves associated with the species of interest.

Examples of particle growth as a function of RH are shown in Fig. 2.17 for ammonium sulfate and bisulfate. D/D_o refers to the ratio of the particle diameter after it has absorbed water, divided by the diameter of the dry particle. Typically, inorganic salts take up and release water abruptly at RH values referred to as the deliquescent and crystallization points. For instance, upon exposing a dry ammonium sulfate particle to increased RH, it will not take up water until about $RH = 80\%$. At that point it spontaneously absorbs and grows by a factor of about 1.5. It then continues to grow in size as shown in Fig. 2.17. If the particle RH environment is then reduced, particle size will decrease continuously until the crystallization point, at which the salt will spontaneously give up all its absorbed water and return to its crystallization state. For ammonium sulfate this occurs at about $RH = 37\%$. In many cases, once the particle has absorbed water, it will not crystallize because of impurities in the aerosol mixture but will continue to lose water on a continuous basis as the RH is reduced. When water is retained below the crystallization point, the aerosol is said to be in a metastable state. It is this metastable D/D_o curve that is shown for both ammonium sulfate and bisulfate in Fig. 2.17.

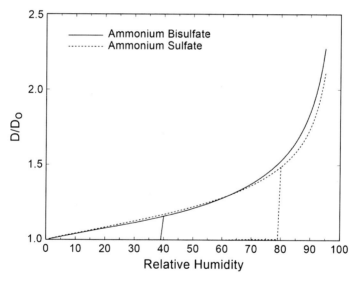

FIGURE 2.17   Deliquescent and metastable D/D_o curves for ammonium sulfate and bisulfate.

In Fig. 2.18, two $f(RH)$ curves were calculated by employing the metastable curves shown in Fig. 2.17, assuming $D_g = 0.2$ μm and 0.5 μm and $\sigma_g = 1.7$ for ammonium bisulfate, and one $f(RH)$ curve was calculated for ammonium sulfate, assuming $D_g = 0.2$ μm and $\sigma_g = 1.7$. First, notice the significant difference between the $f(RH)$ curves for ammonium bisulfate with $D_g = 0.2$ μm and 0.5 μm. At $RH = 90\%$, the difference between the two curves is nearly a factor of 2, implying that the same particle species will scatter twice as much light if its dry size distribution starts with a mass mean diameter of 0.2 μm rather than 0.5 μm. This, of course, is because the aerosol with the smaller particle size distribution grows into a size that is more efficient at scattering light, compared to the aerosol with larger initial dry size. Also notice that, given the same size distribution, the ammonium sulfate $f(RH)$ curve is less than or lower than its bisulfate counterpart. In general, more acidic particles tend to have $f(RH)$ enhancement factors that are greater than for more neutralized particles.

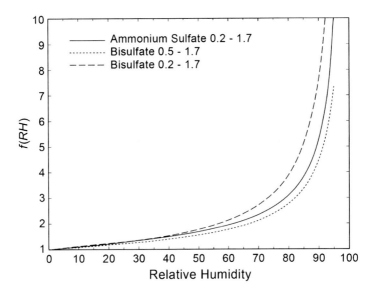

FIGURE 2.18   Metastable $f(RH)$ curves for ammonium bisulfate and sulfate, assuming two different size distributions. $D_g = 0.2$ μm and $\sigma_g = 1.7$ for ammonium bisulfate and sulfate and $D_g = 0.5$ μm and $\sigma_g = 1.7$ for ammonium bisulfate.

The contributions of light scattering and absorption by particles to atmospheric extinction can be estimated from measurements of species mass concentrations using

$$b_{ext} = \alpha_{AS}[(NH_4)_xSO_4]f_{AS}(RH) + \alpha_{AN}[NH_4NO_3]f_{AN}(RH)$$
$$+ \alpha_{POM}[POM]f_{POM}(RH) + \alpha_{soil}[Soil] + \alpha_{sea\ salt}[sea\ salt]f_{sea\ salt}(RH) \quad (2.21)$$
$$+ \alpha_{CM}[CM] + \alpha_{LAC}[LAC].$$

This formulation assumes that contributions to total ambient light scattering are from ammoniated sulfate $[(NH_4)_xSO_4]$, ammonium nitrate, particulate organic carbon (POM), soil, sea salt, and coarse mass (CM). Each species has a corresponding dry mass scattering efficiency α (or absorption efficiency in the case of LAC). The terms in the brackets correspond to mass concentrations of species in $\mu g/m^3$. The effects of the uptake of water by hygroscopic species are estimated by the growth factor $f(RH)$ as a function of relative humidity.

Eq. 2.21 gives a reasonable approximation of the ambient atmospheric extinction coefficient. By its form, Eq. 2.21 suggests that the total extinction coefficient can be budgeted to the specific aerosol components that caused it. However apportioning extinction to individual species in this way is conceptually problematic and has become known as the apportionment problem.

THE APPORTIONMENT PROBLEM

Assigning a fraction of extinction to a species is problematic if the particles are internally mixed. A simple example illustrating the problem is shown in Fig. 2.19 (Malm and Kreidenweis, 1997). The insensitivity of the total computed scattering to the mixing assumption is presented in Fig. 2.19a. Two aerosol species with the same total volume, size distribution, and index of refraction, but with densities differing by a factor of 2, are externally mixed. They have mass scattering efficiencies of 3 and 6 m^2/g, respectively. The externally mixed aerosol contains 10 $\mu g/m^3$ of species 1 and 5 $\mu g/m^3$ of species 2, with each species contributing 30 Mm^{-1} to the total aerosol scattering of 60 Mm^{-1}.

If species 1 and species 2 are assumed to be internally mixed (see the lower portion of Fig. 2.19a), the specific mass scattering efficiency for the mixed particle is now required for calculation of the total aerosol scattering. The specific mass scattering efficiency can be computed as the mass-weighted average of the efficiencies of the individual species (4 m^2/g), assuming several conditions are met. The optical properties of the mixture must be similar to those for the individual species (which must also be similar; this condition is met, for example, by a mixture of ammonium sulfate and organics, with real refractive indexes of 1.53 and 1.5, respectively). Volume conservation must be invoked, and the size distribution of the internal mixture must be similar to the size distributions of the external mixtures. By applying the specific mass scattering efficiency, a total aerosol b_{sp} for the sample is again computed as 60 Mm^{-1} (see Fig. 2.19a).

However, the apportionment of scattering to species 1 and 2 is different under these two models. Using the externally mixed model, species 1 and 2 each contributes 50% of the total scattering. But in the internally mixed case, species 1 contributes 66% of the scattering budget (because it makes up 66% of the total mass), while species 2 contributes 33%. This discrepancy arises because when the species are internally mixed, the same scattering efficiency is implicitly assigned to both species,

		a		b	
		species 1	species 2	species 1	species 2
mass	external case	10 µg/m³	5 µg/m³	10 µg/m³	species 2 removed
aerosol type	external case	◯	●	◯	species 2 removed
species density	external case	2 g/cm³	1 g/cm³	2 g/cm³	species 2 removed
specific mass scattering efficiency	external case	3 m²/g	6 m²/g	3 m²/g	species 2 removed
particle scattering	external case	30 Mm⁻¹	30 Mm⁻¹	30 Mm⁻¹	species 2 removed
total scattering		60 Mm⁻¹		30 Mm⁻¹	
mass	internal case	15 µg/m³		10 µg/m³	
aerosol type	internal case	◯		◯	
species density	internal case	1.5 g/cm³		2 g/cm³	
specific mass scattering efficiency	internal case	4 m²/g		2 m²/g	
total scattering		60 Mm⁻¹		20 Mm⁻¹	

FIGURE 2.19　(a) Contrasting mass scattering efficiencies for internally versus externally mixed particles. (b) Partial scattering efficiencies for internally and externally mixed particles.

although independently they have different efficiencies. This discrepancy can only be resolved if the specific mass scattering efficiency of the mixed aerosol is prorated to its chemical constituents, based on their relative densities. The apportionment of scattering to specific species is therefore independent of whether an externally or internally mixed model is assumed.

This simple example demonstrates that the contribution of each species to total extinction is sensitive to the assumed microscopic structure of the aerosol. This issue has become known as the apportionment problem, distinct from the problem of computing total extinction from estimates of size-dependent chemical compositions that was demonstrated to be similar for the two cases in Fig. 2.19a. For the externally mixed case, the total extinction for the mixture is calculated by computing the contribution of each species separately, and the species contribution to the total is well defined if we know the size distribution of each species. For the internally mixed case, it may be misleading to assign a percentage of the total extinction to each species on a mass-weighted basis, as demonstrated in Fig. 2.19a.

Issues also arise when apportioning scattering to species that have different hygroscopic properties. Consider an external mixture of hygroscopic and nonhygroscopic aerosol species with similar masses. The mass scattering efficiencies of the dry nonhygroscopic and hygroscopic species are each 3 m^2/g, while that of the wet hygroscopic species is 8 m^2/g because of associated water. When internally mixed, the wet specific mass scattering efficiency will be 3–8 m^2/g, depending on the hygroscopic properties of the mixed aerosol. Several authors discuss aerosol growth characteristics as a function of RH and have successfully predicted the scattering characteristics of the mixed particles (Sloane, 1983, 1986; Anderson et al., 1994; Malm and Pitchford, 1997; Malm et al., 2000, 2003, 2009). However, the apportionment of scattering to a particular species is still problematic. Typically, one growth curve is developed for the mixed species, and scattering due to water is apportioned proportionally among all aerosol components, even though only one of the species may be hygroscopic. An example would be a mixture of ammonium sulfate and a weakly hygroscopic organic species.

REMOVAL OF A SPECIES FROM THE ATMOSPHERE

From a regulatory standpoint, the change in total scattering as a function of the removal or addition of a species to the atmosphere may be of interest. Estimating the change in extinction due to the removal or addition of a single species is different from calculating each species' contribution to extinction and then summing them to yield total atmospheric extinction. Taking an ambient mix of particles and removing a fraction of one species or another can change the atmospheric chemistry and result in varying the relative composition of multicomponent aerosols as well as particle size.

Calculating a mass extinction efficiency resulting from the removal of a chemical species or mixture of species is referred to as a "partial mass extinction efficiency" ($\alpha_{ext,part}$):

$$\alpha_{ext,part} = \left(\frac{\partial b_{ext}}{\partial m_j} \right). \tag{2.22}$$

Total extinction can be computed for the mixture, and for the case when one species has been removed from the mixture, the difference between these values provides the contribution of that species to total extinction. As an example, we compute the partial scattering efficiency for species 2 under the assumptions of externally mixed particles. The partial scattering efficiency is obtained by computing the total aerosol scattering assuming only species 1 is present (30 Mm^{-1}; see Fig. 2.19b). Subtracting this value from the total scattering obtained for the mixture (60 Mm^{-1}; see Fig. 2.19a), we deduce that species 2 contributed 30 Mm^{-1} (half) to the mixed aerosol scattering. The derived partial scattering efficiency of 6 m^2/g for species 2 [$\Delta bsp/\Delta mass$ = (60 − 30 Mm^{-1})/(5 μg/m^3)] is the same as its mass scattering efficiency from Fig. 2.19a.

Investigating the same scenario but for the internally mixed case (see the lower part Fig. 2.19b), we assume that the removal of one species from the internal mixture would conserve particle number concentrations but reduce particle size. This size reduction alters the specific scattering efficiency of the mixture, which is now assumed to be 2 m^2/g. The total aerosol scattering is thus 20 Mm^{-1}, which leads to the conclusion that the partial scattering efficiency of species 2 is 8 m^2/g [$\Delta bsp/\Delta mass$ = (60 − 20 Mm^{-1})/(5 μg/m^3)].

White (1990) discusses some real-world particle mixing possibilities:

- Illustrated in Fig. 2.20a, the removal of species 2 permits additional formation of species 1. The partial scattering efficiencies in this case can be small. An example would be the coexistence of sulfates and nitrates in which sulfuric acid takes the cation from ammonium nitrate and liberates nitric acid to the gas phase. The effect of such competition is to attenuate the benefits of sulfate control in nitrate-rich environments (Pilinis, 1989).
- Species 2 may be an active nucleus that depresses the equilibrium vapor pressure of species 1. Under unsaturated conditions, the partial removal of species 2 then eliminates species 1 in the same proportion. In this manner, the scattering efficiencies of hygroscopic species such as sulfuric acid and its ammonium salts are inflated by water. That is, the scattering decrements to be expected from sulfur controls are leveraged by reduced liquid water concentrations (Garland, 1969; Covert et al., 1972) (see Fig. 2.20b).
- The equilibrium vapor pressure of species 1 may be so low that its particulate-phase abundance is limited by gas-phase production rather than the availability of nuclei. The scattering efficiency of species 2 is then diminished because the partial removal of species 2 concentrates the condensation of species 1 on fewer nuclei, increasing the mean particle size of the remaining aerosol. The gas-phase production of sulfuric acid in hazy air probably illustrates this phenomenon. Under such conditions, increasingly tight controls on primary emissions may have little impact on total scattering (McMurry and Friedlander, 1979; White and Husar, 1980) (see Fig. 2.20c).
- Species 2 may physically coat the more-volatile species 1. Under unsaturated conditions, contributions from the imprisoned species then augment the scattering efficiency of the coating. Husar and Shu (1975) and Chang and Hill (1980) presented electron micrograph evidence suggesting that pollutant films may retard the evaporation of coastal fogs in the Los Angeles basin, and Rubel and Gentry (1984) directly demonstrated stabilization of water droplets by partially

evaporated organic coatings in the laboratory. The mass of the coating is a tiny fraction of the total, and control of the species involved could thus yield vastly disproportionate reductions in scattering (see Fig. 2.20d).

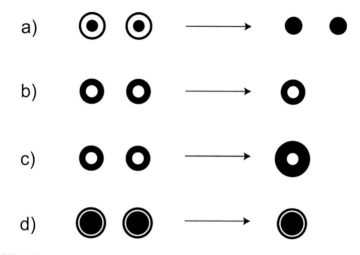

FIGURE 2.20   Response of an aerosol to the removal of a constituent. (a) Species mutually antagonistic in the particulate phase; (b) condensate and nuclei, nuclei limited; (c) condensate and nuclei, condensate limited; and (d) volatile species coated by nonvolatile species.

These examples suggest that the scattering decrement produced by a given control strategy depends not only on the microstructure of the particulate phase but also on the relationship of the particulate phase to the gas phase.

Another example of nonlinearity associated with Eq. 2.22 is highlighted in a paper by Nguyen et al. (2015). They demonstrate that reductions in atmospheric sulfate concentrations led to reductions in aerosol liquid water because of the hygroscopic nature of sulfate salts, which in turn resulted in less formation of biogenic secondary organic aerosol (SOA). Here is an example of where control of anthropogenic emissions, in this case sulfur dioxide, results in an improvement in visibility that is greater than what might have been expected.

Fig. 2.21 exemplifies the discussion above. The red and blue graphs represent temporal plots of ammonium sulfate and *POM* mass concentrations at Great Smoky Mountains National Park (GRSM). Over the course of 10 years, sulfates have decreased by about a factor of 3.5 because of sulfur dioxide emission reductions, while there has been a commensurate decrease in *POM* of about a factor of 1.7. Most *POM* at GRSM is secondary

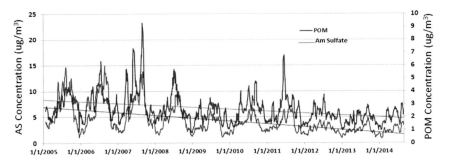

FIGURE 2.21   Temporal plots of ammonium sulfate and *POM* at Great Smoky Mountains National Park along with linear trend lines for the two particle species.

biogenic organic material, and its precursor, biogenic volatile organic compound gas emissions, has not decreased. So it appears that the mechanism for *POM* reduction may be linked to the reduction of aerosol liquid water associated with hygroscopic sulfates.

WHAT DO WE KNOW ABOUT REAL-WORLD MASS EXTINCTION EFFICIENCIES?

Methods for Deriving Mass Scattering Efficiencies

The methods typically used to estimate mass scattering efficiencies depend on the type of measurements and data available. In this section, the term "mass scattering efficiency" will be used to refer to estimates that correspond to a single aerosol component such as dry ammonium sulfate, and although one could argue that species such as POM and soil correspond to a combination of chemical components, we will use this term to correspond to those efficiencies as well. The term "specific mass scattering efficiency" is used to refer to aerosols composed of more than one species, for example, a "fine-mode specific mass scattering efficiency." This term will also be used to refer to aerosols that include water mass; for example, if an efficiency for an inorganic salt particle with associated water mass is reported, we refer to it as a "specific mass scattering efficiency."

The simplest method for computing efficiencies is with measured mass concentrations from filter samples and measured light scattering coefficients from nephelometry. For a particle population, a specific mass scattering efficiency α_{sp_spec} can be defined as the ratio of a light-scattering coefficient corresponding to the mass concentration (M) of that population:

$$\alpha_{sp\_spec} = \frac{b_{sp}}{M}. \tag{2.23}$$

The average specific mass scattering (or absorption) efficiency can be estimated by dividing the average scattering (absorption) coefficient by the average mass concentration for a given set of measurements, or a linear regression can be performed on the data and the slope of the linear fit can be interpreted as the specific mass scattering efficiency. A specific mass scattering efficiency derived in this manner represents an average particle that could be changing due to variations in RH, size distribution, and composition during the sampling period, as, for example, would be the case when RH changes the size and composition of particles composed of hygroscopic materials. Mass concentrations used in this method could be either gravimetric mass or the sum of chemically analyzed masses or be computed from integrated volume distributions using an average assumed density.

A number of investigators have taken advantage of the form of Eq. 2.17 ($b_{ext} = \sum_i \alpha_i M_i$) to construct a multilinear regression (MLR) model with b_{ext} as the independent variable and the measured aerosol mass concentrations for each species $i(M_i)$ as the dependent variables (White and Roberts, 1977). The regression coefficients are then interpreted as mass extinction (or scattering or absorption) efficiencies. The assumptions required in this formulation are that all the components contributing to extinction are included, Eq. 2.17 is a reasonable approximation of the relationship between extinction and the various particle species, the number of samples is large enough to give stable results, and the concentrations of the species are uncorrelated.

Regression-derived efficiencies are vulnerable to a variety of systematic and random errors. At the most basic level, there is the issue that meteorologically driven fluctuations of the existing particle concentrations may cause particle mass concentrations of different species to be collinear. The high degree of collinearity commonly found among particle species can make regression results very sensitive to the choice of species to be included in the analysis. A more practical difficulty is that regression results can be biased by random uncertainties in the measurements. Standard regression models tend to overpredict the coefficients of species that have lower uncertainties (e.g., sulfates) and underpredict the coefficients of species with larger measurement uncertainties (e.g., organics) (White and Macias, 1987).

The theoretical approach described at the beginning of this section (Eq. 2.10) is often used to compute mass scattering efficiencies from mass (or number) distributions of particle species from impactor measurements or mobility or optical measurements. However, this method requires assumptions of the measured species' chemical form and mixing properties of the particles, as well as its optical properties. Estimates from this method often assume an externally mixed particle. However, if the size distribution measurement is obtained by mobility or optical measurements (number size distribution), an internally mixed particle is assumed, but perhaps for separate size modes. This method has been shown to accurately reconstruct light scattering. For example, measurements of size distributions have been incorporated with estimates of particle optical properties to reconstruct measured scattering (and estimate specific mass scattering efficiencies) on fine time scales (Hand et al., 2002; Malm et al., 2005; McMeeking et al., 2005). These estimates have the added advantage of allowing for the investigation of the functional dependence of efficiencies on particle size, composition, and humidity.

The theoretical method provides the most accurate estimate of efficiencies of those discussed in this section. A disadvantage to this method is that size distribution measurements are more analytically intensive and not often available for routine monitoring, so computing mass scattering efficiencies by this method is not routinely possible.

Estimating mass scattering efficiencies as the change in b_{ext} resulting from the removal (or addition) of a single species is different from assigning a fraction of measured extinction to a chemical species or a mixture of species. The change in extinction resulting from a change in particle species concentration was described previously. As species are added to or removed from a particle, the efficiency is dependent on change in composition as well as on change in particle size. Computing partial mass scattering efficiencies in this context requires a model to account for the effects of removal of mass on light scattering coefficients, such as the Elastic Light Scattering Interactive Efficiencies (ELSIE) model (Sloane, 1983; Sloane et al., 1991; Lowenthal et al., 1995; Omar et al., 1999). Among the assumptions in the ELSIE model are those associated with removing particle mass. Mass can be removed or added by assuming that the particle size changes but not the particle number or that the particle number changes but not the particle size (Lowenthal et al., 1995). Unless the option is chosen to keep the size constant, the efficiencies derived from this method cannot be compared to efficiencies from other methods. Also, assumptions must be made regarding how to treat water mass associated with a removed species. Although this technique is useful in a regulatory context, in practice it is not possible to measure the change in light scattering due to a change in mass concentration, so more empirical approaches were adopted to estimate mass scattering efficiencies.

Summary of Estimated Mass Extinction Efficiencies

A survey of mass scattering efficiencies from ground-based measurements reported since 1990 was presented by Hand and Malm (2007). The mass scattering values are summarized for different aerosol species and by the four methods used to derive them as discussed above. All four methods were averaged for mixed fine-mode, coarse-mode, and total-mode specific mass scattering efficiencies under both dry and ambient conditions, depending on what is reported. For species-dependent mass scattering efficiencies, only three methods are averaged (the measurement method is not included) and correspond to dry conditions ($RH \approx 0\%$).

The estimates of α_{sp} from the various methods are also separated into geographical regions (urban, remote/rural continental, and ocean/marine) and averaged as a function of species. The study site location was used to determine the geographical region; however, it is impossible to determine the degree of aging and the length and time of transport the aerosols underwent that were measured at these locations, so these distinctions are approximate. Also, remote/rural and ocean regions may be associated with both pristine and polluted (or a combination of) conditions. For example, it is possible that aerosols measured in remote continental locations were influenced by urban sources or regional hazes, and measurements performed on board cruise ships were influenced by continental outflow.

A summary of the survey of α_{sp} as a function of aerosol species for the four methods is reported in Table 2.3, and average values of α_{sp} for each aerosol species as a function of region are summarized in Table 2.4. In Tables 2.3 and 2.4, the number of observations in each average is noted in parentheses. For the references to the various observations, refer to Hand and Malm (2007).

In Table 2.4, the value listed first is the average α_{sp} from all the methods, followed by α_{sp} corresponding to an individual method, noted with T, MLR, M, or P for the theoretical, multilinear regression, measurement, or partial scattering method, respectively. The number of observations can vary considerably for each geographical region, and estimates for a given region can be dominated by values derived by a single method. The number of study locations and range of periods varies; obviously, the same sites and measurement time periods are not available for every method.

Dry, fine-mode ammonium sulfate mass scattering efficiencies are surprisingly similar when compared over method and region. The overall grand average of 93 observations is 2.5 ± 0.6 m²/g where the variability is one standard deviation of the mean. This fairly low variability is notable given the variation in seasons, locations, and different experimental and analysis methods used to derive these values, not to mention differences in the aerosol microphysical properties. This narrow range in values may reflect the high level of understanding of the optical and hygroscopic properties of sulfate aerosols, as well as the lower uncertainties associated with sulfate measurements.

Some differences were observed in dry ammonium sulfate mass scattering efficiencies as a function of location. Estimates in ocean regions were somewhat lower (2 m²/g) compared to remote/rural continental (2.7 m²/g) or urban locations (2.6 m²/g), and lower estimates (as low as 2.0 m²/g) were also observed in fairly clean and arid locations (e.g., the

TABLE 2.3   Average and One Standard Deviation of Mass Scattering Efficiencies of Fine, Coarse, and Total Correspond to the Size-Range Mode of the Aerosol

Species/Mode	Theoretical (m²/g)	Measurement (m²/g)	MLR (m²/g)	Partial (m²/g)	All Methods (m²/g)
Fine mixed	4.3 ± 0.7 (26)	3.4 ± 1.2 (54)	3.1 ± 1.4 (16)	3.4 ± 1.6 (2)	3.6 ± 1.2 (98)
Coarse mixed	1.6 ± 1.0 (21)	0.40 ± 0.15 (4)	0.7 ± 0.4 (26)		1.0 ± 0.9 (51)
Total mixed	2.2 ± 1.0 (9)	1.7 ± 1.1 (11)			1.9 ± 1.1 (20)
Fine sulfate	2.1 ± 0.7 (34)		2.8 ± 0.5 (53)	2.2 ± 0.7 (6)	2.5 ± 0.6 (93)
Fine nitrate			2.8 ± 0.5 (42)	2.3 ± 0.5 (6)	2.7 ± 0.5 (48)
Fine POM	5.6 ± 1.5 (19)		3.1 ± 0.8 (39)		3.9 ± 1.5 (58)
Coarse POM	2.6 ± 1.1 (19)				2.6 ± 1.1 (19)
Total POM	3.8 ± 0.5 (7)		1.4 (1)		3.5 ± 1.0 (8)
Fine dust	3.4 ± 0.5 (19)		2.6 ± 0.4 (4)		3.3 ± 0.6 (23)
Coarse dust	0.7 ± 0.2 (20)		0.40 ± 0.08 (2)		0.7 ± 0.2 (22)
Total dust	1.2 ± 0.3 (9)	0.9 ± 0.8 (5)	0.7 ± 0.2 (3)		1.1 ± 0.4 (12)
Fine sea salt	4.5 ± 0.7 (22)		3.7 ± 1.7 (3)		4.5 ± 0.9 (25)
Coarse sea salt	1.0 ± 0.2 (19)		0.72 ± 0.02 (2)		1.0 ± 0.2 (21)
Total sea salt	2.2 ± 0.5 (8)		1.8 ± 0.3 (2)		2.1 ± 0.5 (10)

"Mixed" refers to a mixed-composition aerosol. Sulfate efficiencies correspond to dry ammonium sulfate, nitrate entries correspond to dry ammonium nitrate, POM efficiencies have been normalized to an Roc value of 1.8, and sea salt efficiencies have been adjusted to a dry state. The final column gives the overall average for all methods for the mixed-composition aerosols, and the average of three methods (theoretical, MLR (multilinear regression), and partial) for the remaining species. The number of observations is given in parentheses. Estimates are for visible wavelengths (near 550 nm).

southwestern United States) compared to more humid regions, suggesting that aqueous aerosol processing could result in larger size distributions and higher mass scattering efficiencies.

The average theoretical estimate of dry ammonium sulfate mass scattering efficiency is 2.1 ± 0.7 m²/g, while the MLR average is 2.8 ± 0.5 m²/g. This difference between methods may in part be due to MLR tending to

TABLE 2.4 Mass Scattering Efficiencies from Earlier Tables Separated into Urban, Remote, and Ocean Regions

Species/ Mode	Urban (m²/g)	Remote/Rural Continental (m²/g)	Ocean/Marine (m²/g)
Fine mixed	3.2 ± 1.3 (32) (M) 3.4 ± 1.4 (22) (MLR) 2.3 ± 0.8 (6) (P) 3.4 ± 1.6 (2)	3.1 ± 1.4 (24) (T) 4.0 ± 0.5 (3) (M) 2.9 ± 1.3 (14) (MLR) 3.3 ± 1.7 (7)	4.1 ± 0.8 (42) (T) 4.3 ± 0.7 (23) (M) 3.8 ± 0.8 (18) (MLR) 4.3 (1)
Coarse mixed	(MLR) 0.6 ± 0.3 (6)	0.7 ± 0.4 (24) (T) 0.6 (1) (M) 0.40 ± 0.15 (4) (MLR) 0.7 ± 0.5 (19)	1.6 ± 1.0 (21) (T) 1.7 ± 1.0 (20) (MLR) 0.6 (1)
Total mixed	1.7 ± 1.0 (14) (T) 1.4 ± 0.7 (3) (MLR) 1.7 ± 1.1 (11)		(T) 2.5 ± 1.0 (6)
Fine sulfate	2.6 ± 0.7 (9) (MLR) 2.8 ± 0.5 (5) (P) 2.3 ± 0.8 (4)	2.7 ± 0.5 (56) (T) 2.5 ± 0.3 (6) (MLR) 2.8 ± 0.5 (48) (P) 2.05 ± 0.07 (2)	(T) 2.0 ± 0.7 (28)
Fine nitrate	2.2 ± 0.5 (6) (MLR) 2.1 ± 0.3 (2) (P) 2.2 ± 0.6 (4)	2.8 ± 0.5 (42) (MLR) 2.8 ± 0.5 (40) (P) 2.50 ± 0.14 (2)	
Fine POM	(MLR) 2.5 (1)	(MLR) 3.1 ± 0.8 (38)	(T) 5.6 ± 1.5 (19)
Coarse POM			(T) 2.6 ± 1.1 (19)
Total POM			3.5 ± 0.9 (8) (T) 3.8 ± 0.5 (7) (MLR) 1.4 (1)
Fine dust		(MLR) 2.6 ± 0.4 (4)	(T) 3.4 ± 0.5 (19)
Coarse dust		(MLR) 0.5 ± 0.2 (3)	(T) 0.7 ± 0.2 (19)
Total dust		(MLR) 0.71 (1)	1.1 ± 0.4 (11) (T) 1.2 ± 0.3 (9) (MLR) 0.7 ± 0.3 (2)
Fine sea salt		(MLR) 1.8 (1)	4.6 ± 0.7 (24) (T) 4.5 ± 0.7 (22) (MLR) 4.71 ± 0.07 (2)
Coarse sea salt			0.96 ± 0.18 (21) (T) 0.99 ± 0.17 (19) (MLR) 0.72 ± 0.02 (2)
Total sea salt			2.1 ± 0.5 (10) (T) 2.2 ± 0.5 (8) (MLR) 1.8 ± 0.3 (2)

Mixed fine-, coarse-, and total-mode specific mass scattering efficiencies are averages of all four methods for both dry and ambient conditions. The mass scattering efficiencies for individual species are averages of only the theoretical (T), multilinear regression (MLR), and partial (P) scattering methods under dry conditions ($RH = 0\%$). The average for a specific method is listed for the T, measurement (M), MLR, or P methods. Estimates are for visible wavelengths (near 550 nm but could vary depending on individual study). The number of observations in the average is in parentheses.

overestimate regression coefficients and thus mass scattering efficiencies for variables that are measured with more accuracy relative to others. However, 34 of the 53 estimates from the MLR method were carried out assuming that the fine mode was internally mixed and formed one variable in the regression equation, thus minimizing or eliminating this problem.

Fine-mode POM mass scattering efficiencies ranged from 3.1 ± 0.8 to 5.6 ± 1.5 m^2/g, with an average of the three methods being 3.9 ± 1.5 m^2/g for 58 observations. POM efficiencies were normalized assuming a molecular weight per carbon weight ratio of 1.8 and were assumed to be nonhygroscopic. The high value was obtained by the theoretical method and the low value was computed using the MLR method. Again, it is possible that the MLR estimates of POM efficiencies are artificially low due to biases in the MLR technique for data with higher uncertainties. Assuming an internal mixture by combining POM data with other data of less uncertainty (e.g., sulfate data) in the regression may help to avoid this issue. Much higher values of POM efficiencies were obtained over the ocean (5.6 ± 1.5 m^2/g) compared to rural areas (3.1 ± 0.8 m^2/g); however, many of the ocean measurements were influenced by large continental outflow that probably included combustion sources. All of the ocean estimates were computed with the theoretical method, compared to the MLR method for the more rural regions. Perhaps the sources and processing of the organic aerosols also contribute to these differences, especially since higher POM efficiencies have been observed in regions influenced by biomass burning. The much larger range in POM efficiencies (compared to sulfate) may reflect the greater uncertainty in organic aerosol chemical and optical properties, size distributions, possible hygroscopic properties, and varying sources and atmospheric processes, as well as greater uncertainties in the measurements themselves.

Fine-mode, dry ammonium nitrate mass scattering efficiencies ranged from 2.3 ± 0.5 to 2.8 ± 0.5 m^2/g, with an average value of 2.7 ± 0.5 m^2/g for 48 observations. These values are similar to the estimates for dry ammonium sulfate and in fact are often assumed to be the same (Malm et al., 1994; Malm and Hand, 2007). These estimates were derived with only two methods, with the high value associated with the MLR technique and the low value associated with the partial scattering method. Estimates were somewhat higher in remote/rural locations (2.8 ± 0.5 m^2/g) compared to urban regions (2.2 ± 0.5 m^2/g). The estimates for the urban locations were computed predominantly with the partial scattering method, while the rural estimates were computed predominantly by the MLR method. Because the MLR technique depends on bulk mass measurements, the assumption is made that nitrate is associated with the fine mode. The partial scattering method relies on size distributions, so the assumption of fine-mode nitrate is more easily justified. It is possible that a given observation nitrate is not associated with fine-mode ammonium nitrate, as we have assumed in our normalization, but instead it could be associated with coarse-mode nitrate species.

Fine-mode, mixed aerosol specific mass scattering efficiencies ranged from 3.1 ± 1.4 to $4.3 \pm 0.7 \, \text{m}^2/\text{g}$, with an average value of $3.6 \pm 1.2 \, \text{m}^2/\text{g}$ for 98 observations, using all four methods. These estimates correspond to a mixture of aerosol species and have not been normalized to the same composition or to dry conditions, so they undoubtedly include some water mass. The high value was obtained with the theoretical method, and the low value was obtained with the MLR method. The range of estimates is still fairly low given the range of aerosol physio-chemical properties and the variety of conditions under which these measurements were performed. The lowest estimates ($\sim 3.2 \, \text{m}^2/\text{g}$) were observed in urban and rural/remote regions, compared to the highest estimates obtained over the ocean ($4.3 \pm 0.7 \, \text{m}^2/\text{g}$), most likely due to water associated with sea salt in the fine mode.

Mixed aerosol, coarse-mode specific mass scattering efficiencies ranged from 0.40 ± 0.15 to $1.6 \pm 1.0 \, \text{m}^2/\text{g}$, with an average value of $1.0 \pm 0.9 \, \text{m}^2/\text{g}$ for 51 observations, using all methods. As with the fine-mode specific mass scattering efficiencies, these estimates correspond to a range of composition and RH values. The high estimates were derived using the theoretical method, and most of these observations were made over the ocean, so it is likely that water due to hygroscopic sea salt is included in these values. Lower estimates of coarse mode α_{sp} were derived using the MLR and measurement methods and were associated mostly with rural/remote regions and probably dust aerosols. Variations in coarse-mode size distributions and composition can have a significant impact on coarse aerosol specific mass scattering efficiencies, more so than with the fine-mode specific mass scattering efficiencies, as evidenced by the range in variability.

Fine dust mass scattering efficiencies ranged from 2.6 ± 0.4 to $3.4 \pm 0.5 \, \text{m}^2/\text{g}$, with an average value of $3.3 \pm 0.6 \, \text{m}^2/\text{g}$ for 23 values. Estimates were obtained using two methods, with the theoretical method corresponding to the high value (all obtained over the ocean) and the MLR corresponding to the low value (in remote/rural regions). It is likely that the fine dust estimates correspond to the fine tail of a coarse-mode size distribution extending into the fine-mode size range. Given the different assumptions in computing dust composition and the variety of different source regions that contributed to these estimates, the range in values is fairly narrow. Coarse-mode dust mass scattering efficiencies ranged from 0.40 ± 0.08 (two observations with the MLR method) to $0.7 \pm 0.2 \, \text{m}^2/\text{g}$ (20 observations), with the majority of estimates obtained using the theoretical method over ocean regions. The lower values compared to fine dust estimates reflect the less-efficient light-scattering capability of the larger particles at visible wavelengths.

Dry, fine-mode sea salt mass scattering efficiencies ranged from 3.7 ± 1.7 for the MLR method (3 observations) to $4.5 \pm 0.7 \, \text{m}^2/\text{g}$ obtained with the theoretical method (22 observations), with an average value of $4.5 \pm 0.9 \, \text{m}^2/\text{g}$, predominantly obtained over the ocean. Similar to dust, these estimates correspond to the fine tail of the coarse-mode size distribution. The coarse-mode sea salt mass scattering efficiency ranged from 0.72 ± 0.02 (2 observations) to 1.0 ± 0.2 m^2/g (19 observations, all obtained over the ocean). The data used in these estimates were collected in a variety of conditions all over the globe and so reflect average global sea salt mass scattering efficiencies. Normalization to dry conditions decreased the variability originally observed in the estimates.

Mass Absorption Efficiencies

A compilation of estimates of mass scattering and absorption efficiencies of light absorbing carbon (LAC), as well as of other species, is presented in Table 2.5 (Bond and Bergstrom, 2006; Hand and Malm, 2006). Efficiencies listed in this table have been derived using all of the methods described above, which are noted in the first column of the table. Typically, mass absorption efficiencies are computed by the measurement method, by dividing a light-absorption coefficient (measured with an aethalometer, particle soot absorption photometer, or photo acoustic instrument, by the mass of LAC from thermal analyses (see Chapter 7 on measurement methods)). The theoretical method can be used if the size distribution of LAC is measured or by applying a complex refractive index to number or volume size distributions in the Mie calculation. In Table 2.5, the values reported for a wavelength other than 550 nm have been corrected to 550 nm, assuming a λ^{-1} relationship (Bohren and Huffman, 1998). It is well known that large uncertainties surround the values of mass absorption efficiencies by LAC, as is evidenced by the range of values reported in the table. Much of the uncertainty stems from the fact that estimates of LAC mass concentrations are method dependent, and light absorption efficiencies are dependent on the wavelength at which the measurements are performed, due to the spectral dependence of light absorption. Uncertainties also arise due to the wide range in optical properties (e.g., complex refractive index) of LAC applied in theoretical methods.

Reconstructed extinction equations typically treat LAC as externally mixed (see Eqs. 2.4 and 2.5), as well as assume that it is the only component of the aerosol that is absorbing solar radiation in the visible wavelengths. This formulation is based on the assumption that LAC refers only to "soot" particles; however, there is substantial evidence that some of the carbonaceous aerosol, especially biomass smoke, being characterized

TABLE 2.5   Summary of Measured and Derived Mass Absorption Efficiencies

Method	$\alpha_{sp(LAC)}$ (m^2/g)	$\alpha_{ap(LAC)}$ (m^2/g)	$\alpha_{ap(Soil)}$ (m^2/g)	$\alpha_{ap(POM)}$ (m^2/g)
M		9.4 ± 5.2		
MLR	3.6	10.7 ± 3.1	2.8 ± 1.3	5.6 ± 3.2
T	3.1 ± 1.4	12.9 ± 6		

as LAC by thermal methods is not soot and in fact is probably refractory organic carbon. Others have reported that LAC and a significant fraction of organic carbon (OC) from biomass smoke have similar volatility and combustion temperatures, causing the split between LAC and OC to be poorly defined (and method dependent) and leading to an overestimation of LAC and an underestimation of OC for biomass smoke (e.g., Novakov and Corrigan, 1995; Gelencsér et al., 2000; Mayol-Bracero et al., 2002; Formenti et al., 2003; Guyon et al., 2003). A study by Malm et al. (1994) in the northwestern United States showed that light-absorption coefficients (as measured independently) correlated well with higher temperature fractions of OC as well as LAC. Analyses of single particle data from Yosemite National Park (Hand et al., 2005) showed that both the measured LAC and OC from thermal optical reflectance (TOR) analysis (Chow et al., 2007) corresponded to a type of nongraphitic refractory carbonaceous particle called tar balls, which are efficient at both scattering and absorbing light. Caution should be taken in always interpreting data from TOR analyses only as externally mixed OC and LAC particles, because TOR analyses are based on the refractory properties of carbon particles and not necessarily on their light absorbing characteristics.

Theoretical calculations performed by Fuller et al. (1999) suggest a reasonable LAC mass absorption efficiency of 6.5 m^2/g; however, the average value of the empirical values listed in Table 2.5 is 9.7 ± 4.3 m^2/g. In Bond and Bergstrom's (2006) review, they recommend an absorption efficiency of 7.5 ± 1.2 m^2/g. Estimates summarized in Table 2.5 suggest an average fine-mode LAC mass scattering efficiency of approximately 3.3 m^2/g. Clearly, more work is needed to understand the absorption and scattering characteristics of carbonaceous aerosols.

Some Recommendations

Part of the difficulty in assessing reported mass scattering efficiencies is the wide variety of data and methods used to derive them. While microphysical and chemical differences undoubtedly exist between these aerosols, it is more likely that the methods themselves contribute to the large range of variability, especially given what is known about biases associated with different methods. Correct interpretation of a reported value requires that the authors of the study carefully describe the many assumptions required in computing them.

This review of dry ($RH \approx 0\%$) mass scattering efficiencies in the visible wavelength range provides a basis for recommending values to be used for visibility and climate forcing calculations:

- Ammonium sulfate: An α_{sp} of 2.5 m^2/g is appropriate for average conditions; however, we observed lower values (~2 m^2/g) in dry, more pristine environments compared with higher values (~3 m^2/g) in more polluted environments.

- Ammonium nitrate: An α_{sp} value of 2.7 m^2/g is appropriate for remote/rural locations but may be lower for urban regions. We caution that the chemical forms (and size distribution) of nitrate may vary considerably, which would affect its mass scattering efficiency.
- POM: An appropriate average α_{sp} for POM (assuming a molecular weight to carbon weight ratio of 1.8) is 3.9 m^2/g, but much higher values (~6 m^2/g or greater) were observed for aerosols influenced by industrial and biomass combustion sources.
- Mixed fine mode: An α_{sp} of 3.6 m^2/g is reasonable; however, this value can vary considerably due to RH conditions, aerosol composition, and size distribution.
- Mixed coarse mode: An α_{sp} of 1.0 m^2/g falls within the range of observed values but can vary due to RH conditions, aerosol composition, and size distribution.
- Fine dust: An α_{sp} estimate of 3.3 m^2/g is appropriate if the application takes into account only fine-mode aerosols. For fine-mode data influenced by coarse mass, a much lower value near 1.0 m^2/g is recommended.
- Coarse dust: An α_{sp} near 0.6 m^2/g is recommended.
- Fine-mode sea salt: A mass scattering efficiency of 4.5 m^2/g is recommended; however, as discussed above for fine dust, we caution that this value only be applied to data that include only fine mass. A much lower value of about 1.0–1.3 m^2/g is recommended if fine sea salt mass is derived from a nonideal sampler.
- Coarse-mode sea salt: An α_{sp} of 1.0 m^2/g is recommended.

It is also recommended that the following conditions always be reported: the assumed chemical form of the species, including the refractive index and density (if used in the calculation), the wavelength at which the nephelometer measurement or Mie calculations were performed, the RH of the measurement, including any assumptions regarding hygroscopic growth, and a clear description of the method used to calculate efficiency. Spurious interpretations of mass scattering efficiencies can arise when values are reported without a clear description of the measurement and analysis method used. All measurement details (including RH and wavelength) and all assumptions of aerosol composition, size, and hygroscopicity should be routinely reported so that the values can be properly interpreted.

An Example of Real-World Implementation of Recommendations

For a few years prior to about 1988, optical and particle data were collected at about 20 national parks throughout the United States. The optical data were collected using transmissometers and integrating nephelometers, while 24-h-duration mass concentrations for PM_{10} and $PM_{2.5}$, as well as most of the $PM_{2.5}$ component concentrations, were collected on a one-day-in-three schedule (Malm et al., 1994). Using these data, Malm et al. suggested using Eq. 2.21 in the following form as a way to estimate atmospheric extinction from particle mass concentrations measurements.

$$b_{ext} \approx 3 \times f(RH) \times [(NH_4)_2SO_4] + 3 \times f(RH) \times [NH_4NO_3] \\ + 4 \times [POM] + 10 \times [LAC] + 1 \times [Soil] + 0.6 \times [CM] + 0.01. \quad (2.24)$$

This equation assumes constant mass scattering efficiencies for all mass concentrations, that sulfates are always in the form of ammonium sulfate, that sulfates and nitrates have the same mass scattering efficiencies and $f(RH)$ curves, and that $POM = 1.4 \times OC$ where OC is measured carbon mass concentration.

Around 1988 the Interagency Monitoring of Protected Visual Environments (IMPROVE) was established, which resulted an expansion of the visibility monitoring effort from the original 20 sites to about 160 sites. At 21 of those monitoring sites, hourly averaged nephelometer RH data are also routinely available. Data from these sites have been key to evaluating the performance of the above described equation, as well as to the development and performance evaluation of various alternative revised algorithms.

Lowenthal and Kumar (2005) found that $PM_{2.5}$ mass scattering efficiencies increased with increasing levels of particle light scattering and mass concentration. This was attributed to the growth of the dry particle size distribution into size ranges with higher scattering efficiencies as particle mass increases, possibly because of cloud processing during transport. Malm and Hand (2007) analyzed a 10-year span of IMPROVE data and showed a strong relationship between increasing mass and increased mass scattering efficiencies.

Because atmospheric extinction derived from particle mass measurements is used to track progress toward improving visibility under the U.S. 1999 Regional Haze Rule (40 CFR 51, 64 Federal Register 126, Regional Haze Regulations; Final Rule, 1999), it was necessary to redo Eq. 2.24 such that it accounted for the mass scattering efficiency dependence on particle mass concentrations. To that end, the IMPROVE steering committee established a subcommittee of atmospheric scientists to develop a revised algorithm that reduces biases in light extinction estimates associated with Eq. 2.24 and that is consistent with current scientific literature and illustrated in Tables 2.3, 2.4, and 2.5, while constrained by the need to use only those data that are available from the routine particle monitoring programs (Pitchford et al., 2007).

The consensus equation is given by

$$
\begin{aligned}
b_{ext} \approx\ & 2.2 \times f_s(RH) \times [small\ sulfate] + 4.8 \times f_L(RH) \times [large\ sulfate] \\
& + 2.4 \times f_s(RH) \times [small\ nitrate] + 5.1 \times f_L(RH) \times [large\ nitrate] \\
& + 2.8 \times [small\ POM] + 6.1 \times [large\ POM] \\
& + 10 \times [LAC] \\
& + 1 \times [soil] \\
& + 1.7 \times f_{xx}(RH) \times [sea\ salt] \\
& + 0.6 \times [CM] \\
& + Ray \\
& + 0.33 \times [NO_2\ (ppb)],
\end{aligned}
\tag{2.25}
$$

where

$$
[large] = [conc^2]/20 \text{ and } [small] = [large] - [conc] \text{ for } [conc] \le 20.0, \\
\text{and } [large] = [conc] \text{ and } [small] = 0.0 \text{ for } [conc] > 20.0.
\tag{2.26}
$$

Large and small species concentrations are linearly scaled between 0.0 and 20 $\mu g/m^3$ based on the ambient concentration of a given species as given by Eq. 2.26. The mass scattering efficiencies along with $f(RH)$ curves were calculated using the Mie theory, based on an assumed log-normal mass size distribution with geometric mean diameter, geometric diameter standard deviation, index of refraction, and density, given in Table 2.6. Diameter, standard deviation, index of refraction, and density are given by d, σ, n, and ρ, respectively, while L and S refer to the large and small mode, respectively. The $f(RH)$ curves were again calculated using the Mie theory and D/D_o crystallization growth curves predicted by the Aerosol Inorganics Model (AIM) thermodynamic equilibrium model (Clegg et al., 1998). Eq. 2.25 also differs from Eq. 2.24 in that the Roc factor for POM changed from 1.4 to 1.8, and sea salt and NO2 light absorbing gas were added to the equation. For additional details the reader is referred to Pitchford et al. (2007).

The revised algorithm reduces the biases compared to measurements at the high and low extreme mass concentrations but has somewhat reduced precision. Most of the reduction of bias associated with the new algorithm is attributed to the use of the split-component extinction efficiency

TABLE 2.6   Geometric Mean, Standard Deviation, Index of Refraction, and Density Used in Calculating Mass Scattering Efficiencies and Associated $f(RH)$ Curves for Different Species and for the Large and Small Mass Size Distributions

Species	d_{gS}	d_{gL}	σ_{gS}	σ_{gS}	σ_{gS}	σ_{gS}	n	ρ
(NH₄)₂SO₄	0.2 μm	0.5 μm	2.2	1.5	2.2 m²/gm	4.8 m²/gm	1.55 + 0i	1.77 g/m³
NH₄NO₃	0.2 μm	0.5 μm	2.2	1.5	2.4 m²/gm	5.1 m²/gm	1.53 + 0i	1.73 g/m³
POM	0.2 μm	0.5 μm	2.2	1.5	2.8 m²/gm	6.1 m²/gm	1.55 + 0i	1.4 g/m³
Sea Salt	2.5 μm	na	2.0	Na	1.7 m²/gm	na	1.55 + 0i	1.9 g/m³

method for sulfate, nitrate, and OC components, which permitted variable extinction efficiencies depending on the component mass concentration. Though not subject to explicit performance testing, the revised algorithm also contains specific changes to the original algorithm that reflect more recent scientific literature (e.g., the change to 1.8 from 1.4 for the organic compound mass to carbon mass ratio), a more complete accounting for contributors to haze (e.g., sea salt and NO_2 terms), and the use of site-specific Rayleigh scattering to reduce elevation-related bias.

It is also worth noting that in Eqs. 2.24 and 2.25 $f(RH)$ for POM is set equal to one. There have been few direct measurements of ambient POM $f(RH)$ curves, although Malm et al. (2005) show that as the ratio $[POM]/[(NH_4)_2SO_4]$ approaches 12, the $f(RH)$ for RH between 80% and 95% approaches about 1.2, implying an $f(RH)$ for $POM \approx 1.2$ for high RH levels. Field measurements of $f(RH)$ for smoke outside North America, summarized by Reid et al. (2005a,b) and Day et al. (2006), show there is a broad range of $f(80\%)$ varying from 1 to 2.1. Kotchenruther and Hobbs (1998) found that fresh smoke had an $f(80) = 1.1$, which increased with plume aging to 1.35. This can be explained by the formation of secondary salts in the aged plume. However, measured $f(RH)$ in the African savanna (Magi et al., 2003) decreased as the plume aged, defying a simple explanation of the evolution of smoke plume hygroscopicity. Hand et al. (2010) showed that almost all of the smoke hygroscopicity could be explained by internally mixed inorganic salts, suggesting that the carbonaceous component of an internally mixed salt/carbon aerosol, even though water soluble, does not absorb significant amounts of water.

The above measurements were made on ambient aerosols. Lowenthal et al. (2015) summarize growth factor (GF) $D(RH)/D(RH \approx 0)$ laboratory measurements of water-soluble organic carbon (WSOC) extracts from bulk collection of POM on filter substrates. $D(RH \approx 0)$ is the particle diameter at RH approximately equal to 0, while $D(RH)$ is particle diameter at some elevated relative humidity. They conclude that GF is, on average, about 1.14 at $RH = 90\%$ in the rural areas of the eastern United States, which corresponds to an $f(RH) \approx 1.5$. Without reference to any specific publication, the interested reader can find many more studies of aerosol growth factors in the literature.

References

Anderson, T.L., Charlson, R.J., White, W.H., McMurry, P.H., 1994. Comment on: Light scattering and cloud condensation nucleus activity of sulfate aerosol measured over the northeast Atlantic Ocean. J. Geophys. Res. 99, 25947–25949.

Bohren, C.F., Huffman, D.R., 1998. Absorption and Scattering of Light by Small Particles. Wiley, New York.

Bond, T.C., Bergstrom, R.W., 2006. Light absorption by carbonaceous particles: an investigative review. Aerosol Sci. Technol. 40, 27–67.

Born, M., Wolf, E., 1999. Principles of Optics Electromagnetic Theory of Propagation, Interference and Diffraction of Light, seventh ed. Oxford University Press, Oxford.

Brasseur, G.P., Prinn, R.G., Pszenny, A.A.P. (Eds.), 2003. Atmospheric Chemistry in a Changing World. Springer-Verlag, Berlin, Germany, p. 300.

Chang, D.P.Y., Hill, R.C., 1980. Retardation of aqueous droplet evaporation by air pollutants. Atmos. Environ. 14, 803–807.

Chow, J.C., Watson, J.G., Chen, L.-W.A., Chang, M.C.O., Robinson, N.F., Trimble, D., et al., 2007. The IMPROVE-A temperature protocol for thermal/optical carbon analysis: maintaining consistency with a long-term database. J. Air Waste Manage. Assoc. 57, 1014–1023.

Clegg, S.L., Brimblecombe, P., Wexler, A.S., 1998. A thermodynamic model of the system H^+-NH_4^+-Na^+-SO_4^{2-}-NO_3^--Cl^--H_2O at 298.15 K. J. Phys. Chem. 102, 2155–2171.

Covert, D.S., Charlson, R.J., Ahlquist, N.C., 1972. A study of the relationship of chemical composition and humidity to light scattering by aerosol. J. Appl. Meteorol. 11, 968–976.

Day, D.E., Hand, J.L., Carrico, C.M., Engling, G., Malm, W.C., 2006. Humidification factors from laboratory studies of fresh smoke from biomass fuels. J. Geophys. Res. Atmos. 111 (D22), Art. No. D22202.

Formenti, P., Elbert, W., Maenhaut, W., Haywood, J., Osborne, S., Andreae, M.O., 2003. Inorganic and carbonaceous aerosols during the South African Regional Science Initiative (SAFARI 2000) experiment: chemical characteristics, physical properties, and emission data for smoke from African biomass burning. J. Geophys. Res. 108 (D13), 8488.

Fuller, K.A., Malm, W.C., Kreidenweis, S.M., 1999. Effects of mixing on extinction by carbonaceous particles. J. Geophys. Res. Atmos. 104, 15941–15954.

Garland, J.A., 1969. Condensation on ammonium sulfate particles and its effect on visibility. Atmos. Environ. 3, 347–354.

Gebhart, K.A., Malm, W.C. An investigation of the size distributions of particulate sulfate concentrations measured during WHITEX, In: Mathai, C.V. (Ed.), Visibility and Fine Particles: Transactions of the Air & Waste Management Association. Air & Waste Management Assoc., Pittsburgh, PA, 1990.

Gelencsér, A., Hoffer, A., Molnár, A., Krivácsy, Z., Kiss, Gy., Mészáros, E., 2000. Thermal behaviour of carbonaceous aerosol from a continental background site. Atmos. Environ. 34, 823–831.

Guyon, P., Graham, B., Roberts, G.C., Mayol-Bracero, O.L., Maenhaut, W., Artaxo, P., et al., 2003. In-canopy gradients, composition, sources, and optical properties of aerosol over the Amazon forest. J. Geophys. Res. 108 (D18), 4591.

Hand, J.L., Day, D.E., McMeeking, G.R., Levin, E.J.T., Carrico, C.M., Kreidenweis, S.M., et al., 2010. Measured and modeled humidification factors of fresh smoke particles from biomass burning: role of inorganic constituents. Atmos. Chem. Phys. 10, 6179–6194.

Hand, J.L., Kreidenweis, S.M., Sherman, D.E., Collett, Jr., J.L., Hering, S.V., Day, D.E., et al., 2002. Aerosol size distributions and visibility estimates during the Big Bend Regional Aerosol Visibility and Observational Study (BRAVO). Atmos. Environ. 36, 5043–5055.

Hand, J.L., Malm, W.C., 2006. Review of the IMPROVE equation for estimating ambient light extinction coefficients, Report Colorado State University-Appendix 2.

Hand, J.L., Malm, W.C., 2007. Review of aerosol mass scattering efficiencies from ground-based measurements since 1990. J. Geophys. Res. 112.

Hand, J.L., Malm, W.C., Laskin, A., Day, D.E., Lee, T., Wang, C., et al., 2005. Optical, physical, and chemical properties of tar balls observed during the Yosemite Aerosol Characterization Study. J. Geophys. Res. Atmos. 110 (D21).

Husar, R.B., Shu, W.R., 1975. Thermal analyses of the Los Angeles smog aerosol. J. Appl. Methodol. 14, 1558–1565.

Kotchenruther, R.A., Hobbs, P.V., 1998. Humidification factors of aerosols from biomass burning in Brazil. J. Geophys. Res. 103 (D24), 32,081–32,089.

Lowenthal, D.H., Kumar, N., 2005. Variation of mass scattering efficiencies in IMPROVE. J. Air Waste Manage. Assoc. 108, 4279.

Lowenthal, D.H., Rogers, C.F., Saxena, P., Watson, J.G., Chow, J.C., 1995. Sensitivity of estimated light extinction coefficients to model assumptions and measurement errors. Atmos. Environ. 29, 751–766.

Lowenthal, F., Zielinska, B., Samburova, V., Collins, D., Taylor, N., Kumar, N., 2015. Evaluation of assumptions for estimating chemical light extinction at U.S. national parks. J. Air Waste Manage. Assoc. 65 (3), 249–260.

Magi, B.I., Hobbs, P.V., Schmind, B., Redmann, J., 2003. Vertical profile of light scattering, light absorption, and single-scattering albedo during the dry, biomass burning season in southern Africa and comparisons of in situ and remote sensing measurements of aerosol optical depth. J. Geophys. Res. 108 (D18), 8504.

Malm, W.C., Cahill, T., Gebhart, K.A., Waggoner, A., 1986. Optical characteristics of atmospheric sulfur at Grand Canyon, Arizona, presented at the Air Pollution control Association International Specialty Conference: Visibility Protection—Research and Policy Aspects, Grand Teton National Park, Wyoming, Sept 7–10, 1986.

Malm, W.C., Day, D.E., Carrico, C.M., Kreidenweis, S.M., Collett, Jr., J.L., McMeeking, G., et al., 2005. Inter-comparison and closure calculations using measurements of aerosol species and optical properties during the Yosemite Aerosol Characterization Study. J. Geophys. Res. 110 (D14), Art. No. D14302.

Malm, W.C., Day, D.E., Kreidenweis, S.M., 2000. Light scattering characteristics of aerosols as a function of relative humidity: Part 1—A comparison of measured scattering and aerosol concentrations using the theoretical models. J. Air Waste Manage. Assoc. 50, 686–700.

Malm, W.C., Day, D.E., Kreidenweis, S.M., Collett, Jr., J.L., Lee, T., 2003. Humidity-dependent optical properties of fine particles during the Big Bend Regional Aerosol and Visibility Observational Study. J. Geophys. Res. 108 (D9), 4279.

Malm, W.C., Gebhart, K.A., Latimer, D., Cahill, T., Eldred, R., Pielke, R., Stocker, R., Watson, J.G., 1989. Final Report—WHITEX (Winter Haze Intensive Tracer Experiment), Chapter 5: Light Extinction Budgets, available at http://vista.cira.colostate.edu/improve/Studies/WHITEX/Data/Chapter5_5-1.pdf.

Malm, W.C., Hand, J.L., 2007. An examination of the physical and optical properties of aerosols collected in the IMPROVE program. Atmos. Environ. 41, 3407–3427.

Malm, W.C., Kreidenweis, S.M., 1997. The effects of models of aerosol hygroscopicity on the apportionment of extinction. Atmos. Environ. 31, 1965–1976.

Malm, W.C., McMeeking, G.R., Kreidenweis, S.M., Levin, E., Carrico, C.M., Day, D.E., et al., 2009. Using high time resolution aerosol and number size distribution measurements to estimate atmospheric extinction. J. Air Waste Manage. Assoc. 59, 1049–1060.

Malm, W.C., Pitchford, M.L., 1997. Comparison of calculated sulfate scattering efficiencies as estimated from size-resolved particle measurements at three national locations. Atmos. Environ. 31 (9), 1315–1325.

Malm, W.C., Schichtel, B.A., Pitchford, M.L., Ashbaugh, L.A., Eldred, R.A., 2004. Spatial and monthly trends in speciated fine particle concentration in the United States. J. Geophys. Res. 109, 1–22.

Malm, W.C., Sisler, J.F., Huffman, D., Eldred, R.A., Cahill, T.A., 1994. Spatial and seasonal trends in particle concentration and optical extinction in the United States. J. Geophys. Res. 99 (D1), 1347–1370.

Mayol-Bracero, O.L., Guyon, P., Graham, B., Roberts, G., Andreae, M.O., Desesari, S., et al., 2002. Water-soluble organic compounds in biomass burning aerosols over Amazonia 2: apportionment of the chemical composition and importance of the polyacidic fraction. J. Geophys. Res. 107 (D20), 8091.

McMeeking, G.R., Kreidenweis, S.M., Carrico, C.M., Collett, Jr., J.L., Day, D.E., Malm, W.C., 2005. Observations of smoke-influenced aerosol during the Yosemite Aerosol Characterization Study: 2. Aerosol scattering and absorbing properties. J. Geophys. Res. 110 (D18209).

McMurray, P.H., Friedlander, S.K., 1979. New particle formation in the presence of an aerosol. Atmos. Environ. 13, 1635–1651.

Nguyen, T.K.V., Capps, L.S., Carlton, A.G., 2015. Decreasing aerosol water is consistent with OC trends in the southeast U.S. Environ. Sci. Technol 49, 7843–7850.

Novakov, T., Corrigan, C.E., 1995. Thermal characterization of biomass smoke particles. Mikrochim. Acta. 119, 157–166.

Omar, A.H., Biegalski, S., Larson, S.M., Landsberger, S., 1999. Particulate contributions to light extinction and local forcing at a rural Illinois site. Atmos. Environ. 33 (17), 2637–2646.

Ouimette, J.R., Flagan, R.C., 1982. The extinction coefficient of multicomponent aerosols. Atmos. Environ. 16, 2405–2419.

Pilinis, C., 1989. Numerical simulation of visibility degradation due to particulate matter: model development and evaluation. J. Geophys. Res. 94 (D7), 9937–9946.

Pitchford, M., Malm, W.C., Schichtel, B.A., Kumar, N., Lowenthal, L., Hand, J., 2007. Revised algorithm for estimating light extinction from IMPROVE particle speciation data. J. Air Waste Manage. Assoc. 57, 1326–1336.

Reid, J.S., Eck, T.F., Christopher, S.A., Kippmann, R., Dubovik, O., Eleuterio, D.P., Holben, B.N., Reid, E.A., Zhang, J., 2005a. A review of biomass burning emissions part III: intensive optical properties of biomass burning particles. Atmos. Chem. Phys. 5, 827–849.

Reid, J.S., Koppmann, R., Eck, T.F., Eleuterio, D.P., 2005b. A review of biomass burning emissions part II: intensive physical properties of biomass burning particles. Atmos. Chem. Phys. 5, 799–825.

Rubel, G.O., Gentry, J.W., 1984. Measurement of the kinetics of solution droplets in the presence of absorbed monolayers: determination of water accommodation coefficients. J. Phys. Chem. 88, 3142–3148.

Seinfeld, J.H., Pandis, S.N., 1998. Atmospheric Chemistry and Physics. Wiley, New York.

Sloane, C.S., 1983. Optical properties of aerosols: comparison of measurements with model calculations. Atmos. Environ. 17, 409–416.

Sloane, C.S., 1986. Effect of composition on aerosol light scattering efficiencies. Atmos. Environ. 20, 1025–1037.

Sloane, C.S., Watson, J., Chow, J., Pritchett, L., Richards, L.W., 1991. Size-segregated fine particle measurements by chemical-species and their impact on visibility impairment in Denver. Atmos. Environ. Part A Gen. Top. 25 (5–6), 1013–1024.

Tang, I.N., Wong, W.T., Munkelwitz, H.R., 1981. The relative importance of atmospheric sulfates and nitrates in visibility reduction. Atmos. Environ. 15, 2463.

van de Hulst, H.C., 1981. Light Scattering by Small Particles. Dover Publications, New York.

Whitby, K.T., Husar, R.B., Liu, B.Y.H., 1972. The aerosol size distribution of Los Angeles smog. J. Colloid Interface Sci. 39, 177–204.

White, W.H., 1980. Visibility: Existing and Historical Conditions—Causes and Effects, National Acid Precipitation Assessment Program, State of Science and Technology Report 24, Section 6. Government Printing Office, Washington DC.

White, W.H., Husar, R.B., 1980. A Lagrangian model of the Pasadena smog aerosol. In: Hidy et al. (Eds.), Advances in Environmental Science and Technology. John Wiley & Sons, New York.

White, W.H., Macias, E.S., 1987. On measurement error and the empirical relationship of atmospheric extinction to aerosol composition in the non-urban West. In: Bhardwaja (Ed.), Visibility Protection: Research and Policy Aspects. Air Pollution Control Assoc., Pittsburgh, PA.

White, W.H., Roberts, P.T., 1977. On the nature and origins of visibility-reducing aerosols in the Los Angeles air basin. Atmos. Environ. 11, 803–812.

CHAPTER

3

How the Transfer of Light (Radiation) through the Atmosphere Affects Visibility

Scattering and absorption phenomena, discussed in Chapter 2, are responsible for the colors of hazes in the sky. The sky is blue because blue photons, with their shorter wavelengths, are nearer the size of the molecules that make up the atmosphere than are their green, orange, and red counterparts. Thus, blue photons are scattered more efficiently by air molecules than red photons, and as a consequence, the daytime sky looks blue, and sunsets dominated by the less-well-scattered photons have a reddish-orange appearance.

Fig. 3.1 schematically shows what happens when the red, blue, and green photons of white light strike small particles. Only the blue photons are scattered because scattering efficiency is greatest when the size relationship of photon to particle is close to 1:1. The red and green photons pass on through the particles. To an observer standing to the side of the particle concentration, the haze would appear to be blue. Fig. 3.2 shows

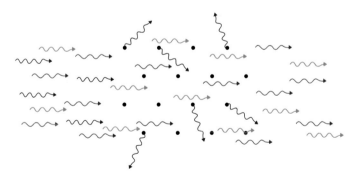

FIGURE 3.1 As a beam of white light (consisting of photons of all colors) passes through a haze made up of small particles, it is predominantly the blue photons that are scattered in various directions.

Visibility: The Seeing of Near and Distant Landscape Features. http://dx.doi.org/10.1016/B978-0-12-804450-6.00003-6

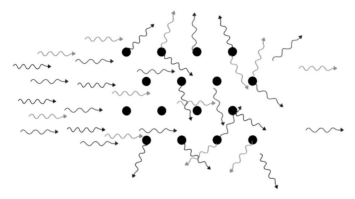

FIGURE 3.2   When particles are nearly the same as or larger than the wavelength of the incident light, photons of all colors are scattered out of the beam path.

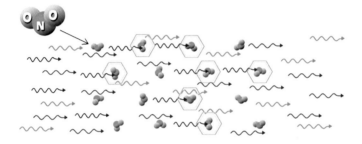

FIGURE 3.3   An atmosphere containing nitrogen dioxide (NO_2) will tend to deplete the number of blue photons through the absorption process. As a result, white light will tend to look reddish or brownish in color after passing through an NO_2 haze.

what happens when the particles are about the same size as the incoming radiation. All photons are scattered equally, and the haze appears to be white or gray. Fig. 3.3 is a similar diagram of white light passing through a concentration of nitrogen dioxide (NO_2) molecules. Blue photons are absorbed, so a person standing in the beam of light would see it as being reddish-brown (i.e., without blue) rather than white.

Fig. 3.4 further exemplifies the relationship of particle size and the color of scattered light. Fig. 3.4a shows the smoke from a lit cigarette held in a strong beam of white light. Notice that the smoke appears to have a bluish tinge to it. One can conclude that these particles must be quite small because they are scattering more blue than green or red photons. Fig. 3.4b is smoke from the same cigarette. However, the smoke in Fig. 3.4b has been changed by being held in a person's mouth for a few seconds. The inside of a person's mouth is humid, and some smoke particles have an affinity for water vapor. These hygroscopic particles tend to grow to sizes that are near the wavelengths of light and thus scatter all wavelengths of

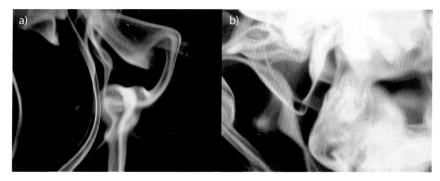

FIGURE 3.4 (a) Photograph showing the color of small particles that have been illuminated by white light. Because the smoke appears blue, it can be concluded that the scattering particles must be quite small, less than the wavelength of visible light. (b) Photograph of similar particles after they have been allowed to grow in a humid environment. Note that as a result of equal scattering of all photon colors, these larger particles appear white instead of blue.

light equally. Scattered photons having wavelengths that extend over the whole visible spectrum are, of course, perceived to be white or gray.

The fact that light scatters preferentially in different directions as a function of particle size is extremely important in determining the effects that atmospheric particulates have on a visual resource. The angular relationship between the sun and the observer in conjunction with the size of particulates determines how much of the sunlight is redistributed into the observer's eye. These concepts are schematically represented in Fig. 3.5. When the atmospheric particle concentration is more than a few $\mu g/m^3$ and the viewing angle is such that the observer is looking in the direction of or away from the sun, Mie scattering will dominate Rayleigh scattering,

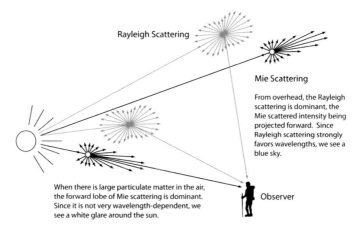

Rayleigh Scattering

Mie Scattering

From overhead, the Rayleigh scattering is dominant, the Mie scattered intensity being projected forward. Since Rayleigh scattering strongly favors wavelengths, we see a blue sky.

When there is large particulate matter in the air, the forward lobe of Mie scattering is dominant. Since it is not very wavelength-dependent, we see a white glare around the sun.

Observer

FIGURE 3.5 Schematic contrasting how the molecular-volume scattering function effects on scattered radiation differ from those of the fine particle scattering function.

as shown in Fig. 1.27. On the other hand, if the observer views a vista such that the scattering angle between the sun and the observer is about 90°, Mie scattering may be on the order of or less than Rayleigh scattering, and in many cases the sky will appear blue.

QUANTIFYING THE TRANSFER OF RADIATION THROUGH THE ATMOSPHERE

As discussed above, the alteration of radiant energy as it passes through the atmosphere is due to scattering and absorption by gases and particles. The effect of the atmosphere on the visual properties of distant objects theoretically can be determined if the concentrations and characteristics of air molecules, particles, and absorbing gases are known throughout the atmosphere and, most importantly, along the line of sight between the observer and the object.

Radiometric and photometric concepts were briefly discussed in Chapter 2 and summarized in Table 2.1 and Appendix 1. Radiometric concepts refer to radiant energy in a single wavelength, while photometric variables are radiant energy weighted in proportion to their ability to stimulate our sense of light.

Notation used in this chapter is similar to that used by Duntley (1948). The basic symbol employed for spectral radiance is N, and the symbol for luminance is B. Other variables used to quantify light energy in the atmosphere are also presented in Appendix 1. The position in the atmosphere is denoted by $\bar{r}$. The direction of any path of sight is specified by zenith angle θ and azimuth angle φ, the photometer being directed upward, $0 \leq \theta < \pi/2$, as in Fig. 3.6; r, θ, and φ are always written as parenthetic attachments to the parent symbol. When postsubscript r is appended to any symbol, it denotes that the quantity pertains to a path of length r. The subscript o always refers to the hypothetical concept of any instrument or observer located at zero distance from the object, as for example, denoting the inherent radiance of a surface. Presubscripts identify the objects: presubscript b refers to background and l to landscape feature. Thus, the monochromatic, inherent spectral radiance of a landscape feature at position $\bar{r}$ as viewed in the direction (θ, φ) is $_lN_r(r_l, \theta, \varphi)$. A postsuperscript or postsubscript $*$ is employed as a symbol signifying that the radiometric quantity has been generated by the scattering of ambient light reaching the path from all directions. Thus, $N_r^*(r, \theta, \varphi)$ is the spectral path radiance observed at position r in the indicated direction, and $N_*(r, \theta, \varphi)$ is used to denote the path function, two quantities to be defined later.

Image-forming light is lost by scattering and absorption in each elementary segment of the path of sight, and contrast-reducing air light or path radiance is generated by the scattering of the ambient light that

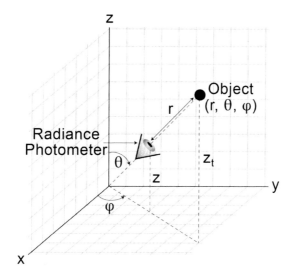

FIGURE 3.6   Illustrative geometry for a sight path.

reaches the segment from all directions, both of which reduce the contrast of landscape features as viewed against some background. The loss
of image-forming light due to attenuation by scattering and absorption
within any path segment is proportional to the amount of image-forming
light present; the coefficient of proportionality is $b_{ext}(\vec{r})$, the attenuation or
extinction coefficient at position $\vec{r}$. $b_{ext}(\vec{r})$ is a function of position within
the path of sight. It does not depend upon the image transmission direc-
tion; it is independent of the manner in which the path segment is lit by
the sun or sky; and it is a physical property of the atmosphere alone. Absorption refers to any thermodynamically irreversible transformation of
radiant energy, including conversion of light into heat but also fluores-
cence and photochemical processes.

Attenuation by scattering results from any change of direction sufficient to cause the radiation to fall outside the summative radius of the
observer detector system, such as the human eyeball.

As defined in Chapter 2,

$$b_{ext}(\vec{r}) = b_{scat}(\vec{r}) + b_{abs}(\vec{r}) \tag{3.1}$$

where $b_{scat}(\vec{r})$ and $b_{abs}(\vec{r})$ are the scattering and absorption coefficients at
position $\vec{r}$. b_{scat} can be further subdivided into scattering by gases b_{sg} and
scattering by particles b_{sp}, as b_{abs} can be expressed as the sum of absorption
by gases b_{ag} and absorption by particles b_{ap}.

Also defined in Chapter 2 are the volume scattering function $\sigma(\bar{r},\beta)$ and the normalized volume scattering function $\sigma'(\bar{r},\beta)=\sigma(\bar{r},\beta)/b_{scat}(\bar{r})$. These two variables are a measure of the atmosphere's ability to scatter light in a given direction. β refers to the scattering angle.

Schematics of how direct sunlight, reflected sunlight, and diffuse radiation affect the seeing of landscape features are shown in Figs. 1.2 and 3.7. Image-forming information is lost by the scattering of imaging radiant energy out of the sight path and absorption within the sight path, while ambient light scattered into the sight path adds radiant energy to the observed radiation field. This process is described by

$$\frac{dN_r(\theta,\varphi,r)}{dr} = \underset{(loss)}{-b_{ext}N_r(\theta,\varphi,r)} + \underset{(gain)}{N_*(\theta,\varphi,r)} \qquad (3.2)$$

where $N_r(\theta,\varphi,r)$ is the apparent radiance at some vector distance $\bar{r}$ from a landscape feature, $N_*(\theta,\varphi,r)$ (referred to as the path function) is the radiant energy gain within an incremental path segment, and $b_{ext}N_r(\theta,\varphi,r)$ is radiant energy lost within that same path segment. Although not explicitly stated, it is assumed that each variable in, and each variable derived from,

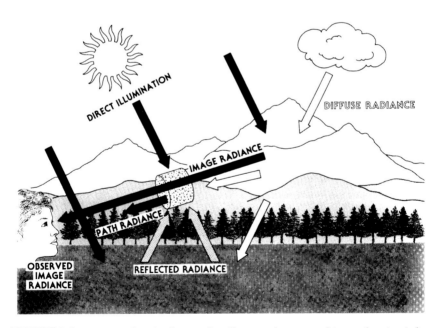

FIGURE 3.7 Diagram showing how path radiance and attenuated image-forming information combine to form the observed image at the eye of an observer. Direct, diffused, and reflected radiance all contribute to illuminating the image, as well as the incremental volumes of atmosphere that sum to form sight path radiance.

Eq. 3.2 is wavelength dependent. The parenthetical variables (θ, φ, r) indicate that N_r and $N_*(\theta, \varphi, r)$ are dependent on both the direction of image transmission and the position within the path segment. For the sake of brevity, the parenthetical variables will be dropped in the following equations.

When N_r has some special value N_q, such that $b_{ext}N_q = N_*$, then $dN_q/dr = 0$. N_q is independent of r and is commonly referred to as equilibrium radiance. The atmosphere is said to be in optical equilibrium when as much light is scattered into as out of a sight path. Therefore, under these circumstances, for every path segment

$$\frac{dN_r}{dr} = -b_{ext}(N_r - N_q).$$

(3.3)

If N_q is constant, Eq. 3.2 can be integrated to yield

$$\frac{N_r - N_q}{N_0 - N_q} = T_r$$

(3.4)

where T_r is the transmittance over path length r and is given by

$$T_r = e^{-\int_0^r b_{ext}(r)dr}.$$

(3.5)

Rearranging Eq. 3.4 yields

$$N_r = N_0 T_r + N_q (1 - T_r)$$

(3.6)

where the first term on the right of Eq. 3.6 is the residual image-forming radiance, while the second term is the path radiance (air light) N_r^*, which results from scattering processes throughout the sight path. The parameter N_r^* is the sky radiance when $r = \infty$:

$$N_\infty^* = N_q (1 - T_\infty).$$

(3.7)

If T_∞ is approximately zero, then $N_q = N_\infty^* = N_s$ and

$$N_r^* = N_s (1 - T_r)$$

(3.8)

where N_s is sky radiance. Eq. 3.8 allows for a simple approximation of N_r^* when N_s is known.

The explicit dependence of N_r^* on illumination and directional scattering properties of the atmosphere is best examined by considering

$$N_r^* = \int_0^r N_* T_r \, dr$$

(3.9)

where

$$N_* = h_s \sigma + \int_{4\pi} N\sigma \, d\Omega$$

(3.10)

where N is the apparent radiance of the sun, sky, clouds, or ground, and $d\Omega$ is an element of solid angle. The parameter h_s is sun irradiance, and σ is the volume scattering function.

CONTRAST TRANSMITTANCE IN REAL SPACE

Any landscape feature can be thought of as consisting of many small pieces, or elements, with a variety of physical characteristics. For instance, the reflectivity of an element as a function of wavelength, along with characteristics of the incident radiation, determines its color and brightness. The brightness of a scenic element at some observing distance and at one wavelength is referred to as monochromatic apparent spectral radiance. The monochromatic apparent spectral radiance of any scenic element is given according to Eq. 3.6 by

$$_l N_r = T_{rl} N_o + N_r^*$$ (3.11)

where N_r^* is substituted explicitly for $N_q(1 - T_r)$. The subscript l indicates that the radiance is associated with a specific uniform scenic landscape feature.

A scenic element is always seen against some background, such as the sky or another landscape feature. The apparent and inherent background radiances are related by an expression similar to Eq. 3.11:

$$_b N_r = T_{rb} N_o + N_r^*.$$ (3.12)

Subtracting Eq. 3.12 from Eq. 3.11 yields the relation

$$[_l N_r - _b N_r] = T_r [_l N_o - _b N_o].$$ (3.13)

Thus, radiance differences are transmitted along any path with the same attenuation as that experienced by each image-forming ray.

The image-transmitting properties of the atmosphere can be separated from the optical properties of the object by the introduction of the contrast concept. The inherent spectral contrast C_o of a scenic element is, by definition,

$$C_o = [_l N_o - _b N_o] / _b N_o.$$ (3.14)

The corresponding definition for apparent spectral contrast at some distance r is

$$C_r = [_l N_r - _b N_r] / _b N_r.$$ (3.15)

Contrast defined in the form of Eqs. 3.14 and 3.15 is referred to as universal contrast. Another form of contrast known as modulation contrast is defined as

$$m_o = [_l N_o - _b N_o]/[_l N_o + _b N_o] \tag{3.16}$$

where m_o refers to the inherent modulation contrast. Contrast as defined by Eq. 3.16 will be discussed in more detail in the section on equivalent contrast.

If Eq. 3.13 is divided by the apparent radiance of the background $_b N_r$ and combined with Eqs. 3.14 and 3.15, the result can be written as

$$C_r = C_o \frac{_b N_o}{_b N_r} T_r. \tag{3.17}$$

Substituting Eq. 3.12 for $_b N_r$ and rearranging yields

$$C_t \equiv C_r / C_o = 1/[1 + N_r^* / _b N_o T_r]. \tag{3.18}$$

The right-hand member of Eq. 3.18 is an expression for the contrast transmittance C_t of the path of sight. Eq. 3.18 is the law of contrast reduction by the atmosphere, expressed in the most general form. It should be emphasized that Eq. 3.18 is completely general and applies rigorously to any path of sight, regardless of the extent to which the scattering and absorbing properties of the atmosphere or the distribution of lighting exhibit nonuniformities from point to point.

If the background is the sky and $_s N_o/_s N_r = 1$, then Eq. 3.17 becomes $C_r = C_o T_r$, which is known as the familiar Koschmieder relationship (Koschmieder, 1924). An alternative derivation to the Koschmieder equation can be found in Middleton (1968).

EQUIVALENT CONTRAST

The above discussion and derivations were formulated in real space or in an x, y, z coordinate system. The relationships between contrast of specific scenic elements and atmospheric extinction and volume scattering functions work well when one is interested in the perceptibility of isolated scenic elements or in estimating parameters such as visual range. However, if one is interested in human responses to changes in scenic quality of an entire scene consisting of a complex mixture of scenic elements with varying degrees of edge sharpness as haze increases or decreases, linear systems theory may be more appropriate (Henry, 1977). This approach incorporates human response to variations of spatial frequencies (size and shape effects) of various landscape features. There are many linear system theory models available that describe how humans respond to image quality of electronic displays such as television or computer monitors. On the other hand, very little work has been carried out to apply these formalisms to the visibility issue. However, the fundamentals of this approach will be described here.

It could be argued that this section would be better discussed in Chapter 4 on visibility metrics or Chapter 5 on the human perception of the effects of haze on landscape features. It is presented here to develop a more-general form of contrast transmittance, which is the atmospheric modulation transfer function. A first step is to develop a quantitative descriptor of the scene itself.

Although a typical landscape projected on the human retina is two-dimensional, for purposes of simplicity the following discussion will be carried out in one dimension. The generalization to two dimensions is straightforward.

It has been shown that for discrimination tasks the variation of a mean square radiance field is a significant psychophysical variable (Cohen, 1978). Consider the simple sine wave superimposed on a background radiance field:

$$N(x) = \frac{N_m + N_b}{2} + \frac{N_m - N_b}{2}\cos(2\pi f x) \qquad (3.19)$$

where N_m and N_b are radiance values associated with the peak and trough of a sine wave, respectively. The mean square radiance associated with Eq. 3.19 can be shown to equal

$$\overline{N^2} = \overline{N}^2 + \frac{1}{2}m^2\overline{N}^2 \qquad (3.20)$$

where m is the modulation contrast, $m = (N_m - N_b)/(N_m + N_b)$, $\overline{N}^2$ is the average image radiance squared, and $\frac{1}{2}m^2\overline{N}^2$ is the mean square radiance fluctuation.

A more complex, one-dimensional "slice" of a scene or image can be represented as weighted sums of light and dark bars of various frequencies and intensities. Mathematically,

$$N_r(x) = a_o + \sum_{n,n\neq 0}^{\infty} a_n e^{in2\pi f_0 x} \qquad (3.21)$$

where $N_r(x)$ is the one-dimensional image radiance field at an observer image distance r, a_o is average image radiance, and a_n are the weighting factors associated with various sinusoid frequencies in increments of $2\pi f_o$.

The second term, $\sum_{n,n\neq 0}^{\infty} a_n e^{in2\pi f_o x}$, is the modulation of average image radiance a_o. The modulation contrast of a given sinusoid is just a_n/a_o. Calculation of the mean square radiance associated with Eq. 3.21 yields

$$\overline{N_r^2} = a_o^2 + \frac{1}{N'} \sum_{n,n\neq 0} |a_n|^2 \tag{3.22}$$

where N' is total number of pixels in the one-dimensional slice. Comparison of Eq. 3.20 to Eq. 3.22 suggests that the second term in each of these equations can be set equal to each other to yield

$$\frac{1}{2} m^2 \overline{N_r}^2 = \frac{1}{N'} \sum_{n,n\neq 0} |a_n|^2. \tag{3.23}$$

Solving for m gives

$$C_{eq,i} \equiv m = \frac{\sqrt{2 \frac{1}{N'} \sum_{n,n\neq 0} |a_n|^2}}{\overline{N_r}} \tag{3.24}$$

where C_{eq} is defined to be the average image equivalent contrast. C_{eq} is essentially the average sine wave modulation contrast associated with the amplitude of the various spatial frequencies that make up the image. C_{eq} is sensitive to changes in both the edge sharpness and size of the image structure.

For the discriminative tasks such as determination of visibility impairment, it is convenient to somewhat modify Eq. 3.24 to better reflect the workings of the human visual system. There is considerable evidence that the visual system behaves as if it were Fourier decomposing the scene into channels or adjacent bands of spatial frequency and processing these bands independently (Campbell and Robson, 1964; Campbell and Kulikowski, 1966; Campbell et al., 1968). If the intent is to fully understand the response of the human visual system to image modification, then Eq. 3.24 can be rewritten as

$$C_{eq,i} \equiv m = \frac{\sqrt{2 \frac{1}{N'} \sum_{\Delta n,n\neq 0} |a_n|^2}}{\overline{N_r}} \tag{3.25}$$

where i refers to the ith human visual system channel, and the sum over Δn under the square root sign is carried out for those frequencies that correspond to ith human channel.

If the frequency dependence of the equivalent contrast is not of interest, mean square radiance and thus average equivalent contrast can be calculated using

$$\overline{N_r^2} = \sum_i \left(\overline{N_r} - N_{r,i}\right)^2 / N' \tag{3.26}$$

where $\overline{N_r^2}$ is the mean square radiance calculated on a pixel by pixel basis and N' is the total number of pixels. A pixel is a small area of an actual scene or photograph. Average equivalent contrast is then

$$C_{eq} = \frac{\sqrt{2\sum_i \left(\overline{N_r} - N_{r,i}\right)^2 / N'}}{\overline{N_r}}. \tag{3.27}$$

In summary, a scene can be decomposed into varying light/dark sine waves of various spatial frequencies and intensities whose brightness changes are proportional to a sine wave function (essentially a Fourier decomposition of the image radiance field). Fig. 3.8 shows brightness sine waves of varying frequencies and intensities that make up a decomposition of any two-dimensional pictorial representation of landscape features.

Equivalent contrast C_{eq} is just the average contrast of those sine waves, and $C_{eq,i}$ is the equivalent contrast within specified frequency ranges consistent with the human visual system. C_{eq} can then be used in human visual system models to estimate the probability that a human observer will notice a change in the appearance of a landscape feature as aerosols are added or removed from the atmosphere.

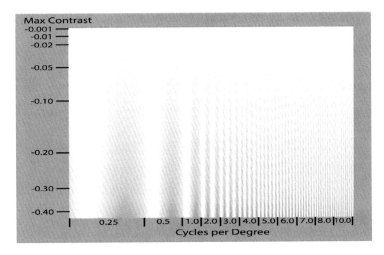

FIGURE 3.8　Brightness sine waves of varying frequencies and intensities.

CONTRAST TRANSMITTANCE IN SPATIAL FREQUENCY SPACE (MODULATION TRANSFER FUNCTION)

The modulation contrast $m = a_n/a_o$ is modified in accordance with Eq. 3.11. The effect of an increase in atmospheric aerosol concentration on the average radiance field is the reduction of the average radiance by a factor equal to the atmospheric transmittance T_r while adding path radiance N_r^*:

$$a_o' = a_o T_r + N_r^* \tag{3.28}$$

where a_o' is the new average radiance. On the other hand, the modulation term $a_n e^{i2\pi n f_o x}$ is only reduced by a factor equal to T_r. This is because radiance differences are attenuated equally by an amount equal to the atmospheric transmittance (see Eq. 3.13); thus the effect of an increase in atmospheric aerosol concentration on Eq. 3.21 is

$$N_r'(x) = a_o T + N_r^* + T_r \sum_n a_n e^{i2\pi n f_o x} \tag{3.29}$$

where $N_r'(x)$ is the one-dimensional radiance field due to an increase in atmospheric particle concentration. The modulation contrast of any sinusoid is then

$$m' = T_r a_n / (a_o T + N_r^*) \tag{3.30}$$

and the transfer of modulation contrast is

$$M_{tf,a} = m'/m = \frac{1}{1 + N_r^*/a_o T}. \tag{3.31}$$

Provided atmospheric turbulence does not interfere with transfer of modulation contrast, Eq. 3.31 is independent of spatial frequency and is the general form for the atmospheric contrast transmittance in spatial frequency space. It is more commonly referred to as the modulation transfer function of the atmosphere (Malm, 1985). Therefore,

$$C_{eq,r} = C_{eq,o} M_{tf,a} \tag{3.32}$$

where $C_{eq,r}$ and $C_{eq,o}$ are the equivalent contrasts at distances r and o.

Comparison of Eqs. 3.18 and 3.31 shows that if $_bN_o = a_o$, then contrast transmittance in real and spatial frequency space is identical. In most cases, the feature within the image of interest is small compared with its surroundings, and average radiance a_o is very nearly the same as background radiance $_bN_o$. This is a very satisfying result. Whether one is interested in using psychophysical spatial frequency models to examine how much aerosol can be introduced into the atmosphere before it is noticed or how image contrast is changed as a function of aerosol load, the calculation is reduced to understanding the dependence of the atmospheric modulation transfer function, or contrast transmittance, on aerosol chemical and physical properties. This result is highlighted in Chapter 5 on perception.

DEPENDENCE OF CONTRAST TRANSMITTANCE (C_t) ON ATMOSPHERIC OPTICAL VARIABLES

Contrast transmittance C_t is the one variable that contains all the information required to describe how various physical descriptors of scenic landscape features are modified as a function of atmospheric particle loading, illumination, and observer–vista geometry. Therefore, it is of interest to examine how sensitive C_t is to changes in atmospheric extinction coefficient (atmospheric particle mass concentration). Two different illumination conditions that correspond to geometries that are conducive primarily to forward- and backscattering are considered.

Fig. 3.9 shows $S \equiv |\Delta C_t / \Delta b_{ext}|$, where $\Delta b_{ext} = 0.01$ km^{-1} as a function of b_{ext}. Fig. 3.9a corresponds to a typical fine particle aerosol mass size distribution, scattering angle $\theta_s = 15°$ and $_bN_o = 0.13 N_s$, where N_s is the Rayleigh sky radiance. Fig. 3.9b corresponds to the same aerosol but with $\theta_s = 125°$ and $_bN_o = 0.5 N_s$. An immediately evident trend shown in the figure is that there is a distance where S is maximum. S decreases to zero as $R \to 0$ and $R \to \infty$. Second, the distance at which S is maximum increases as $_bN_o$ increases (brighter landscapes). In a forward-scattering situation in which landscapes are in a shadow ($C_o \approx -0.87$), S is maximum in the 5–10 km range. Although not shown in Fig. 3.9a, in a backscatter geometry ($\theta_s = 125°$), the most sensitive distance is still around 5–10 km if the landscape is dark. However, the maximum sensitivity drops by about a factor of two and is not nearly as sensitive to distance. On the other hand, Fig. 3.9b shows that when the landscape is more reflective and illuminated ($C_o \approx 0.50$ and $\theta_s = 125°$), the distance of maximum sensitivity increases, is quite sensitive to background b_{ext}, and remains sensitive to changes in b_{ext} long after dark targets have lost their sensitivity (dark targets will have disappeared, while bright targets can still be seen).

Fig. 3.10a and b examines in more detail the relative contributions of N_r^* and T_r to S. Fig. 3.10a shows contributions of N_r^* and T to S at $R = 10$ km (forward scattering, sulfate aerosol, and dark target). Changes in N_r^* are primarily responsible for changes in $M_{tf,a}$ as aerosol is added or subtracted from a clean atmosphere. As background aerosol loading is increased (larger b_{ext}), the relative importance of T_r to S increases to a point where T_r dominates the effect on S. However, it should be emphasized that this only occurs after C_t has increased to a point where landscape features would be barely visible. Fig. 3.10b shows N_r^* and T_r contributions to S at

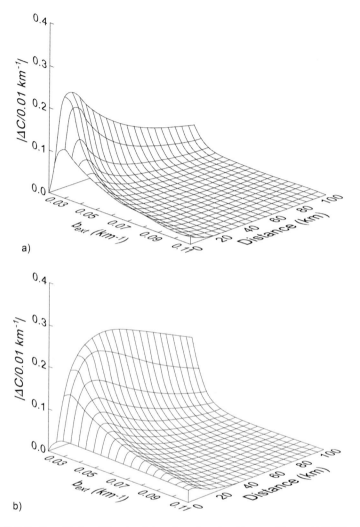

FIGURE 3.9 Sensitivity of the absolute value of contrast transmittance $|\Delta C_t / \Delta b_{ext}|$ plotted as a function of extinction coefficient and distance to a landscape feature. (a) corresponds to a shadowed vista with a scattering angle of 15° and $_bN_o = 0.13N_s$, and (b) corresponds to an illuminated vista with a scattering angle of 125° and $_bN_o = 0.5N_s$.

$R = 70$ km (backscattering, sulfate aerosol, and bright target). With this geometry, attenuation of image-forming information T_r is responsible for much of the change in $M_{tf,a}$. In fact, N_r^* can decrease as b_{ext} increases and can compensate slightly (contribute to cause C_t to increase) for decreases in T_r.

The foregoing discussion shows that increasing b_{ext} (aerosol concentration) for a scattering aerosol in almost all situations causes C_t to decrease. However, in forward-scattering situations in which targets tend to be dark,

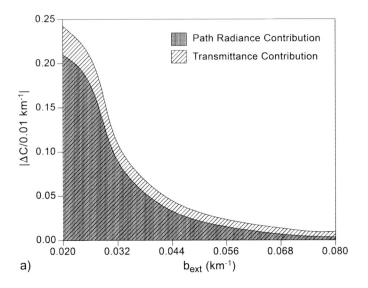

a)

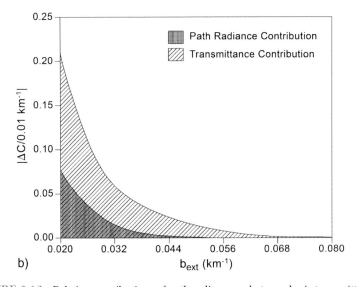

b)

FIGURE 3.10   Relative contributions of path radiance and atmospheric transmittance to the sensitivity of the absolute value of contrast transmittance $|\Delta C_t/\Delta b_{ext}|$ as a function of atmospheric extinction. In (a) the scattering angle is 15° and $_bN_o = 0.13N_s$, implying a shadowed vista, while in (b) the scattering angle is 125° and $_bN_o = 0.5N_s$, implying an illuminated vista.

N_r^* dominates changes in C_t. On the other hand, when the observer is looking at brightly colored landscape features with the sun behind his back (backscatter), the relative importance of N_r^* to visibility becomes smaller, and changes in C_t as a result of increased b_{ext} are more dependent on image-forming radiance being attenuated over the sight path. However, for a specific scene under static illumination conditions, contributions of N_r^* and T_r to changes in C_t as a function of aerosol concentration tend to track each other.

The relative contributions of path radiance and image attenuation as a function of sun angle–landscape–observer geometry are further exemplified in Figs. 3.11–3.14. These images were created using image-processing techniques, described in Chapter 6, that make it possible to isolate the effects of path radiance and atmospheric transmittance on the observed scene. Fig. 3.11a and b shows a shadowed, morning Grand Canyon scene under near-Rayleigh conditions and with $b_{ext} = 0.1$ km^{-1} and a volume scattering function corresponding to a lognormal mass size distribution with

FIGURE 3.11   (a) Grand Canyon scene during morning, shaded, forward-scattering conditions under a near-Rayleigh atmosphere. The distant features are about 30 km distant. (b) The same scene but with a particle atmospheric extinction coefficient equal to 0.9 km^{-1}. The scattering angle β is 30°.

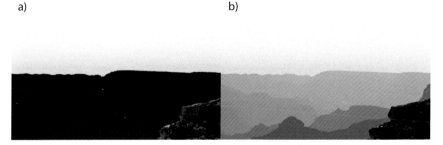

FIGURE 3.12   (a) shows that the already low image radiance is nearly attenuated to zero (black), while (b) shows the added path radiance, which competes with image-forming information, "washing out" the inherent scenic quality, of which there is little, in the shadowed landscape feature.

$D_g = 0.4$ μm and $\sigma_g = 2$. The scattering angle β is equal to 30°. Fig. 3.12 independently shows the contribution of attenuation to visibility impairment in Fig. 3.12a and path radiance in Fig. 3.12b. Because there is little image-forming information to attenuate and the sun–observer angle is conducive to forward scattering where particle scattering is accentuated over molecular scattering, path radiance contributions to contrast transmittance or visibility impairment are significantly greater than attenuation effects. Notice that the attenuated image radiance is near zero (black).

Figs. 3.13 and 3.14 show the same sequence of photographs under more direct illumination conditions. Here, the landscape features are illuminated and the sun–observer scattering angle of 150° corresponds to backscattering where particle scattering is minimized relative to total scattering. In this instance, image attenuation plays a more significant role in contrast transmittance reduction, although it is apparent that path radiance still contributes significantly to the haziness of the scene. Even for a directly illuminated scene, image attenuation is nearly complete in that the attenuated image is black for the distant features and quite dark for nearby landscape features.

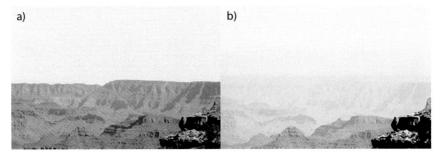

FIGURE 3.13 (a) Grand Canyon scene during afternoon backscatter conditions under a near-Rayleigh atmosphere. The distant features are about 30 km distant. (b) The same scene but with a particle atmospheric extinction coefficient equal to 0.9 km^{-1}. The scattering angle β is 150°.

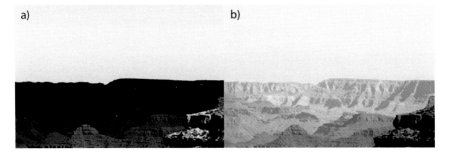

FIGURE 3.14 (a) shows that image radiance is nearly attenuated to zero (black), while (b) shows the added path radiance, which competes with image-forming information, "washing out" the inherent scenic quality of the landscape features.

Because most research to date has focused on apportionment of b_{ext}, and therefore T, to aerosol species, it is fortunate that, for fine particle ($<PM_{2.5}$) scattering aerosols, an understanding of this relationship yields significant insight into how aerosols affect visibility under a wide range of viewing conditions. However, under many circumstances, the major cause of visibility degradation can be associated with path radiance, and path radiance explicitly requires knowledge of the volume scattering function in addition to b_{ext}. Little effort has been expended on examining how path radiance is affected as a function of aerosol characteristics or on apportioning path radiance to aerosol species. Large particles and particles that absorb light contribute to path radiance differently than do fine particles that only scatter light. Furthermore, the contribution of scatterers and absorbers to the angular dependence of path radiance is not additive. Conversely, the effect of total scattering and absorption to b_{ext} is additive. Therefore, when appreciable concentrations of coarse and light absorbing particles or gases are present, knowledge of just b_{ext} (attenuation) may not be adequate to describe changes in visibility.

The concepts discussed above are summarized in Fig. 3.15. Those variables enclosed in the box on the left side of Fig. 3.15 are dependent on illumination–observer geometry, while those on the right are not. Path radiance, a geometry-dependent variable, is combined with atmospheric transmittance, a geometry-independent parameter, and average scene luminance to yield contrast transmittance or modulation transfer function.

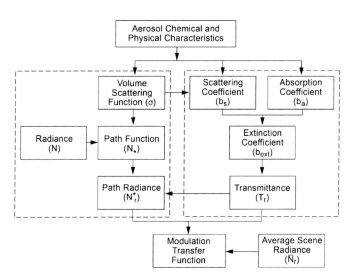

FIGURE 3.15 Flow diagram showing how particle physio-chemical properties relate to the optical variables required to specify the atmospheric modulation transfer function/contrast transmittance.

References

Campbell, F.W., Kulikowski, J.J., 1966. Orientational selectivity of the visual cell of the cat. J. Physio. London 187, 437.

Campbell, F.W., Robson, J.G., 1964. Application of Fourier analysis to modulation response of the eye. J. Opt. Soc. Am. 54, 581.

Campbell, F.W., Cleveland, B., Cooper, G.F., Enroth-Cogell, C., 1968. The spatial selectivity of the visual cell of the cat. J. Physiol. London 198, 237.

Cohen, R., 1978. Applying psychophysics to display design. Photogr. Sci. Eng. 22 (2), 56–59.

Duntley, S.Q., 1948. The reduction of apparent contrast by the atmosphere. J. Opt. Soc. Am. 38 (2), 179.

Henry, R.C., 1977. The application of the linear system theory of visual acuity to visibility reduction by aerosols. Atmos. Environ. 11, 697–701.

Koschmieder, H., 1924. Theorie der horizontalen Sichtweite. Beits. Phys. zur Atmos. 12, 33–55.

Malm, W.C., 1985. An examination of the ability of various physical indicators to predict judgment of visual air quality, Proceedings of the 78th Annual Meeting and Exhibition of the Air Pollution Control Association, paper number 85-10.1, Detroit, 16–21 June, 1985.

Middleton, W.E.K., 1968. Vision through the Atmosphere, corrected edition, University of Toronto Press, Toronto, 250 pp.

Visibility Metrics

Visibility metrics fall into two broad categories: those that are scene dependent and those that are not. For the purposes of this discussion, those that are not scene dependent are referred to as universal metrics, while those that depend on scene characteristics are scene-dependent metrics. Table 4.1 is a list of the visibility metrics that will be discussed, as well as the symbol used to represent each metric, the relationship each metric has to fundamental physical quantities, and at least one reference for each variable.

UNIVERSAL METRICS

Atmospheric Extinction Coefficient

The atmospheric extinction coefficient (b_{ext}), which is the sum of scattering and absorption by particles and gases, is a measure of the alteration of radiant energy as it passes through the atmosphere. It is discussed in some detail in Chapter 2. The effect of the atmosphere on the visual properties of distant objects theoretically can be determined if the concentration and characteristics of air molecules, particles, and absorbing gases are known throughout the atmosphere and, most importantly, along the line of sight between the observer and the object. This connection to the causes of visibility impairment makes b_{ext} particularly useful for apportioning haze to the pollutants that are responsible, though as described in Chapter 2, it can be challenging to do that correctly. Also, as shown below, b_{ext} is fundamental to the definition of all of the other universal metrics. The extinction coefficient has not been a popular visibility metric for the general public, who prefers the use of the seemingly easier-to-understand "visual range," which has the additional advantage of using simple length units (e.g., miles or kilometers), while b_{ext} uses reciprocal length units.

Visibility: The Seeing of Near and Distant Landscape Features. http://dx.doi.org/10.1016/B978-0-12-804450-6.00004-8

TABLE 4.1 Universal and Scene-Dependent Visibility Metrics

Universal Visibility Metrics

Name	Symbol	Units	Relationship to Other Variables	Use	Reference
Atmospheric extinction	b_{ext}	km^{-1}	b_{ext} proportional to particle optical cross section.	Indicator of haze	Trumpler (1930)
Visual range	V_r	km	$V_r = 3.912/b_{ext}$	Farthest distance at which an object can be seen	Bennet (1930) Koschmieder (1924)
Standard visual range	SVR	km	$V_{SVR} = 3.912\Big/(b_{sp} + b_{ap} + b_{ag} + 0.01)$	Normalized visual range	Malm and Molenar (1984)
Unimpaired visual range	V_{ur}	km	$V_{ur} = 0.17/b_{ext}$	Distance seen without noticing haze	Malm and Pitchford (2012)
Deciview	dv	None	$dv = 10\left(\ln \dfrac{b_{ext}}{0.01 \text{ km}^{-1}} \right)$	Haze index	Pitchford and Malm (1994)

Scene-Dependent Visibility Metrics

Name	Symbol	Units	Relationship to Other Variables	Use	Reference
Apparent contrast	C_r	None	$C_r = (_t N_r - _b N_r)/_b N_r$	Contrast of feature against background	Duntley (1948)
Edge detection	Sobel-like operators (G)	None	Proportional to radiance difference $(N_2 - N_1)$ of adjacent pixels	Edge detection	Luo et al. (2002)

Average scene contrast	$\bar{C}$	None	$\bar{C} = \dfrac{\sum_i \left|(\bar{N}_r - N_{r,i})\right|/N'}{\bar{N}_r}$	Average contrast	Malm et al. (2015)
Equivalent contrast	C_{eq}	None	$C_{eq} = \sqrt{\dfrac{2\sum_i \left(\bar{N}_r - N_{r,i}\right)^2/N'}{\bar{N}_r}}$	Proportional to average contrast	Malm (1985)
Root mean square radiance	RMS	None	$RMS = \sqrt{\sum_i \left(\bar{N}_r - N_{r,i}\right)^2/N'}$	Indicator of image quality	Pokhrel and Lee (2011)
Color difference	ΔE	None	$\Delta E = \sqrt{(L_1 - L_2)^2 + (a_1 - a_2)^2 + (b_1 - b_2)^2}$	Color difference between two images	Ibraheem et al. (2012)
Color contrast	C_c	None	$C_{eq,c} = \sqrt{C_{eq,r}^2 + C_{eq,g}^2 + C_{eq,b}^2}$	Average color contrast of an image	Malm et al. (1980)
Just noticeable change/difference	JNC/JND	None	$\Delta C = kC_i$	Suprathreshold. Change in contrast that can be noticed	Ross and Murray (1996)

A complete description of optical variables such as b_{ext}, N_r, and C_{eq} can be found in Chapters 2 and 3 and in the discussion following this table.

Visual Range Concept

Visual range has historically been defined in the context of how far away a black object has to be such that it is just noticeable or visible. This distance was especially relevant for military operations that relied primarily on the human eye to see or detect an enemy target such as a warship or aircraft. However, a landscape feature at a distance equal to its visual range has lost all of its inherent scenic qualities; it is just barely noticeable as some distant object. A more relevant visual range is the distance at which an observer can see and appreciate a landscape feature that is unimpaired by haze other than the naturally occurring background haze associated with air molecules. The distance at which a landscape feature can just be detected will be referred to as the visual range (V_r), while the greatest distance at which an unimpaired landscape feature can be observed will be referred to as unimpaired visual range (V_{ur}). Additional discussion concerning perception of landscape features will be presented in Chapter 5.

Start with the contrast reduction equation found in Chapter 3:

$$C_r = C_o \frac{{}_b N_o}{{}_b N_r} e^{-\bar{b}_{ext} r} = C_o \gamma e^{-\bar{b}_{ext} r} \tag{4.1}$$

where $\gamma = {}_b N_o / {}_b N_r$, and $\bar{b}_{ext}$ is the average extinction over sight path r. Assuming the background is the sky (${}_b N_o = {}_s N_o$ and ${}_b N_r = {}_s N_r$) and further assuming the sky radiance at the observer and landscape feature are the same ($\gamma = 1$), Eq. 4.1 reduces the familiar Koschmieder relationship (Koschmieder, 1924):

$$C_r = C_o e^{-\bar{b}_{ext} r}. \tag{4.2}$$

Solving Eq. 4.1 for $\bar{b}_{ext}$ yields

$$\bar{b}_{ext} = -\frac{1}{r} \ln C_r / \gamma C_o. \tag{4.3}$$

It should be emphasized that $\bar{b}_{ext}$ is the average extinction coefficient of the atmosphere between the observer and the target when they are separated by a distance equal to r.

Let $V_r \equiv r$ be the distance from a feature at which a threshold contrast (a contrast level that can just be detected) of ε is achieved. Eq. 4.3 can then be written as

$$V_r = \frac{1}{\bar{b}_{ext,V_r}} \ln \frac{|C_o| \gamma}{\varepsilon}. \tag{4.4}$$

This relationship is the defining equation for a "monochromatic" visual range of an object with an inherent contrast equal to C_o. In this equation, $\bar{b}_{ext,V_r}$ is the average extinction coefficient between the observer and a landscape feature that is at a distance sufficient to reduce its apparent contrast to ε. It is the same $\bar{b}_{ext}$ as determined by Eq. 4.3 if $C_r = \varepsilon$ and r is equal to the visual range. For a black object, $C_o = -1(|C_o| = 1)$, Eqs. 4.3 and 4.4 become

$$\bar{b}_{ext} = -\frac{1}{r}\ln(C_r/\gamma) \tag{4.5}$$

and

$$V_r = \ln(\gamma/\varepsilon)/\bar{b}_{ext,v_r}. \tag{4.6}$$

In addition, if the earth is assumed to be flat, the atmospheric particle distribution is horizontally homogeneous, the object is viewed at a zenith angle of 90°, and the object is viewed under a cloudless sky (or uniform illumination), then $\gamma = 1$ (the sky radiance at the target and sky radiance at the observation point are equal) and $\bar{b}_{ext} = \bar{b}_{ext,v_r} = b_{ext}$.

If these assumptions are met, Eq. 4.6 yields

$$V_r = -\ln(\varepsilon)/b_{ext}. \tag{4.7}$$

Sometimes, this equation is further simplified by ignoring the absorption component of the extinction coefficient or assuming that it is equal to zero. Then,

$$V_r = -\ln(\varepsilon)/b_{sp}. \tag{4.8}$$

Much of the early visibility perception research concentrated on quantifying ε, the contrast between an object and its background that is "just noticeable" or visible. These early experiments are discussed in more detail in Chapter 5.

If ε, the threshold contrast, is taken to be 0.02, then Eq. 4.7 becomes

$$V_r = 3.912/b_{ext} \tag{4.9}$$

where V_r and b_{ext} are expressed in consistent length and inverse length units, respectively. Eq. 4.9 allows visual range data to be interpreted in terms of extinction and, vice versa, extinction measurements to be interpreted in terms of visual range.

Standard Visual Range

In Eq. 4.9, b_{ext} is the sum of Rayleigh and particle scattering and extinction, and Rayleigh scattering varies with altitude and temperature. b_{sp}, b_{ap},

b_{ag}, and b_{sg} are used to denote scattering and absorption by particles and scattering and absorption by gases, respectively. Because at times it is desirable to compare visual range estimations across monitoring networks with sites at much different elevations, it is convenient to define a visual range that is normalized to some constant b_{sg}. This can be especially important under near-pristine circumstances where Rayleigh scattering is a substantial fraction of the total atmospheric extinction, so site elevation would distort the apparent spatial gradient of visual range. This normalized visual range is referred to as standard visual range (SVR) and is defined as

$$V_{SVR} = \frac{3.912}{(b_{sp} + b_{ap} + b_{ag} + 0.01)} \tag{4.10}$$

where b_{sg} has been set equal to 0.01 km^{-1}, which corresponds to Rayleigh scattering at about 1800 m above sea level (Malm and Molenar, 1984).

Unimpaired Visual Range

Visual range has been misinterpreted by some as that as long as scenic elements of interest are viewed from a point that is within the visual range, there is not any visibility degradation. In general this is not the case. The sensitivity of changes to a landscape feature's appearance to changes in b_{ext} increases with the sight path distance. While visual range is the distance at which a landscape feature is no longer visible, unimpaired visual range (V_{ur}) is the distance at which an observer can just begin to see a change in the appearance of the landscape feature as a result of a specified change in b_{ext}. The change in the appearance of a landscape feature beyond the V_{ur} will be noticeable.

The relationship between extinction and unimpaired visual range can be derived assuming contrast reduction takes the form of Eq. 4.2. In an atmosphere free of atmospheric particles, the contrast of landscape features as a function of distance and atmospheric extinction is given by

$$C_r = C_o e^{-b_{ray} r} \tag{4.11}$$

where b_{ray} is the Rayleigh scattering coefficient. The question is at what distance an observer can see no change in contrast under conditions of an increase in atmospheric extinction equal to b_{ext}. The ensuing contrast of landscape features as a function of distance under this increase in extinction is

$$C_r = C_o e^{-(b_{ray} + b_{ext}) r}. \tag{4.12}$$

Although the farthest distance an observer can see a landscape feature is a threshold problem, the unimpaired visual range question presents itself in the context of a suprathreshold issue in that it becomes one of how much contrast change from some base contrast can be noticed. The simplest relationship for the increase or decrease in contrast that can be just noticed is Weber's law:

$$\Delta C = kC_i \tag{4.13}$$

where $\Delta C = C_f - C_i$, C_f and C_i are given by Eqs. 4.11 and 4.12, and k is a constant determined empirically. There are a number of Weber-type equations that can be found in the literature, but for the purpose of this derivation, Eq. 4.13 will be assumed (Ross and Murray, 1996).

Substituting Eqs. 4.11 and 4.12 into Eq. 4.13 and solving for r defines the unimpaired visual range as

$$V_{ur} = \text{const}/b_{ext} \tag{4.14}$$

where const $= -\ln(1 - k)$. The choice of k is situation dependent and is discussed in more detail below. For a value of $k = 0.16$, const $= 0.17$ and the unimpaired visual range is given by

$$V_{ur} = 0.17/b_{ext}. \tag{4.15}$$

Fig. 4.1 is a plot of visual and unimpaired visual range. In a particle-free atmosphere, the visual range is 391 km, while the unimpaired visual range is infinite in that it is defined relative to a Rayleigh atmosphere. It should be kept in mind that even though $V_{ur} = \infty$, objects farther than

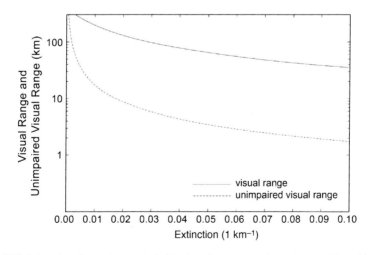

FIGURE 4.1   Plot of visual range (solid line) and unimpaired visual range (dotted line) as a function of particle extinction.

391 km could not be seen. At atmospheric particle concentrations equal to Rayleigh or about 0.01 km^{-1} (total b_{ext} = 0.02 km^{-1}), V_r = 196 km and V_{ur} = 17 km, while at b_{ext} = 0.10 km^{-1}, the two visual ranges are 35 and 1.7 km, respectively. Under the above assumptions, the increase in particle scattering that would cause a just noticeable change in landscape visual air quality would occur at a value that is about 8% greater than Rayleigh scattering, or about 0.0008 km^{-1}.

Deciview Scale

Because neither extinction nor the visual ranges defined above are proportional to perceived changes in visual air quality, a scale was derived such that a change in the index would be perceived to be the same under any visual air quality (haziness) condition. Weber's law as defined above can be written in terms of extinction and distance r to a landscape feature as

$$C_f - C_i = \Delta C = kC_i. \tag{4.16}$$

$$C_o e^{-(b_{ext}+\Delta b_{ext})r} - C_o e^{-(b_{ext})r} = kC_o e^{-(b_{ext})r} \tag{4.17}$$

where Δb_{ext} is an incremental increase in extinction from some baseline extinction.

Eq. 4.17 can be solved to yield

$$-\Delta b_{ext} r = \text{const} \tag{4.18}$$

where const = $-\ln(1 + k)$.

If r is an inverse function of b_{ext} such that the visual range ($r \equiv V_r = 3.912/b_{ext}$) or the distance at which a change in extinction evokes the largest contrast change ($r = 1/b_{ext}$), then Eq. 4.18 becomes

$$\frac{\Delta b_{ext}}{b_{ext}} = \text{const.} \tag{4.19}$$

Eq. 4.19 states that a fractional or percent change in extinction equal to const will be just noticed and, furthermore, that this percent change is perceived to be constant under any atmospheric extinction level as long as there is a landscape feature at a distance proportional to $1/b_{ext}$. Mathematically

$$dp = L\frac{db_{ext}}{b_{ext}} \tag{4.20}$$

where dp is some incremental perceived change in an observed landscape. Integrating Eq. 4.20,

$$P = L\{\ln(b_{ext})\} + K. \tag{4.21}$$

K, a constant, can be determined from the boundary condition that at some reference $b_{ext} = b_{ag}$ and $P = 0.0$. Then $K = -L\{\ln(b_{as})\}$ and Eq. 4.21 can be written as

$$P = L\left(\ln \frac{b_{ext}}{b_{sg}} \right). \tag{4.22}$$

If L is chosen to be equal to 10 and $b_{sg} = 0.01$ km^{-1}, then P is defined as deciviews and Eq. 4.22 becomes

$$dv = 10\left(\ln \frac{b_{ext}}{0.01 \text{ km}^{-1}} \right). \tag{4.23}$$

Eq. 4.23 defines the deciview scale (Pitchford and Malm, 1994). One deciview corresponds to about a 10% change in extinction, and one deciview will be perceived to be approximately constant under the assumptions of atmospheric and landscape feature conditions stated above. The question of whether a one deciview change, or a 10% change in extinction, can be noticed will be discussed later.

SCENE-DEPENDENT VISIBILITY METRICS

Digital or photographic cameras can be used to collect two-dimensional arrays, referred to as pixels, of film densities or digitized voltages, in three color channels, that are proportional to the image radiance field. The response of film or camera detector arrays to scene radiance is nonlinear, and to derive true radiance levels, the camera systems must be calibrated. The calibration process will be discussed in Chapter 7 on monitoring techniques.

The overarching goal of any scene-dependent metric is to relate human judgments of the visual air quality of a scene under varying haze conditions to some basic atmospheric optical variable such as atmospheric extinction. Relationships between extinction and various scene-dependent metrics will also be discussed in Chapter 7.

The following discussion assumes calibrated radiance levels, although the calculations could be carried out using the reported or uncalibrated digital values for each collected image. The problem then, of course, would be that the calculations done by different individuals using different film or camera systems would not be intercomparable.

Contrast

Eqs. 4.24 and 4.25 define universal and modulation contrast, respectively:

$$C_r = ({}_{l1}N_r - {}_{l2}N_r)/{}_{l2}N_r \tag{4.24}$$

$$m_r = ({}_{l1}N_r - {}_{l2}N_r)/({}_{l1}N_r + {}_{l2}N_r) \tag{4.25}$$

where *l*1 and *l*2 are two distinct landscape features or one is the background sky. Either of these contrast equations can be used to describe the visibility of a contrast edge either of contiguous landscape features or of landscape features as seen against a background sky. However, the advantage of Eq. 4.24 is that it, with a number of assumptions, relates to atmospheric extinction in a simple and straightforward way (see Eq. 4.2).

Sobel and Similar Indexes

Given an image, Eq. 3.13 states that the radiance difference between adjacent pixels representing landscape features at the same distance is attenuated in direct proportion to the atmospheric transmittance over the path between the landscape features represented by the pixels and the observer. The radiance difference between pixels can be estimated by a gradient operator applied to an image of interest. The average radiance difference associated with any given image then becomes an index that is proportional to the amount of haze present when the photo was taken.

One of the more popular gradient operators is the Sobel filter. Others such as the Roberts, Scharr, and Prewitt operators are similar, but all essentially identify the areas of an image with a change or gradient in the image radiance field. In order to perform edge detection with the Sobel operator, the original image is convolved with the following two kernels:

$$G_y = \begin{bmatrix} -1 & -2 & -1 \\ 0 & 0 & 0 \\ 1 & 2 & 1 \end{bmatrix} \quad \text{and} \quad G_x = \begin{bmatrix} -1 & 0 & 1 \\ -2 & 0 & 2 \\ -1 & 0 & 1 \end{bmatrix}. \tag{4.26}$$

Let $N(x,y)$ be a point in the original image, $G_y(x,y)$ be a point in an image formed by convolving with the first kernel, and $G_x(x,y)$ be a point in an image formed by convolving with the second kernel. The gradient can then be defined as

$$\nabla N(x,y) = G(x,y) = \sqrt{G_x^2 + G_y^2}. \tag{4.27}$$

The edges identified by these various operators and associated with various landscape features in the image are attenuated in proportion to the amount of haze in the atmosphere and the distance between the camera and landscape features corresponding to the edge at the time the photo was taken. The visibility index is usually taken to be the average of the pixel values of the filtered image $G(x, y)$.

Fast Fourier Transform

Another approach to identifying radiance difference edges is through the use of the fast Fourier transform (FFT). The FFT transforms the spatial variations in a radiance field into a frequency domain according to

$$F(u, v) = \sum_{i=0}^{N-1} \sum_{j=0}^{N-1} f(i, j) \exp\left[-i2\pi\left(\frac{ui}{N} + \frac{vj}{N}\right)\right] \quad (4.28)$$

where $f(i, j)$ is the radiance associated with each pixel i and j, and u and v are frequencies associated with the i and j spatial domain. By truncating $F(u, v)$ to higher spatial frequencies and then inverting Eq. 4.28 back to the spatial domain, it is possible to extract only the edges of the embedded landscape features, essentially accomplishing the same thing as the gradient filters discussed above. Again, the visibility index is taken to be the average value of the pixel values of the inverted filtered image. The advantage of the FFT approach is that the analyst, by selecting the frequency domain he wishes to consider, has some control over the "sharpness" of the edges used in the analysis.

Root Mean Square Radiance

The mean squared error (MSE) is a routinely used metric to quantify the visual difference between two images:

$$MSE = \sum_{i} (_2N_i - _1N_i)^2 / N' \quad (4.29)$$

where $_2N_i$ and $_1N_i$ are the derived radiance values for each pixel in respective images and N' is the total number of pixels in the images. The peak signal-to-noise ratio (PSNR), another difference metric, is defined as

$$PSNR = 10 \log \frac{m^2}{MSE} \quad (4.30)$$

where m is the maximum radiance that an image can have. MSE measures image difference while PSNR is more closely aligned with image fidelity.

Cohen (1978) has argued that a metric akin to the MSE, the mean square radiance fluctuation (MSRF), given by Eq. 3.26, is the most relevant psychophysical variable for describing perception of image radiance fields. It is defined as

$$\overline{N_r^2} = \sum_i (\overline{N_r} - N_{r,i})^2 / N' \tag{4.31}$$

where $\overline{N_r^2}$ is the mean square radiance calculated on a pixel-by-pixel basis, $\overline{N_r}$ is the average radiance across all pixels in the image, $N_{r,i}$ is radiance associated with an individual pixel i, and N' is the total number of pixels. r refers to the distance to the landscape feature represented by pixel i. Pokhrel and Lee (2011) used the square root of Eq. 4.31,

$$RMS = \sqrt{\overline{N_r^2}}, \tag{4.32}$$

as an indicator of image quality.

Average Scene Contrast

Another possible contrast visibility index is the overall average image contrast, defined as

$$\overline{C} = \frac{\sum_i \left| (\overline{N_r} - N_{r,i}) \right| / N'}{\overline{N_r}} \tag{4.33}$$

where $\overline{N_r}$ and $N_{r,i}$ are the average scene radiance and the pixel-by-pixel radiance of each image element of the scene, respectively, and N' are the total number of pixels.

Equivalent Contrast

Universal or modulation contrasts in and of themselves are somewhat limited because they cannot account for the human visual system response to size and shape of various scenic landscape features. Any two-dimensional scene can be Fourier decomposed into sums of sinusoidal patterns of varied frequencies and amplitudes. The average contrast of these sinusoids is referred to as equivalent contrast and is given by Eq. 3.25. Average equivalent contrast is just the square root of the mean square radiance fluctuation divided by the average scene radiance:

$$C_{eq} = \frac{\sqrt{2 \sum_i (\overline{N_r} - N_{r,i})^2 / N'}}{\overline{N_r}} \tag{4.34}$$

where $\overline{N_r}$ is the average radiance over the entire image.

The advantage of this approach over simple contrast is that the average equivalent contrast incorporates all radiance variability across a scene, whether it is in the form of sharp edges or slow spatial variability of radiance values associated with receding landscape features, such as rolling hills represented by altitude changes of landscape features that occur gradually and change with distance.

If the power spectra of a Fourier-transformed image are known, the equivalent contrast associated with various spatial frequencies can be calculated. For a one-dimensional "slice" of the general image radiance field,

$$C_{eq,\Delta n} \equiv m = \frac{\sqrt{2\frac{1}{N'}\sum_{\Delta n, n \neq 0}|a_n|^2}}{\overline{N}_r} \tag{4.35}$$

where $C_{eq,\Delta n}$ is the equivalent contrast in the frequency range associated with Δn, and the sum of the power spectra is only over those frequencies (Malm, 1985). The use of equivalent contrast in spatial-frequency-specific bands is helpful in discrimination studies in which the sensitivity of the human visual system is a function of spatial frequencies. It is well established that the human visual system (HVS) is most sensitive to spatial frequencies around 3 cycles per degree (cpd) and decreases for frequencies greater or less than that value.

The advantage to using contrast or equivalent contrast as formulated in Eq. 4.34 over gradient indexes is that, by dividing mean square radiance or radiance differences by background radiance or average radiance, the index becomes a ratio of radiance values as opposed to absolute differences. By taking ratios of radiance values, the effects of varying exposures and calibrations on absolute pixel values tend to cancel out.

Image Color Characteristics

The perception of color is neither purely physical nor purely psychological. It is the evaluation of radiant energy in terms that correlate with visual perception. The study of how the human observer responds to radiant energy of different wavelengths has been ongoing for some 200 years. Ibraheem et al. (2012) is a fairly easy-to-read review of various color models, including human visual system, color difference, color printing, display models, and more. Basically, they all weight red, blue, and green color channels using some algorithm to produce the sensation of the full color spectrum.

Any of the indexes discussed above can be calculated using the red, green, and blue color channels that make up any color photograph or digital image. These indexes can be combined in a variety of ways. A color model, referred to as color contrast, was proposed by Malm et al. (1980) and also by Horvath et al. (1987):

$$C_{eq,c} = \sqrt{C_{eq,r}^2 + C_{eq,g}^2 + C_{eq,b}^2} \qquad (4.36)$$

where $C_{eq,r}$, $C_{eq,g}$, and $C_{eq,b}$ are the red, green, and blue color channel equivalent contrasts, respectively.

Delta E and Chromaticity

Although many theories of color vision have been proposed, the Commission Internationale de l'Eclairage (CIE), over a period of four decades, has set colorimetric standards that form the basis of the so-called CIE method of color specification. While these standards are adequate for the evaluation of color in a controlled laboratory setting, they may not accurately quantify color in natural settings, even though they have been widely used. The delta E (ΔE) color difference model and the chromaticity difference model have been used in visibility perception studies and actually have been included in Environmental Protection Agency modeling guidance documents (Latimer et al., 1978; Latimer and Ireson, 1980).

The CIE chromaticity diagram is one way to quantify the concept of color. In a chromaticity diagram, the spectral distribution of light is first weighted with three functions corresponding to the spectral response of the human eye. For any color of light, these three values define a coordinate value that defines a point in space. The projection of all possible points onto a unit plane ($x + y + z = 1$) defines a two-dimensional shape called a chromaticity diagram (see Fig. 4.2). Monochromatic light, or light at one wavelength, defines the outer edges of the diagram, and white light is located in the center. Any color can thus be represented by its coordinates (x,y) on the diagram.

Since the chromaticity diagram does not distinguish between differences in intensity (e.g., between yellow and brown or white, gray, and black), chromaticity coordinates (x,y) must be used in conjunction with a descriptor of light intensity for a complete specification of color. Thus, a color solid can be formed by taking the two-dimensional chromaticity diagram and adding a third dimension perpendicular to that plane to represent brightness.

Fig. 4.3 is a drawing of a color solid. The brightness in such a coordinate system is usually specified by a value L, while color parameters a and b are related to the x and y values, respectively, shown in the chromaticity diagram. The Munsell color system is the most widely used means of specifying colors. In this system, colors are arranged by brightness (or value, which is $L^*/10$), hue (the shade of color, for example, yellow, red, green,

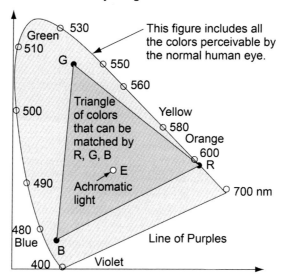

FIGURE 4.2 Commission Internationale de l'Eclairage (CIE) 1931 color space chromaticity diagram. The outer curved boundary is the spectral (or monochromatic) locus, with wavelengths shown in nanometers.

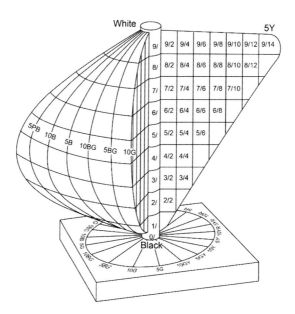

FIGURE 4.3 Munsell color system, showing circles of hues; the neutral values from 0 to 10; and the chromas or saturation.

or blue), and chroma or saturation (the degree of departure of a given hue from a neutral gray of the same value).

In 1976, the CIE adopted two color-difference formulas for calculating the perceived magnitude of color differences. Color differences are specified by a parameter, ΔE, given by

$$\Delta E = \sqrt{(L_1 - L_2)^2 + (a_1 - a_2)^2 + (b_1 - b_2)^2} \ . \tag{4.37}$$

ΔE is a function of the change in brightness or value (ΔL) and the change in chromaticity (Δx and Δy, which are proportional to a and b, respectively, in Eq. 4.37). The objective of the ΔE parameter is to provide a univariate measure of the difference between two arbitrary colors as perceived by humans. This parameter allows us to make quantitative comparisons of the perceptibility of two plumes, even though one may be a reddish discoloration viewed against a blue sky, while the other may be a white plume viewed against a dark forest canopy.

ΔE is related to the distance between two colors in a color space (e.g., Fig. 4.3), with appropriate weights given to the sensitivity of the human eye–brain system to the specific changes in chroma and saturation involved. In this way, color changes having equal ΔE values can be said to be equally perceptible.

The ΔE formalism has been used extensively (Latimer et al., 1978; Latimer and Ireson, 1980) to predict the perceptibility of plumes viewed against a variety of backgrounds. However, the color space associated with ΔE, developed under laboratory conditions using "color chips" and uniform backgrounds, may not be directly applicable to perception of color differences under natural conditions. Matamala and Henry (1990) found that the eye–brain system may be much more sensitive to color differences in a natural setting than predicted by the ΔE formalism. Furthermore, the ΔE formalism does not account for the change in plume perceptibility resulting from changes in plume size, shape, and edge sharpness. To that end, Henry (1986) introduced a new color difference parameter, which he referred to as ΔF, which incorporated the dependence of the human visual system to varying spatial frequencies inherent to different landscape features and layered hazes.

Kim and Kim (2005) defined a color space in terms of color, hue, and saturation differences based on the uncalibrated digital values of the three color channels of a digital camera and related them to independently measured b_{ext}.

All colorimetric schemes have serious shortcomings when applied to measurements of natural landscape colors. A major limitation associated with using a chromaticity diagram is deciding on the proper "white point," while the CIE color space requires a definition of Y_m, the Y tristimulus value of some reference "white object" color stimulus (MacAdam, 1942, 1943).

Both problems are centered on the inability to establish the chromatic adaptation state of the eye–brain system in a natural environment.

THE JUST NOTICEABLE CHANGE (JNC)

One concept that has evolved as a possible visibility metric to characterize visibility impairment is the just noticeable difference (JND) index (Henry, 1977, 1979; Carlson and Cohen, 1978). JND refers to the minimum change in visual stimuli that causes the physiological and psychophysical response in the human visual system that is just noticeable when two stimuli are viewed side by side. The JND concept has been used as a tool to assess the perceptual impact of distortions introduced in the production, distribution, and display of video or screen images. A variety of JND models have evolved over the years, with the more recent models incorporating the human visual system's sensitivity to color and brightness contrast as a function of spatial frequencies of image details, luminance levels, and flicker (Geisler, 1989; Sarnoff Corporation, 2001; Wang et al., 2004; Chen and Liu, 2014).

To differentiate between a JND associated with distortion in a video display and a JND in landscape features due to haze, Malm and Pitchford (1989) introduced the idea of a just noticeable change (JNC) when referring to a change in visibility impairment that is just noticeable.

The JNC concept is readily applicable to layered-haze thresholds in which haze forms a contrast edge within the scene itself (Ross et al., 1988, 1990) to determine whether one can see the haze or not. It is also applicable under uniform haze conditions in which a contrast edge is at its threshold value. The question becomes one of how much additional haze is required to make that contrast edge drop below the perception threshold, or conversely, how much of a decrease in haze would allow some contrast edge that is below the detection threshold to become visible. The threshold or JNC for these types of hazes that have a discernible edge is usually expressed as the contrast between the haze and its background or contrast of some scenic landscape feature as viewed against some background.

Applying the JNC concept to changing landscape features as a function of increasing or decreasing haze levels is not straightforward and is paradigm dependent. In a real-world situation, comparison of side-by-side scenes is not possible. An observer would have to compare some reference scene to an impaired scene at some earlier increment of time, possibly days, weeks, or months. Under these circumstances, the observer would have to remember what the unimpaired scene looked like and judge whether the current appearance of the scene is different, possibly at a different time of day with different lighting conditions, cloud cover, or weather conditions.

Studies have shown, discussed in Chapter 5, that each scenic element is unique, and judgments of visual air quality inherently include the variability of these unique features. Observers cannot separate the singular effects of haze from other ever-changing landscape feature characteristics. Therefore, establishing a JNC associated with haze under varying lighting and changing meteorological conditions is somewhat of an ill-defined problem.

However, from a hypothetical construct of a side-by-side comparison of scenes, the concept of a JNC associated with uniform haze is useful to set a lower limit or threshold on the amount of haze that could be noticed.

Weber's-Law-Type JNC

A very simple suprathreshold model, known as Weber's law, was discussed above in the context of unimpaired visual range. Basically, Weber's law states that the change in contrast that is just noticeable is a constant fraction of the initial contrast. For the most part, this relationship works well for initial contrasts well above the threshold contrast. A relatively simple, in concept, JNC model that incorporates contrast thresholds near the threshold contrast was proposed by Carlson and Cohen (1978).

The basic hypothesis of the model is that, in order to perceive a change in the appearance of a video display screen with a given error rate, the change in the mean square luminance per unit frequency, integrated over any single human visual system channel, must be a constant fraction of the interfering signal. Mathematically,

$$C_{eq,f}(v)^2 - C_{eq,i}(v)^2 = C_{eq,T}(v)^2 + k(v)C_{eq,i}(v)^2 \qquad (4.38)$$

where $C_{eq,f}(v)^2$ and $C_{eq,i}(v)^2$ are the frequency-dependent final and initial equivalent sine wave contrasts, respectively, $C_{eq,T}(v)$ is the threshold contrast, and $k(v)$ is a constant fraction at each frequency. $C_{eq,f}(v)$, $C_{eq,i}(v)$, and $C_{eq,T}(v)$ are modulation contrasts defined as

$$C_{eq}(v) \equiv \frac{\{N_1(v) - N_2(v)\}}{\{N_1(v) + N_2(v)\}} \qquad (4.39)$$

where N_1 and N_2 are the maximum and minimum radiance values, respectively, of a brightness field that is varied in accordance with a sine wave response. Eq. 4.38 is in the form of a contrast discrimination model and can be used to predict JNCs in a scenic brightness structure. Furthermore, notice that when $C_{eq,i} \gg C_{eq,T}$, Eq. 4.38 reduces to a Weber's-law-type equation. The frequency-dependent equivalent contrast can be calculated using Eq. 4.35.

The value of the proportionality constant $k(v)$ is obtained from contrast discrimination experiments under conditions where the initial contrast is much greater than the threshold contrast (Carlson and Cohen, 1978). The constant $k(v)$ as a function of spatial frequency is shown in Fig. 4.4. Notice that $k(v)$ increases as spatial frequency increases. $k(v)$ slowly varies with changes in experimental condition and procedure. In some experiments, display luminance was varied over four orders of magnitude, with the resulting change in contrast required for detection varying by less than a factor of 3.

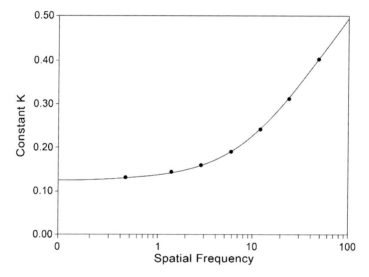

FIGURE 4.4   Proportionality constant that defines the contrast change that is perceptible, shown as a function of spatial frequency.

Substituting $C_{eq,f} = C_{eq,i} M_{tf,a}$ into Eq. 4.38 and solving for M_{tf} yields

$$M_{tf,jnc} = \frac{C_{eq,T}^2(v)}{C_{eq,i}^2(v)} + k(v) + 1 \qquad (4.40)$$

where $M_{tf,jnc}$ is the modulation transfer function required to evoke one JNC, given an initial equivalent contrast of $C_{eq,i}$.

Fig. 4.5 is a three-dimensional plot of Eq. 4.32 showing the change in the modulation transfer function required for one JNC as a function of initial contrast and spatial frequency.

Studies investigating eye fixation and eye motion as observers look at pictures show that pictorial areas with little modulation receive very

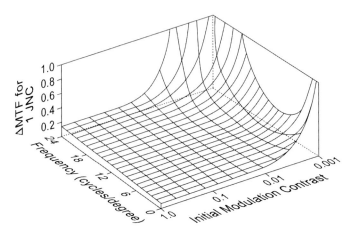

FIGURE 4.5   Change in modulation transfer function required to evoke one JNC as a function of initial contrast and spatial frequency.

little attention, while higher-modulated scenic features receive more (Buswell, 1935). Because sharp contrast edges are quite sensitive to changes in atmospheric modulation transfer, they are good patterns for predicting the relationship between JNCs in scenic appearance and increases in atmospheric aerosol load.

An exemplary application of Eq. 4.40 is a calculation of the percent change in extinction required to evoke a JNC for a dark landscape feature viewed against a background sky. First, for a dark background viewed against a background horizon sky,

$$M_{tf,a} = \frac{1}{1 + N_r^* / a_{oo} T_r} \approx T_r \tag{4.41}$$

is true if the path radiance can be approximated by $N_r^* = N_s(1 - T_r)$ and $a_{oo} \approx N_s$. Therefore ΔM_{tf} is equal to $\Delta T_r = \exp(-\Delta b_{ext} R)$. For small or low contrasts, $C \approx 2C_{eq}$ and C_{eq} for a contrast edge is $C_{eq} = 0.14 C_{edge}$. C_{edge} is the modulation contrast between two adjacent landscape features or between a landscape feature and the sky. With these assumptions, Eq. 4.40 can be written as

$$JNC = \frac{\Delta b_{ext}}{b_{ext}} = \frac{1}{\tau} \ln \left(C_T^2 e^{2\tau} + k + 1 \right)^{\frac{1}{2}} \tag{4.42}$$

where $\tau = b_{ext} R$ is the optical depth. Eq. 4.42 expresses the percent change in extinction required for one JNC as a function of threshold universal contrast, constant k, and optical depth. Eq. 4.42 is only applicable to universal contrast and is so written because universal contrast for a black-shadowed landscape feature is approximately equal to T_r.

Eq. 4.42 is used in Fig. 4.6 to plot the percent change in extinction as a function of optical depth. Here, k was assumed to be 0.158 and $C_T = 0.02$. Note that the smallest percent change in extinction of about 4% occurs at an optical depth of about 2.4. If the atmosphere is free of particles and the background extinction coefficient is near Rayleigh at 0.01 km^{-1}, then the distance to the landscape feature that would first show a JNC in contrast would be at 238 km ($R = \tau/b_{ext} = 2.38/0.01$ km^{-1}), while in an atmosphere where $b_{ext} = 0.1$ km^{-1}, that distance would be 23.8 km. Also, notice that at an optical depth of 3.9 where a dark landscape feature is at a distance from the observer equal to the visual range, it would take a 10% change in extinction to evoke one JNC. This is the condition under which the deciview is defined, in that one deciview is about one JNC for objects at the visual range.

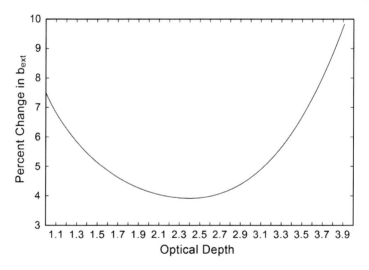

FIGURE 4.6   Percent change in extinction required to evoke a one JNC as a function of optical depth τ.

The above calculation can be repeated for every contrast edge in a scene in a recursive way to estimate the total number of JNCs that would occur for a given reduction/increase in extinction.

References

Bennet, M.G., 1930. The physical conditions controlling visibility through the atmosphere. Quart. J. Roy. Meteorol. Soc. 56, 1–29.

Buswell, G.T., 1935. How People Look at Pictures: A Study of the Psychology of Perception in Art. The University of Chicago Press, Chicago, IL.

Carlson, C.R., Cohen, R.W., July 1978. Image Descriptors for Displays: Visibility of Displayed Information. RCA Laboratories, Princeton, NJ.

Chen, Z., Liu, H. JND modeling: Approaches and applications, presented at the 19th International Conference on Digital Signal Processing, Hong Kong, 20–23 August 2014.

Cohen, R.W., 1978. Applying psychophysics to display design, photographic design and engineering. Photographic Sci. Eng. 22 (2), 56–59.

Duntley, S.Q., 1948. The reduction of apparent contrast by the atmosphere. J. Opt. Soc. Am. 38 (2), 179.

Geisler, S.G., 1989. Sequential ideal-observer analysis of visual discriminations. Psych. Rev. 96 (2), 267–314.

Henry, R.C., 1977. Application of the linear system theory of visual acuity to visibility reduction by aerosols. Atmos. Environ. 11, 697–701.

Henry, R.C., 1979. The human observer and visibility: Modern psychophysics applied to visibility degradation. In: Proceedings: View on Visibility—Regulatory and Scientific. Air Pollution Control Association, Pittsburgh.

Henry, R.C., 1986. Improved predictions of plume perception with a human visual system model. J. Air Pollut. Control Assoc. 36, 1353–1356.

Horvath, H., Gorraiz, J., Raimann, G., 1987. The influence of the aerosol on the color and luminance differences of distant target. J. Aerosol Sci. 18, 99–112.

Ibraheem, N.A., Hasan, M.M., Khan, R.Z., Mishra, P.K., 2012. Understanding color models: review. ARPN J. Sci. Technol. 2 (3).

Kim, K.W., Kim, Y.J., 2005. Perceived visibility measurement using the HIS color difference method. J. Korean Phys. Soc. 46 (5), 1243–1250.

Koschmieder, H., 1924. Theorie der horizontalen Sichtweite. Beitr. Phys. Freien. Atmos. 12, 33–55.

Latimer, D.A., Ireson, R.G., 1980. Workbook for Estimating Visibility Impairment, EPA-450/4-80-031, U.S. Environmental Protection Agency, Office of Air Quality Planning and Standards, Research Triangle Park, NC.

Latimer, D.A., Bergstrom, R.W., Hayes, S.R., Lui, M.K., Seinfeld, J.H., Whitten, G.Z., et al., 1978. The Development of Mathematical Models for the Prediction of Anthropogenic Visibility Impairment, EPA-450/3-78-110a,b,c, U.S. Environmental Protection Agency, Office of Air Quality Planning and Standards, Technical Support Division, Research Triangle Park, NC.

Luo, C.H., Liu, S.H., Yuan, C.S., 2002. Measuring atmospheric visibility by digital image processing. Aerosol Air Quality Res. 2 (1), 23–29.

MacAdam, D.L., 1942. Visual sensitivities to color differences in daylight. J. Opt. Soc. Am. 32, 247.

MacAdam, D.L., 1943. Specification of small chromaticity differences. J. Opt. Soc. Am. 33, 18.

Malm, W.C. An examination of the ability of various physical indicators to predict judgement of visual air quality, paper #85-10.1, presented at the 78th annual meeting of Air Pollution Control Association, Detroit, 16 June 1985.

Malm, W.C., Molenar, J.V., 1984. Visibility measurements in National Parks in the Western United States. J. Air Pollut. Control Assoc. 34, 899–904.

Malm, W.C., Pitchford, M.L., 1989. The use of an atmospheric quadratic detection model to assess change in aerosol concentrations to visibility, paper #89-67.3, presented at the 82nd Annual Meeting of the Air & Waste Management Association, Anaheim.

Malm, W.C., Pitchford, M.L. A review of old visibility metrics and a proposal of a new metric, presented at the Air & Waste Management Association specialty conference, Aerosol and Atmospheric Optics: Visibility and Air Pollution, Whitefish, Montana, September 24–28, 2012.

Malm, W.C., Leiker, K.K., Molenar, J.V., 1980. Human perception of visual air quality. J. Air Pollut. Control Assoc. 30 (2), 122–131.

Malm, W.C., Cismosky, S., Schichtel, B.S. Use of webcam images for quantitative character-ization of haze, presented at the Air & Waste Management Association Annual Meeting, Raleigh, NC, June 2015.

Matamala, L.V., Henry, R.C., 1990. Initial results of a color matching study in the Grand Canyon. In: Mathai, C.V. (Ed.), Transactions: Visibility and Fine Particles. Air and Waste Management Association, Pittsburgh.

Pitchford, M.L., Malm, W.C., 1994. Development and applications of a standard visual index. Atmos. Environ. 28 (5), 1049–1054.

Pokhrel, R., Lee, H., 2011. Algorithm development of a visibility monitoring technique using digital image analysis. Asian J. Atm. Environ. 5 (1), 8–20.

Ross, H.E., Murray, D.J. (Ed. and Transl.), 1996. E.H. Weber on the Tactile Senses, second ed. Erlbaum (UK) Taylor & Francis, Hove, East Sussex, UK.

Ross, D.M., Malm, W.C., Iyer, H.K., Loomis, R.J., 1988. Human detection of layered haze using natural scene slides with a signal detection paradigm. In: Proceedings of the 81st Annual Meeting of the Air Pollution Control Association, Pittsburgh.

Ross, D.M., Malm, W.C., Iyer, H.K., Loomis, R.J., 1990. Human visual sensitivity to layered haze using computer generated images. In: Mathai, C.V. (Ed.), Transactions of Visibility and Fine Particles. Air and Waste Management Association, Pittsburgh.

Sarnoff Corporation. JND: A Human Vision System Model for Objective Picture Quality Measurements, Sarnoff Corporation, Princeton, NJ, June 2001.

Trumpler, R.J., 1930. Preliminary results on the distances, dimensions and space distribution of open star clusters. Lick Observatory Bull. 14 (420), 154–188.

Wang, Z., Bovik, A.C., Sheikh, H.R., Simoncelli, E.P., 2004. Image quality assessment: from error visibility to structural similarity. IEEE Trans. Image Process. 13 (4), 600–612.

Human Perception of Haze and Landscape Features

The rationale for visibility perception research is rooted in the 1977 Clean Air Act Amendments (1977 CAAA, Public Law 95-95) that specifically identified visibility as an air-quality-related value of certain protected geographic areas. The amendments also charged the responsible federal land managers (FLMs) with the duty of protecting the visibility of those areas from adverse impairment. However, adverse impairment was not defined in a quantifiable manner that could be used effectively by the FLMs to carry out this mandate. The Environmental Protection Agency (EPA) attempted to address this situation by defining visibility impairment as "any humanly perceptible change in visibility (visual range, contrast, coloration) from that which would have existed under natural conditions." Consequently, visibility took on a substantially different meaning from how it had been interpreted prior to 1977, and the focus of many visibility studies was directed toward how haze affects the visual quality of scenic landscape features and on whether haze from a single or multiple sources could be seen or detected.

The overriding goal of visibility perception studies was to link judgments of visual air quality and the perceptibility of haze to emissions of particles and gases in the atmosphere such that these emissions could be effectively managed for the purpose of protecting the visibility of scenic landscape features. The link between human perception of haze and aerosol physical/chemical/optical properties is outlined in Fig. 5.1.

Visibility: The Seeing of Near and Distant Landscape Features. http://dx.doi.org/10.1016/B978-0-12-804450-6.00005-X

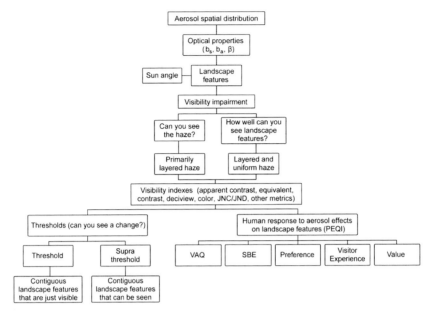

FIGURE 5.1 Flow diagram of how spatial distribution of aerosols relates to haze perceptibility and perceived environmental quality of visually degraded landscape features.

Emissions of gases and particles into the atmosphere result in an aerosol spatial distribution that can manifest itself as either layered or uniform haze. The scattering characteristics of haze, as discussed in Chapter 2, are described by the scattering and absorption coefficients b_{scat} and b_{abs} and the volume scattering function σ, which describes the preferential scattering of radiant energy in a given direction as represented by the scattering angle β. Given an aerosol spatial distribution and its scattering and absorption properties, the sun's zenith and azimuth angles, and the landscape reflectance characteristics, one can estimate the radiance field arriving at the eye of an observer or a photometric instrument. This is represented in Fig. 5.1 as the box labeled "Visibility impairment."

Then, from a human perception perspective, the distribution or variability of the radiance field determines whether an observer can see the haze itself, as well as describes the degradation of the scenic quality of landscape features observed through the haze, whether it is uniform or layered. It is possible that an observer may not even be aware of a haze unless it has a spatial distribution that results in some sort of achromatic or colored contrast edge.

Because of the spatial distribution of radiance values, there have been many and varied visibility metrics proposed to describe the perceptual characteristics of haze as well as the degradation of the visual quality of landscape features seen through the haze. These metrics, discussed in Chapter 4, describe different visual air quality (VAQ) characteristics. Indexes such as extinction coefficient, visual range, unimpaired visual range, and deciview (dv) are all transforms of one another and should be viewed as "local" indicators of visibility impairment. They are not expected to be universal predictors of how observers judge the appearance of varying levels of haze on diverse scenic landscapes. They are independent of landscape characteristics and therefore do not directly describe the appearance of scenic landscape features.

Other metrics, such as apparent universal or equivalent contrast, landscape color, and other landscape-specific indexes, that incorporate landscape characteristics, the integrated haziness between the landscape and the observer, and possible characteristics of the human eye–brain system have the possibility of predicting how observers value and judge VAQ. The concept of just noticeable difference or change (JND/JNC) may be more applicable to estimating the change in haze required to just notice a change in the appearance of landscape features or the appearance of or change in a haze layer (Malm and Pitchford, 1989; Henry, 2002).

Given this wide variety of visibility indexes, an obvious question is when and where they might be applicable to predicting perception-related issues. There is the threshold problem of whether you can just see or notice the presence of a haze. JND/JNC models are directly applicable to whether an observer can notice a layered haze or not. A simple example of layered haze visibility impairment is plume blight, in which the radiance difference between a plume and a background such as the sky can be noticed. The observer can essentially compare the haze layer edge to its background on a real-time basis. So the calculation of a JNC for haze layers reduces to comparing the radiance field without the haze layer to the field with the haze layer. These types of threshold estimations are discussed in more detail in the last section of this chapter.

Then there is the suprathreshold problem. The estimation of a JNC associated with how much a radiance field must change after the haze layer is visible or how much a uniform haze must change to cause a noticeable difference in the appearance of a landscape is a more difficult question and is context dependent. Typical laboratory studies are done with side-by-side comparisons of a baseline scene or stimulus and some altered or degraded stimulus. This type of paradigm is unrealistic for a real-world setting because changes in haze intensity take place over time periods of hours or days and under a variety of illumination conditions, and an evaluation of VAQ change requires a person to "remember" what the scene looked like before a given change in air pollution took place. However, the side-by-side comparison sets a lower bound of suprathreshold sensitivity. Furthermore, the number of JNCs associated with a change in haze will be scene dependent and therefore is not expected to be universally related to judgments or value assessments of changing VAQ.

It is also of interest to understand how haze affects judgments of visual air quality or some other perceived effect. Studies have focused on perceived environmental quality indexes (PEQI). These include VAQ judgments and scenic beauty estimates (SBE) (Craik and Zube, 1976; Daniel and Boster, 1976; Malm et al., 1981, 1983; Middleton et al., 1983, 1984; Stewart et al., 1984), which use interval scale indexes, acceptability judgments describing the level of haze obscuring landscape features that would be deemed acceptable, ratings of visitor enjoyment as a function of haze levels, and value assessments in terms of willingness to pay for a certain level of visual air quality.

Human judgments of haze effects on landscape features are uniquely different from determinations of whether a landscape feature is just visually degraded or not. VAQ judgments fall into a broader category of using PEQIs as adjuncts to physically based monitoring systems to gauge the state of various aspects of the human environment. VAQ judgments are unique to visibility; however, SBEs have been used extensively to assess the scenic qualities of forest and desert landscapes (Daniel et al., 1973).

Other surveys include visibility preference, visitor experience, and value assessment studies. Visibility preference studies have focused on the levels of haze that were found to be acceptable, primarily in the context of five urban settings (Ely et al., 1991; Pryor, 1996; Abt Associates, 2001; AZ DEQ, 2003; Fajardo et al., 2013). Visitor experience studies have explored the relative importance of visibility to park visitors, visitor awareness of visibility conditions, and the effect of visibility on visitor satisfaction and behavior (Ross et al., 1985, 1987). Value assessment studies, which refer to how much a person is willing to pay to preserve some level of VAQ (Schulze et al., 1983; Loehman et al., 1994; Chestnut and Dennis, 1997; Smith et al., 2005) or how a person might change their behavior as haze levels in their visual environment change (Bell, 1985), have also been carried out.

The following sections of this chapter describe many of the studies on human-observed assessments of the effects of haze on various landscape features, consistent with Fig. 5.1 and the discussion mentioned previously. Topics covered start with the most fundamental colorimetric measurements of scene color and brightness, then the perceived environmental quality studies are presented. Effects of haze on human activity and awareness are next in line and the chapter ends with a discussion of various threshold studies.

COLORIMETRIC MEASUREMENTS
OF HAZE

Two of the more fundamental studies of how haze affects land-scape features were carried out by Henry (2002, 2005). A visual color-imeter was used to assess color differences and perceived brightness changes in a scene as a function of varying haze levels. In these studies an observer looked at a natural landscape feature with one eye while looking through a colorimeter with the other and adjusted the colorim-eter to match the perceived color and/or brightness of the landscape feature.

In one study, Henry (2002) focused on perceived lightness difference. For the difference observations, light- and dark-colored areas of the same distant landscape feature were chosen. The perceived lightness difference as a function of varying extinction was observed and recorded by eight observers. The functional form of the observed perceived light-ness difference versus optical depth was well approximated by the expression $PL = PL_o\exp(-\tau)$, where PL and PL_o are perceived lightness differences at some impaired and reference optical depth τ and τ_o, re-spectively. This result is not unexpected in that it can be shown that radiance differences are attenuated in direct proportion to the transmit-tance of an intervening atmosphere. However, the relationship between perceived lightness difference and radiance difference is not 1:1 because objects tend to be perceived darker than expected.

In a similar study carried out at Great Smoky Mountains National Park, Henry (2005) focused on color shift or alteration as a function of varying haze levels. Observers again used a colorimeter to estimate the perceived color of a red barn and a dark, tree-covered ridge. The per-ceived color of the landscape features was recorded as the amount of red, green, and blue light required to achieve a color match. A telespec-troradiometer was also used to measure the spectrum of light coming from the landscape features. These measurements were then used in a color appearance model to estimate hue and colorfulness, which is analogous to color saturation. It was concluded that the eye–brain sys-tem to some degree compensated for the effects of haze on the color change of the landscape feature as haze levels increased or decreased. As the haze increased, the perceived hue did not change, even though the radiometric measurements showed the hue shifting toward the blue end of the spectrum, while the colorfulness was perceived to decrease. As with perceived lightness difference, the decrease in colorfulness was approximated to be in direct proportion to atmospheric transmittance between the targets and observers, $M = M_o\exp(-\tau)$, where M and M_o are the colorfulness index at some impaired and reference optical depth τ and τ_o, respectively.

VISUAL AIR QUALITY JUDGMENTS

Two study groups focused only on judgments of VAQ (Malm et al., 1981; Middleton et al., 1984); however, many studies included VAQ judgments as part of evaluating other PEQIs such as scenic beauty (Latimer et al., 1981), and some studies focused on establishing the level of VAQ degradation that was judged to be acceptable (Ely et al., 1991; Pryor, 1996; Abt Associates, 2001; AZ DEQ, 2003). The format for all the studies was similar in that participants were asked to rate slides on a 1–10 scale, with 1 or 0 indicating extremely low and 9 or 10 indicating extremely high or good VAQ. A few studies used a 1–7 instead of a 1–10 scale.

Slides representing the best, worst, and intermediate levels of air quality were chosen to correspond to haze levels, sun angles, and cloud and meteorological conditions. Prior to the presentation of slides, participants were read a standard set of instructions. Randomly ordered evaluation slides were preceded by a subset of preview slides. Preview slides were used to orient the observers to the full range of VAQ levels. In the case of the Middleton et al. (1984) Denver VAQ studies, on-site observers made VAQ judgments of actual scenes over the course of the day.

Objectives

The overall objective of all VAQ studies was to link human perceptions of visibility or VAQ to atmospheric aerosol properties and ultimately to their precursor emissions. The design of the Malm et al. (1981) study was uniquely different from those of Middleton (Middleton et al., 1983, 1984) and other VAQ studies referenced previously, as they adopted a stepwise approach in which they investigated the functional relationship between VAQ and various electro-optical parameters for isolated scenic elements. Secondary steps included examining the effects on VAQ judgments of introducing additional scenic elements and investigating the effects of changing illumination conditions and snow and cloud cover on VAQ ratings. Other studies adopted a more indirect approach in which all stimuli representing various levels of haze on highly variable background scenes were employed. Stimuli not only contained photos or scenes of different haze levels but photos taken under varying illumination conditions, cloud cover, and other meteorological conditions such as snow cover, and photos with a variety of scenic elements such as distant mountains and various foreground features. Statistical analyses, such as ordinary least squares (OLS) regressions, were used to relate these various scenic or landscape attributes to measures of air quality such as atmospheric extinction and other variables such as aerosol concentrations.

Comparison of On-Site to Slide Ratings

In order to determine whether slide ratings can be substituted for on-site ratings of real three-dimensional scenes, a visibility perception study, reported on in Malm et al. (1981), was designed to allow observers, upon completion of their rating of the regular evaluation slides, to rate 40 additional slides that corresponded as nearly as possible to the format of an actual on-site, three-dimensional scene that they also rated. There were, out of the approximately 700 on-site VAQ judgments, a number of ratings that corresponded to similar air pollution levels, time of day, and meteorological conditions that existed on the day the slides were taken. Analysis of the dataset showed that there was not a statistical difference between on-site versus slide ratings of VAQ. Other research has consistently shown that color slides are valid representations of actual scenes for obtaining judgments of perceived scenic beauty or landscape preference (Zube, 1974; Daniel and Boster, 1976; Buhyoff and Leuschner, 1978).

Judgments of Visual Air Quality in an Urban Setting

Middleton et al. (1983, 1984) reported on the relationship of emission changes to variations in judgments of VAQ in an urban setting (Denver, Colorado). Although most PEQI studies relating to visibility used slide projections of various scenes and landscape features, the Middleton et al. studies relied on on-site judgments made by trained observers. Mumpower et al. (1981) developed protocols for on-site judgments of VAQ and found that VAQ judgments are strongly related to clarity of objects, color of the air, and the appearance of a border between clear and discolored air.

Middleton et al. (1984) correlated on-site judgments with various aerosol concentrations and with $b_{sp,}$ $b_{ap,}$ and b_{ext}. Note that the Middleton et al. studies did not develop a quantitative relationship between VAQ and physio-chemical indicators but rather presented correlations between these variables. They concluded that the single physio-chemical measure that was found to best satisfy the criteria for a direct indicator of VAQ was b_{sp}. They further concluded that this measure is a good indicator of VAQ, regardless of the direction or location of observation, and that a measure at one site could be used as an indicator of VAQ over a much larger region.

Judgments of Visual Air Quality in National Parks

The key features of VAQ judgment studies in national parks were highlighted by Malm et al. (1981). One vista used in the Malm et al. study was of Mount Trumbull as viewed from Hopi Point, Grand Canyon National Park, and is shown in Fig. 5.2. The 96-km-distant Mount Trumbull

FIGURE 5.2   Photos of Mount Trumbull under near-Rayleigh conditions at 9:00 am (a) and 3:00 pm (b).

represents only 4% of the vista, while 20-km and 10-km foreground canyon walls make up 19% and 77% of the remaining area, respectively.

To isolate visitors' perceptions of air pollution impacts on a single landscape feature, only the Mount Trumbull portion of the scene was shown to participants, as shown in Fig. 5.3a–f. Haze levels varied from near-Rayleigh conditions, in which the Mount Trumbull–sky contrast was near −0.33, to hazy conditions, in which Mount Trumbull was barely visible, with a sky–mountains contrast near 0.02.

Results of these ratings are presented in Figs. 5.4 and 5.5. The error bars around each of the data points show the 95% confidence interval of the ratings, while the line through the data points is the result of a model calculation that is discussed later.

Fig. 5.4 shows VAQ judgments plotted as a function of optical depth τ between Mount Trumbull and the observer ($b_{ext} \times 96$ km), while Fig. 5.5

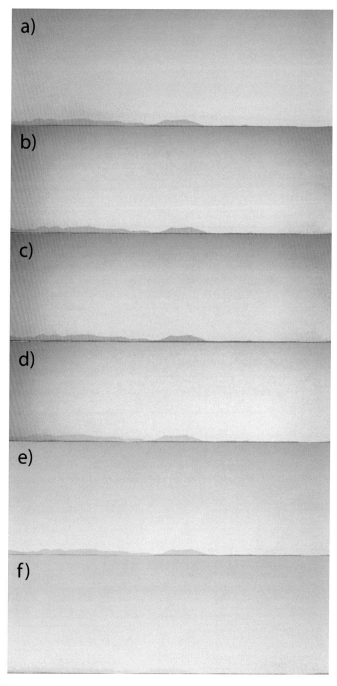

FIGURE 5.3 Six of the 12 slides used in the Mount Trumbull study showing various levels of haze and their effects on sky–mountain contrast.

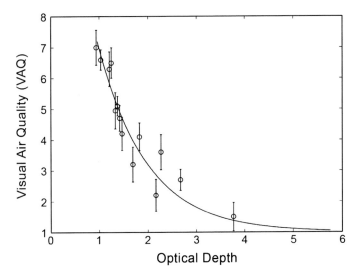

FIGURE 5.4   Average visual air quality (VAQ) ratings of Mount Trumbull as viewed from Hopi Point, Grand Canyon National Park, plotted as a function of optical depth. Error bars correspond to the 95% confidence interval of VAQ ratings. The solid line corresponds to a model calculation discussed later.

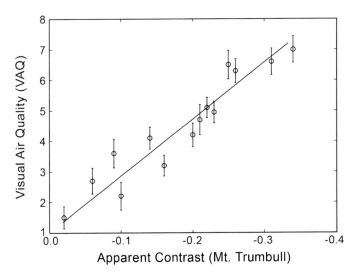

FIGURE 5.5   Average visual air quality (VAQ) ratings of Mount Trumbull as viewed from Hopi Point, Grand Canyon National Park, plotted as a function of apparent contrast of Mount Trumbull (foreground features absent). Error bars correspond to the 95% confidence interval of VAQ ratings. The solid line corresponds to a model calculation discussed later.

shows the same data plotted as a function of the apparent contrast of Mount Trumbull. Although there is some scatter in the data points, the relationship between the VAQ of this isolated scenic element and its apparent contrast is linear, and necessarily, the relationship between VAQ and extinction or optical depth is nonlinear. Even though slides depicting near-Rayleigh conditions were shown to participants, the average highest ratings of this scene were only about 7 on a 10-point scale, suggesting that inherent scenic quality played a role in the VAQ judgments. However, the implication of this experiment is that there is a linear relationship between VAQ and apparent contrast for *any* isolated scenic element.

To investigate the role that other scenic elements play in VAQ judgments, the mask was removed from those images shown in Fig. 5.3 so that the full scene as depicted in Fig. 5.2a and b was visible. Five slides with the dark foreground and six slides with the colored foreground were used. Haze levels were again varied from Rayleigh conditions to levels where Mount Trumbull was just noticeable. Results of this survey are shown in Fig. 5.6. Also shown in Fig. 5.6 are solid and dotted lines through the data points, representing the results of a model calculation presented later.

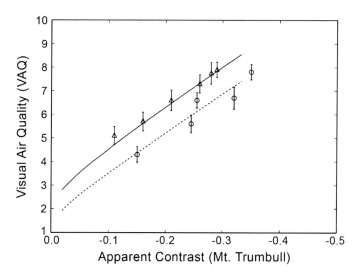

FIGURE 5.6 VAQ judgments of the Mount Trumbull scene with foreground features showing. "o" symbols correspond to the dark foreground feature, while Δ corresponds to the brightly colored foreground. Lines through the data points correspond to a model estimation of VAQ ratings.

The most striking result is the clear separation of VAQ judgments for scenes with different foreground features. The scene with dark-shadowed foreground features was systematically rated lower than those with the brightly colored foreground. With the foreground features present (Fig. 5.2), the maximum ratings went from about 7 when the features were not visible (Fig. 5.3) to about 8 when they were, emphasizing that the inherent scenic quality of the scene plays a role in the VAQ judgments.

Second, the slopes of the VAQ versus apparent contrast curves for the brightly colored and shadowed foreground features do not change, implying that even though the ratings for the more colorful scene were higher than its shadowed counterpart, the sensitivity to an increase or decrease in haze was the same.

Finally, when the scene is represented by only one landscape feature and the appearance of this feature is responsive to changing haze levels, the VAQ versus contrast curve had an intercept term of 1, or the lowest rating possible, corresponding to the whole scene being "hazed" out. On the other hand, Fig. 5.6 shows that when there are foreground landscape elements that are for the most part insensitive to haze levels, the VAQ judgments are well above 1, even though the Mount Trumbull portion of the scene disappears. Although the participants were asked to rate the presented scenes on a 1–10 scale, the lowest they rated the haziest scene was about 5 for the brightly colored foreground scene and 4 for the shadowed foreground. The resultant intercepts are about 3.5 and 2. And while participants were asked to rate visual "air quality," their ratings were sensitive to other features present in the scene, such as landscape features unaffected by haze, and to the inherent scenic characteristics of those landscape features. The context in which varying haze levels are presented is important and plays a role in how VAQ is judged or rated.

Other scenes were also used to investigate the effect that foreground features and other factors such as meteorological conditions had on VAQ ratings for similar levels of air pollution. One of the vistas used in this exploration was the La Sal Mountains as viewed from Island in the Sky, Canyonlands National Park, which is shown in Figs. 5.7–5.9. The La Sal Mountains are 50 km distant, while a foreground red cliff is approximately 4 km away from the observation point. The other feature is a nearby cliff face that is less than 1 km distant. Fig. 5.7 shows the view under near-Rayleigh conditions and under a haze condition in which the La Sal Mountains are just visible. Fig. 5.8 shows the same view but with the foreground features in shadow, and Fig. 5.9 shows a similar scene but with snow covering portions of the mountain. In the Malm et al. (1981) study, the slides representing the snow-covered scene were identical to the scene shown in Fig. 5.7 but with the La Sal Mountains covered with snow.

Even though the apparent contrast of the distant mountains changed by as much as 0.4, the nearby rock cliff and tree changed very little in appearance. Because the relationship between apparent contrast of an

FIGURE 5.7   Photos of the La Sal Mountains taken from Island in the Sky, Canyonlands National Park. (a) was taken under near-Rayleigh conditions and in direct sunlight while (b) was taken under the same lighting conditions but with haze that caused the La Sal Mountain–sky apparent contrast to be reduced to about −0.02.

FIGURE 5.8   Photo of the La Sal Mountains under near-Rayleigh conditions but in shadowed lighting.

FIGURE 5.9   Photo of the La Sal Mountains in direct sunlight but with snow cover.

isolated scenic element and VAQ is linear, it might be anticipated that in the case of the La Sal Mountains vista where the only perceived change in the vista was the contrast of the mountains, the relationship between VAQ and apparent contrast of the mountains would be linear.

A close examination of VAQ ratings of this scene gives some insight into effects of separate scenic elements on the VAQ of the total vista. The results of these ratings, shown in Fig. 5.10, are plotted as a function of the apparent contrast of a tree-covered portion of the 50-km-distant La Sal Mountains for three separate conditions: illuminated, shadow, and snow-covered. The difference in ratings of VAQ for different illumination conditions is of specific interest for this rather unique vista. In Fig. 5.10, the only difference in the slides used to generate the shadow data (■) and those used to create the partial illumination data (□) was the presence of the foreground red cliff. In the partial shadow case, the foreground cliff had about 60% of its face illuminated, while in the shadow case the entire face was in full shadow. In either case, shadow or partial illumination, the rock face did not change

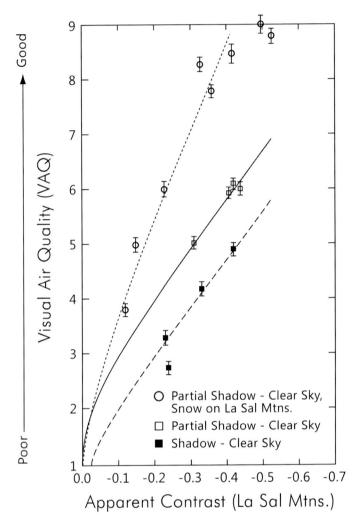

FIGURE 5.10   Average VAQ ratings of the La Sal Mountains vista as viewed from Island in the Sky, Canyonlands National Park, plotted as a function of the apparent contrast of a tree-covered portion of the La Sal Mountains. Error bars correspond to 95% confidence intervals. The lines through the data points correspond to model calculations.

appearance as contrast of the La Sal Mountains changed dramatically. As with the Mount Trumbull scene, the ratings of the more pleasing or prettier, partially illuminated case are about one unit of VAQ higher than the shadowed counterpart. More importantly, the slope of a line drawn through these two sets of data does not differ. The observers are only rating the change they perceive in the slides, the contrast of the La Sal Mountains. An increase in color (change in illumination) causes the VAQ versus contrast curve to shift upward. However, since it is the contrast of the La Sal Mountains that is changing in both cases, the slope of VAQ versus contrast remains constant. The third curve, data represented by ○, shows the effect of adding snow to the La Sal Mountains element of the total scene. In this case, the scenic element whose appearance is changing as a function of air pollution has increased in overall brightness and color (the scene has greater scenic beauty). As a consequence, the slope of the VAQ versus contrast curve for the tree-covered portion of the mountain increased. This indicates that the scene is more sensitive to increases in air pollution. This same increase in sensitivity as a result of increased color was evident in other scenes studied.

A Visual Air Quality Model

Malm et al. (1981) showed that the results presented earlier can be generalized as

$$VAQ = \sum_{i=1}^{n} f_i \left(VAQ_{oi} + const_i \right) e^{-\Delta \tau_i} - const_i \tag{5.1}$$

where VAQ_{oi}, the inherent VAQ of the ith scenic element, is VAQ in the absence of air pollution of the ith scenic element, $\Delta \tau_i = \Delta b_{ext} R_i$, R_i being the distance to the ith landscape element, Δb_{ext} the increment in atmospheric extinction above the clearest or cleanest reference slide, and f_i the fraction of total area of the ith element.

If $\Delta \tau_i = 0$, then

$$VAQ = \sum_{i=1}^{n} f_i VAQ_{oi}. \tag{5.2}$$

Thus, VAQ in the absence of air pollution is the weighted average of the inherent VAQ of each landscape feature.

VAQ_o is context dependent in that it is a function of cloud conditions, sun angle, azimuth angle of the target or scene, geologic structure, etc. Sun angle and azimuth manifest themselves primarily through a change in inherent color of a landscape feature, which may be quantifiable, while geologic structure (vista texture), etc., may not be physically quantifiable. The continuous lines through the data points in Figs. 5.5, 5.6, and 5.10 are a result of applying Eq. 5.1 to the various datasets. Scenes found in other national parks have been studied and are also presented in Malm et al. (1981).

Eq. 5.1 can be expressed in terms of apparent contrast C_R if it is assumed that apparent contrast can be approximated by $C_r = C_o e^{-b_{ext} R}$. Rewriting Eq. 5.1 in terms of C_R yields

$$VAQ = \sum_{i=1}^{n} f_i K_i C_{Ri} - const_i \tag{5.3}$$

where $K_i = \left(VAQ_{oi} + const_i \right) e^{\tau_{ref,i}} / \left(C_{oi} \right).$

C_{oi} is the inherent contrast of the ith landscape feature and $\tau_{ref,i}$ is the optical depth of the reference slide from which $\Delta \tau_i$ is calculated. If the range of haze is such that only a sensitive single vista element is affected by the haze, then Eq. 5.3 is simply

$$VAQ = KC_R - const,\qquad(5.4)$$

a linear relationship between apparent contrast of that landscape element with an intercept of $-const$. This was the case for most scenes discussed earlier.

It is of interest to point out that from a perception perspective Eq. 5.1 implies that no one simple index, such as average scene contrast, average contrast transmittance, average modulation transfer function, or modulation depth, can universally describe or predict VAQ judgments when the scene includes vista elements with varying color and texture and is viewed under varying meteorological conditions. Each scenic element is unique, and judgments of VAQ inherently include the variability of these features. It seems that observers cannot separate the singular effects of haze from other ever-changing landscape feature characteristics.

Clearly, increases in haze affect judgments of VAQ in a very systematic and quantifiable way, but changes in sun angle associated with scene illumination, snow cover, clouds, and other meteorological factors also affect VAQ. It is important to emphasize, however, that haze decreases the VAQ of any scene, some more than others.

Judgments of Scenic Beauty (Scenic Beauty Estimate (SBE))

Latimer et al. (1981) investigated the perceptions of both VAQ and scenic beauty, using several atmospheric optical parameters, including coloration parameters, of several vistas of different landscape types. Although Malm et al. (1981) sought to investigate a fundamental or basic understanding of relationships between judgments of VAQ and various atmospheric optical variables, the approach taken by Latimer et al. was a statistical one, where relationships between VAQ and SBE and visibility metrics were investigated using OLS regressions.

Three hundred and eight photographs of 31 scenic vistas in Arizona were taken under a range of weather, lighting, and visibility (air quality) conditions. Only two or three photographs were taken at some of these vistas; however, 12–20 photographs were taken at each of the 13 intensively sampled sites—eight in Grand Canyon National Park and five on Mount Lemmon near Saguaro National Monument and the city of Tucson. Ninety-nine photographs of six scenic vistas in the eastern United States were also taken. Three of the vistas were in Shenandoah National Park and three were in Great Smoky Mountains National Park. When each photograph was taken, the spectral radiance at four wavelengths (405, 450, 550, and 630 nm) of various elements (targets) of the vista was measured with a Meteorology Research Incorporated multiwavelength telephotometer (see Chapter 7). For some of the photographs taken in the western United States, the light scattering coefficient was measured with nearby integrating nephelometers. The protocol for judging the photographs was similar to those used by Malm et al. (1981) and described previously.

Before regressions were carried out, both perceptual and physical indicators were normalized to a Z score given by

$$Z_{xij} = \left(X_{ij} - \overline{X}_j \right) / \sigma_{xj} \tag{5.5}$$

where Z_{xij} is the standardized Z-transform of variable X for slide i of vista j; X_{ij} is the basic untransformed measurement/judgment for slide i of vista j (e.g., SBE, VAQ, visual range, chromaticity, or luminance); and $\overline{X}_j$ and σ_{xj} are the mean and standard deviation, respectively, for variable x measured at vista j.

The results of OLS regressions, for both western and eastern vistas, of VAQ as a function of visual range, scattering angle, solar zenith angle, cloud cover, sky luminance, and sky chromaticity are summarized in Table 5.1. Not surprisingly, in all cases VAQ increases with increasing visual range, decreasing sky chromaticity (more saturated blue color),

TABLE 5.1　Regression Equations for the Dependence of Z-Transforms for VAQ on Physical Parameters

Vistas	R^2	Regression Equation (95% Confidence Limits on Coefficients Are Shown)
All western vistas	0.39	$ZVAQ = 0.71 + 0.51Z_v - 0.36Z_x - 0.23Z_L - 0.016\alpha$ 　　　　　± 0.10　± 0.11　± 0.011　± 0.007
All Grand Canyon vistas	0.55	$ZVAQ = -1.13 + 0.62Z_v - 0.31Z_x + 0.011\theta$ 　　　　　± 0.14　± 0.14　± 0.003
All Mt Lemmon vistas	0.46	$ZVAQ = 0.39 + 0.50Z_v - 0.20Z_x - 0.23Z_L + 0.017Cu - 0.016\alpha$ 　　　　　± 0.14　± 0.16　± 0.015　± 0.007　± 0.008
All eastern U.S. vistas	Random 0.59	$ZVAQ = 0.01 + 0.51Z_v - 0.36Z_x$ 　　　　　± 0.11　± 0.11
α = Solar zenith angle in degrees θ = Scattering angle in degrees Cu = Percentage of sky covered by cumulus clouds Z_v = Z-transform of visual range	Z_L = Z-transform of sky luminance Z_x, Z_y = Z-transform of sky chromaticity x, y ZVAQ = Z-transform of VAQ ZSBE = Z-transform of SBE	

The ± values below each of the independent variables correspond to the standard error for the coefficient associated with that variable.

decreasing sky luminance (deeper, darker sky color), and either changes in scattering angle or solar zenith angle. For the Mount Lemmon vistas, increases in cumulus clouds also resulted in an increase in VAQ. Notice that the coefficient associated with visual range is larger than those associated with other variables, indicating that visual range or air quality is the biggest driver of human judgments of VAQ.

These results are consistent with the findings of Malm et al. (1981) in that they also found the air quality as represented by landscape feature contrast was the most significant driver of VAQ, but vista color (as represented by sun angle) and cumulus clouds also enhanced or increased VAQ judgments. This study also illustrates that individuals cannot differentiate between the effects of haze and those of sun angle and other meteorological enhancements of the scene.

To that end, the Latimer et al. (1981) study also directly investigated SBEs of the same landscape scenes as were used in their VAQ investigation. They also investigated, using OLS regressions, the relationships between normalized SBE and the same physical variables used in their VAQ portion of the study. These results are shown in Table 5.2.

TABLE 5.2 Regression Equations for the Dependence of Z-Transforms for SBE on Physical Parameters

Vistas	R^2	Regression Equation (95% Confidence Limits on Coefficients Are Shown)
All western vistas	0.27	$ZSBE = -0.56 + 0.26Z_v - 0.27Z_x + 0.011Cu + 0.004\theta$ $\pm 0.10 \quad \pm 0.11 \quad \pm 0.003 \quad \pm 0.003$
All Grand Canyon vistas	0.48	$ZSBE = -0.69 + 0.63Z_v - 0.27Z_y - 0.19Z_L + 0.005\theta + 0.006Cu$ $\pm 0.16 \quad \pm 0.15 \quad \pm 0.15 \quad \pm 0.005 \pm 0.003$
All Mt Lemmon vistas	0.18	$ZSBE = -0.18 - 0.22Z_x + 0.018Cu$ $\pm 0.17 \quad \pm 0.008$
All eastern U.S. vistas	Random 0.57	$ZSBE = 0.53 + 0.48Z_v - 0.606Z_y - 0.011\alpha$ $\pm 0.12 \quad \pm 0.12 \quad \pm 0.008$
α = Solar zenith angle in degrees θ = Scattering angle in degrees Cu = Percentage of sky covered by cumulus clouds Z_v = Z-transform of visual range		Z_L = Z-transform of sky luminance Z_x, Z_y = Z-transform of sky chromaticity x, y ZVAQ = Z-transform of VAQ ZSBE = Z-transform of SBE

The ± values below each of the independent variables correspond to the standard error for the coefficient associated with that variable.

Comparing Tables 5.1 and 5.2, it is evident the judgments of SBE evoke a different psychological response than judgments of VAQ. Visual range, in most cases, is still the variable that is weighted highest in that it is most responsible for variability in SBE. However, as an example, the regression coefficient on visual range for the "all western vistas" case is diminished by about a factor of 2 over that of the VAQ relationship. The effects of sun angle and clouds play a more significant role in SBE judgments.

A graphical comparison of the normalized SBE ratings, ZSBE, and normalized VAQ ratings, ZVAQ, is shown in Fig. 5.11. Each vista has been assigned a number and the mean ZSBE and ZVAQ for that vista are plotted against each other.

Clearly, in most cases, ratings of ZSBE and ZVAQ are very much interrelated. For intensively sampled vistas at the Grand Canyon, at Mount Lemmon, and in the eastern United States, the means are based on 12–22 measurements, whereas the means for some of the other vistas are based on as few as two or three measurements. It is important to note that, though the mean visual ranges measured for all the Grand Canyon vistas were equal (about 200 km), mean ZVAQs for these vistas ranged from −40 to +60 and mean ZSBEs from −20 to +60. Indeed, the linear relationship between the mean ZSBEs and ZVAQs for the views of the Grand Canyon is striking. Thus, the fact that vistas 4, 8, and 9 have mean ZSBEs and

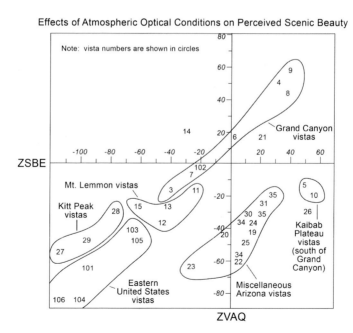

FIGURE 5.11   Average scenic beauty estimates (ZSBE) plotted against visual air quality indices (ZVAQ) for each vista.

ZVAQs so much higher than those of vistas 3 and 7 cannot be ascribed to differences in visibility (air quality) conditions; these differences may rather be due to differences in the foreground, intermediate, and distant landscape features or to the azimuth orientation of the view. For example, vistas 4, 8, and 9 have trees or bushes visible in the foreground, whereas vistas 3 and 7 have only rock visible in the foreground. Moreover, the latter vistas are more-distant views than those of the other vistas, and they are west-looking, so they experience a greater variation in scattering angle with time of day than views from the other vistas. Regardless of the reasons for the differences in mean ZVAQs and ZSBEs between vistas, it is clear that the dependence of perceptions on physical parameters must be treated separately for each vista, because scenic features of each vista affect not only the rating of ZSBE but also the rating of ZVAQ.

Although the analysis methodologies reported in the Malm et al. (1981) and Latimer et al. (1981) studies differed significantly, the conclusions derived from these two approaches are quite similar. First, there is a great deal of agreement between different groups' ratings of color slides for VAQ. The demographic background of an observer has little effect on VAQ ratings. Intercorrelations of separate group means of the evaluation slides within demographic partitions as well as intercorrelation of each individual's ratings with mean ratings for each slide were high. Even groups of ostensibly different professional interests, such as groups from the EPA and from oil companies, are in quite good agreement with one another. These conclusions were also true for SBE ratings.

Second, people's ratings of VAQ and, separately, SBE, differ substantially for photographs of different vista landscapes. People generally rate scenes having colorful foreground features higher than those with minimal or totally shadowed foreground. Because of this difference in ratings among different vista landscapes, the effects of haze on VAQ and SBE ratings must be evaluated by considering inherent differences in ratings of a photograph relative to other photographs of the same vista.

Third, there appears to be differences in sensitivity to haze changes for different landscapes. Visibility impairment may have profoundly different effects on the human aesthetic experience in various scenic areas because of differences in landscape characteristics.

URBAN VISIBILITY PREFERENCE STUDIES

There have been five urban visibility preference studies to date, four carried out in North American cities and one in Beijing, China. A sixth study focused on the validity of the preference study approach to identifying levels of acceptable visibility. The urban visibility preference studies used a focus-group method to estimate the level of visibility impairment that respondents described as "acceptable." In addition to determining the amount of haze deemed to be acceptable, respondents were also asked to make VAQ judgments of the various scenes. The specific definition of acceptable was largely left to each individual respondent, allowing each to identify their own preferences. There are three completed North American studies that used this method and a pilot study (designed as a survey instrument development project) that provides additional information.

The completed studies were conducted in Phoenix, Arizona (AZ DEQ, 2003), two cities in British Columbia, Canada (Pryor, 1996), and Denver, Colorado (Ely et al., 1991). The pilot study was conducted in Washington, D.C. (Abt Associates, 2001). The studies are summarized in Table 5.3. The Beijing study, although discussed later, is not included in Table 5.3 because all of the included information in the table was not available in the published journal article (Fajardo et al., 2013).

In each study, information was collected in a focus-group setting in which slides depicting various visibility conditions were presented. In each study, photographs of a single scene from the study's city were used; each photo included images of the broad downtown area and out to the hills or mountains composing the scene's backdrop. The maximum sight distance under good conditions varied by city, ranging from 8 km in Washington, D.C., to mountains hundreds of kilometers away in Denver. Multiple photos of the same scene were used to present approximately 20 different levels of visibility impairment.

TABLE 5.3 Summary of Urban Visibility Preference Studies

	Denver	Phoenix	British Columbia	Washington, DC
Report date	1991	2003	1996	2001
Session duration		45 min	50 min	2 h
Compensation	None	$50	None	$50
# of focus-group sessions	17	27 total at 6 locations,	4	1
# of participants	214	385	180	9
Age range	Adults	18–65+	University students	27–58
Annual or seasonal	Wintertime	Annual	Summertime	Annual

(Continued)

TABLE 5.3   Summary of Urban Visibility Preference Studies (*cont.*)

	Denver	Phoenix	British Columbia	Washington, DC
Scenes presented	Single scene of downtown with mountains in background at about 80–150 km	Single scene of downtown and mountains, 42-km maximum distance	Single scene from two cities with distant mountains at about 30 km	Single scene of DC Mall with 8-km maximum sight path
# of total visibility conditions	20 levels (+5 duplicates)	21 levels (+4 duplicates)	20 levels (10 each from each city)	20 levels (+5 duplicates)
Source of slides	Actual photos taken between 9 am and 3 pm	WinHaze	Actual photos taken at 1 pm or 4 pm	WinHaze
Medium of presentation	Slide projection	Slide projection	Slide projection	Slide projection
Ranking scale used	1–7 scale	1–7 scale	1–7 scale	1–7 scale
Extinction level presented	0.03–0.55 km^{-1}	0.045–0.331 km^{-1}	0.037–0.122 km^{-1} (Chilliwack) 0.039–0.233 km^{-1} (Abbotsford)	0.025–0.447 km^{-1}
Health issue directions	Ignore potential health impacts; visibility only	Judge solely on visibility, do not consider health	Judge solely on visibility, do not consider health	Health never mentioned, "Focus only on visibility"
Key questions asked	a) Rank VAQ (1–7 scale), b) is each slide "acceptable," c) "How much haze is too much?"	a) Rank VAQ (1–7 scale), b) is each slide "acceptable," c) how many days a year would this picture be "acceptable"	a) Rank VAQ (1–7 scale), b) is each slide "acceptable"	a) Rank VAQ (1–7 scale), b) is each slide "acceptable," c) if this hazy, how many hours would it be acceptable (three slides only), d) valuation question
Mean b_{ext} found "acceptable"	0.067 km^{-1} (19 dv)	0.12 km^{-1} (24.5 dv)	0.10 km^{-1} (~23 dv) (Chilliwack), 0.09 km^{-1} (~22 dv) (Abbotsford)	0.16 km^{-1} (~28 dv)

Actual photographs taken from the same locations were used in the Denver and British Columbia studies to depict various visibility conditions. Photographs prepared using WinHaze software (Air Resource Specialists, Inc., http://www.air-resource.com/resources/downloads.html) were used in the Phoenix and Washington, D.C., studies. WinHaze is a computer-imaging software program that simulates VAQ differences of various scenes, allowing the user to "degrade" an original, near-pristine-condition visibility photograph to create a photograph of each desired VAQ level (see Chapter 6).

Denver, Colorado, Urban Visibility Preference Study

The Denver urban visibility preference study was conducted on behalf of the Colorado Department of Public Health and Environment (CDPHE). A series of focus-group sessions were conducted with 17 civic and community groups, in which a total of 214 individuals were asked to rate slides. The slides depicted varying levels of VAQ, of which three are shown in Fig. 5.12. The extinction coefficient corresponding to each of these images is approximately 0.03 km^{-1} (11 dv), 0.07 km^{-1} (19 dv), and 0.25 km^{-1} (32 dv). This well-known Denver vista includes a broad view of downtown Denver, with the mountains to the west composing the scene's background. The participants were instructed to base their judgments on three factors:

1. The standard was for an urban area, not a pristine national park area where the standards might be stricter.
2. The level of an urban visibility standard violation should be set at a VAQ level considered to be unreasonable, objectionable, and unacceptable visually.
3. Judgments of standard violations should be based on visibility only, not on health effects.

Participants were shown 25 randomly ordered slides of actual photographs. The visibility conditions presented in the slides ranged from 11 to 40 dv, approximating the 10th to 90th percentiles of wintertime visibility conditions in Denver. The participants rated the 25 slides based on a scale of 1 (poor) to 7 (excellent), with five duplicates included. They were then asked to judge whether the slide would violate what they would consider to be an appropriate urban visibility standard (i.e., whether the level of impairment was "acceptable" or "unacceptable"). A level of 19 dv was judged by 50% of the participants to be unacceptable, which corresponds to the slide in Figure 5.12b.

Phoenix, Arizona, Urban Visibility Preference Study

The Phoenix urban visibility preference study was conducted on behalf of the Arizona Department of Environmental Quality. Its focus-group survey process was patterned after the Denver study. The study included 385

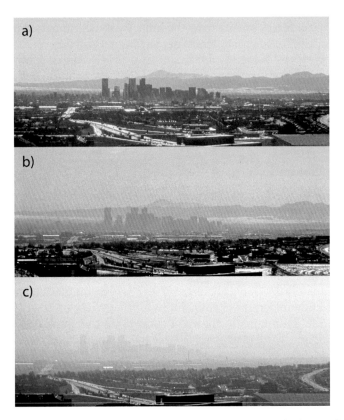

FIGURE 5.12   View of Denver, Colorado. The mountain in the background is about 150 km distant. Haze levels for images (a), (b), and (c) correspond to b_{ext} = 0.03 km^{-1} (11 dv), 0.07 km^{-1} (19 dv), and 0.25 km^{-1} (32 dv), respectively.

participants in 27 separate focus-group sessions. Participants were recruited using random-digit dialing to obtain a sample group designed to be demographically representative of the larger Phoenix population. Focus-group sessions were held at six neighborhood locations throughout the metropolitan area to improve the participation rate. Three sessions were held in Spanish in one region of the city with a large Hispanic population (25%), although the final overall participation of native Spanish speakers (18%) in the study was moderately below the targeted level. Participants received $50 as an inducement to participate.

Participants were shown a series of 25 images, examples of which are shown in Fig. 5.13, of the same vista of downtown Phoenix, with South Mountain in the background at a distance of about 40 km. Photographic slides of the images were developed using WinHaze. The visibility impairment levels ranged from 15 to 35 dv (the b_{ext} extinction coefficient range was approximately 45 Mm^{-1} to 330 Mm^{-1}, or a visual range of 87–12 km).

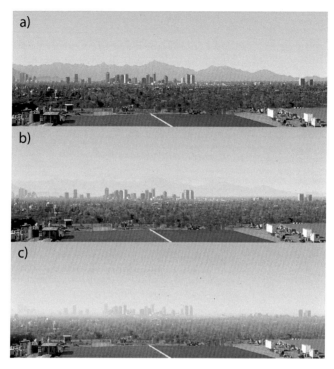

FIGURE 5.13   Examples of photos used in the Phoenix preference study. The most-distant mountainous feature is 40 km distant. The extinction levels of the three photos are 0.045 km^{-1} (15 dv), 0.11 km^{-1} (24 dv), and 0.33 km^{-1} (35 dv), respectively.

The images in Fig. 5.13a, b, and c–correspond to extinction coefficients of 0.045 km^{-1} (15 dv), 0.11 km^{-1} (24 dv), and 0.33 km^{-1} (35 dv), respectively. First, participants individually rated the randomly shown slides on a VAQ scale of 1 (unacceptable) to 7 (excellent). Participants were instructed to rate the photographs solely on visibility and to not base their decisions on either health concerns or what it would cost to have better visibility. Next, the participants individually rated the randomly ordered slides as "acceptable" or "unacceptable," defined as whether the visibility in the slide is acceptable or objectionable. The slide in Fig. 5.13b was judged to be acceptable by 50% of the participants (24 dv).

British Columbia, Canada, Urban Visibility Preference Study

The British Columbia (B.C.) urban visibility preference study was conducted on behalf of the B.C. Ministry of Environment. Focus-group sessions were conducted that were also developed following the methods used in the Denver study. Participants were students at the University of British Columbia who participated in one of four focus-group sessions with between 7 and 95 participants. A total of 180 respondents completed surveys (29 did not complete the survey).

Participants in the study were shown slides of two suburban locations in B.C.: Chilliwack and Abbotsford. The Chilliwack landscape shown in Fig. 5.14a–c corresponds to extinction coefficients of 0.05 km^{-1} (13 dv), 0.96 km^{-1} (23 dv), and 0.17 km^{-1} (28 dv), respectively, while the Abbotsford

FIGURE 5.14 Examples of photos used in the Chilliwack preference study. The most-distant mountainous feature is about 30 km distant. The extinction levels of the three photos are 0.05 km^{-1} (13 dv), 0.96 km^{-1} (23 dv), and 0.17 km^{-1} (28 dv), respectively.

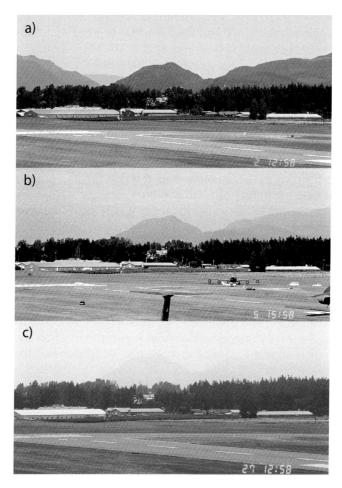

FIGURE 5.15 Examples of photos used in the Abbotsford preference study. The most-distant mountainous feature is about 30 km distant. The extinction levels of the three photos are 0.045 km^{-1} (15 dv), 0.1 km^{-1} (23 dv), and 0.164 km^{-1} (28 dv), respectively.

scene photos, Fig. 5.15a–c, correspond to 0.045 km^{-1} (15 dv), 0.1 km^{-1} (23 dv), and 0.164 km^{-1} (28 dv), respectively. Using the same general protocol as the Denver study, Pryor (1996) found that responses from this study showed the acceptable level of visibility was 23 dv in Chilliwack and 19 dv in Abbotsford. Pryor discusses some possible reasons for the variation in standard visibility judgments between the two locations, including the relative complexity of the scenes, potential bias of the sample population (only university students participated), and the different levels of development at each location. Abbotsford (population 130,000) is an ethnically diverse suburb adjacent to the Vancouver metro area, while Chilliwack (population 70,000) is an agricultural community 100 km east of Vancouver in the Frazier Valley.

The photos in both Figs. 5.14b and 5.15b correspond to the level of haze found to be acceptable by 50% of the participants. The B.C. urban visibility preference study is being considered by the B.C. Ministry of the Environment as a part of establishing urban and wilderness visibility goals in B.C.

Washington, D.C., Urban Visibility Preference Pilot Study

The Washington, D.C., urban visibility pilot study was conducted on behalf of the U.S. EPA and was designed to be a pilot focus-group study, an initial developmental trial run of a larger study. The intent of the pilot study was to study both focus-group method design and potential survey questions. Due to funding limitations, only a single focus-group session was held, consisting of one extended session with nine participants.

The study also adopted the general Denver study method, modifying it as appropriate to be applicable in an eastern urban setting that has substantially different visibility conditions than any of the three western locations of the other preference studies. Washington's (and the entire East) visibility is typically substantially worse than western cities and has different characteristics. Washington's visibility impairment is primarily a uniform, whitish haze dominated by sulfates, relative humidity levels are higher, the low-lying terrain provides substantially shorter maximum sight distances, and many residents are not well informed that anthropogenic emissions impair visibility on hazy days. A single scene, shown in Fig. 5.16a–c, was used. It is a panoramic shot of the Potomac River, Washington mall, and downtown Washington, D.C. The extinction levels shown in the three photos are 0.045 km^{-1} (15 dv), 0.16 km^{-1} (28 dv), and 0.37 km^{-1} (36 dv), respectively. Again, it is the image in Fig. 5.16b that 50% of the participants found to be at the acceptability level.

Fig. 5.17 shows a scatter plot of the percent of observers judging an urban scene to have acceptable VAQ as a function of judged VAQ for each of the studies described previously. Wash, Phx, Chil, Abbt, and Den refer to the Washington, Phoenix, Chilliwack, B.C., Abbotsford, B.C., and Denver studies, respectively. Notice that the shape of the curves representing each of the studies is the same. When asked to judge VAQ, the participants distributed their ratings approximately uniformly across the images they were shown, but when asked to judge acceptability, there tends to be a level of VAQ that is acceptable and one that is not acceptable. Ratings above about 5 are judged to be acceptable and below 2 unacceptable.

It is also interesting to note that the median VAQ of 4 corresponds to about 50% of respondents finding the VAQ acceptable for the Chilliwack image and about 75% of the respondents for the Phoenix scene. Clearly, judging overall VAQ and making a judgment as to what level of haze is acceptable/unacceptable evoked different responses in participants. An underlying goal of eliciting an observer-based response to

FIGURE 5.16 Examples of photos used in the Washington, D.C., preference study. The most-distant feature is about 8 km distant. The extinction levels of the three photos are 0.045 km^{-1} (15 dv), 0.16 km^{-1} (28 dv), and 0.37 km^{-1} (36 dv), respectively.

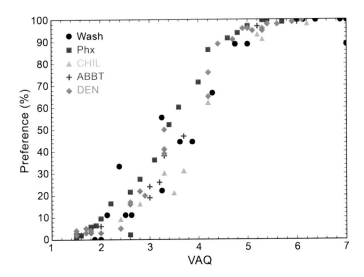

FIGURE 5.17   Scatter plot of the percent of observers judging an urban scene to have acceptable visual air quality (VAQ) as a function of judged VAQ for each of the studies described in the text.

varying haze levels on scenic resources is to develop a physical indicator of response functions, which are in this case visibility preference levels.

Fig. 5.18 shows the same acceptability data as in Fig. 5.17, plotted against the dv levels associated with the various haze levels. A logistical

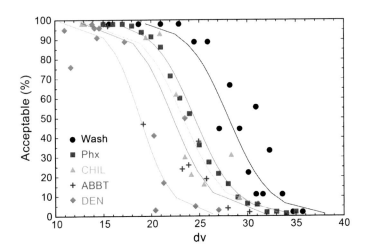

FIGURE 5.18   Percent acceptability levels plotted against deciviews for each of the images used in the various studies.

TABLE 5.4   Percentile Acceptability Levels and Associated Uncertainties, Expressed in Deciviews, for Each of the Scenes Studied

Acceptability	10%	25%	50%	75%	90%
Washington	32.81 ± 0.60	30.44 ± 0.56	28.06 ± 0.51	25.69 ± 0.47	23.31 ± 0.43
Phoenix	29.25 ± 0.22	26.89 ± 0.20	24.53 ± 0.19	22.17 ± 0.17	19.81 ± 0.16
Chilliwack	28.18 ± 0.97	25.76 ± 0.90	23.33 ± 0.83	20.91 ± 0.76	18.48 ± 0.69
Abbotsford	26.82 ± 0.75	24.55 ± 0.70	22.29 ± 0.65	20.02 ± 0.59	17.76 ± 0.54
Denver	22.46 ± 1.36	20.66 ± 1.26	18.86 ± 1.16	17.06 ± 1.06	15.26 ± 0.96

regression model, applied to each dataset, was used to estimate the acceptability levels summarized in Table 5.4.

First, it should be noted that it takes considerably more haze, whether that haze is represented by dv, atmospheric extinction, visual range, or particulate concentration, to cause the Washington, D.C., scene to be judged unacceptable than the Denver scene. The difference in the amount of haze required to create an unacceptable judgment for the Washington and Denver scenes was almost 10 dv or about 36 $\mu g/m^3$ of particulate matter, assuming the particles are not hygroscopic. The amount of haze required in other scenes to be judged as unacceptable is intermediate to Washington and Denver. The dv or extinction level required to reach the 50% level of acceptability is inversely proportional to the distance of the more-distant features in the scene. Closer scenes require higher extinction levels to cause equal amounts of visibility impairment. Extinction or any transforms of extinction are not universal indicators of visibility preference levels.

The same data shown in Fig. 5.18 were plotted against a host of scene- or image-specific variables. They included just noticeable changes (JNCs) and the variants in JNC calculations discussed earlier: contrast, including average contrast of the overall scene, and mean-square fluctuation and its variants. Variables like JNC, average contrast, and mean-square fluctuation are highly dependent on detail and the number of contrast edges in the image and therefore are not good universal indicators of judgments of VAQ. For instance, the number of JNCs relative to a baseline least-haze reference for the 50% acceptance level for the Washington study is 100 JNCs, while for Phoenix it is approximately 50 JNCs.

The best predictor of acceptability level is apparent contrast of a distant, prevalent, but not necessarily dominant, feature (Malm et al., 2011). Fig. 5.19 shows the acceptability levels for each of the studies, plotted against the apparent contrast of the distant feature that is most sensitive to haze. Also shown in Fig. 5.19 is a logistic model curve fit of each of the datasets. Using the logistic equation, the apparent contrast associated

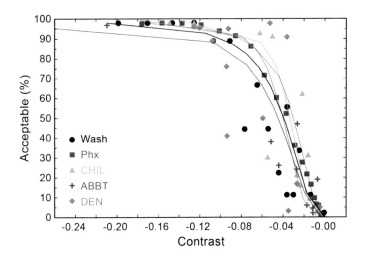

FIGURE 5.19 Percent acceptability levels plotted against apparent contrast of distant landscape features.

with different acceptability percentiles can be estimated. Table 5.5 shows the apparent contrast associated with the 10%, 25%, 50%, 75%, and 90% acceptability levels and estimated uncertainties derived from the logistic model curve fit.

When the feature approximately reaches the visual range or a contrast between about -0.03 and -0.05, about 50% of the observers rated the scene or image as not being acceptable. Referring to Table 5.3, one sees that these features are at about 80–150 km, 42 km, 30 km, and 8 km for the Denver, Phoenix, British Columbia, and Washington, D.C., images, respectively. Because it takes a considerably greater amount of haze/particulate matter to cause an 8-km landscape feature to reach a threshold contrast of -0.03 to -0.05 than a 100-km vista, any

TABLE 5.5   Percentile Acceptability Levels, Expressed as Apparent Contrast, for Each of the Scenes Studied

Acceptability	10%	25%	50%	75%	90%
Washington	0.016 ± 0.0007	0.025 ± 0.0011	0.040 ± 0.0017	0.063 ± 0.0028	0.099 ± 0.0044
Phoenix	0.011 ± 0.0004	0.018 ± 0.0006	0.030 ± 0.0010	0.049 ± 0.0016	0.080 ± 0.0025
Chilliwack	0.017 ± 0.0015	0.024 ± 0.0024	0.033 ± 0.0039	0.046 ± 0.0061	0.065 ± 0.0096
Abbotsford	0.017 ± 0.0007	0.028 ± 0.0011	0.045 ± 0.0018	0.072 ± 0.0028	0.115 ± 0.0044
Denver	0.021 ± 0.0022	0.029 ± 0.0035	0.040 ± 0.0055	0.029 ± 0.0087	0.076 ± 0.0138

Also shown are the uncertainties of each of the reported contrast values.

visibility metric that is proportional to just aerosol concentration, such as b_{ext}, dv, or visual range, without consideration of the distance to and characteristics of landscape features, will not be a general predictor of acceptability or judgments of VAQ.

Beijing, Urban Visibility Preference Study

Fajardo et al. (2013) report on a visibility preference study carried out using Beijing, China, urban scenes. They found that the median preference value for acceptable visibility was between 34.5 dv and 45 dv, considerably higher than the 28 dv median preference level for the Washington, D.C., study. They did not discuss or present the physical landscape features used in any detail, so one can only conjecture what they might have been. In the previous four studies discussed, the most-distant landscape feature varied from 150 km away for the Denver scene to only 8 km away for the Washington, D.C., scene. The closer the landscape features are to the observer, the more particulate matter, as represented by dv levels, it takes to cause the same level of perceived haziness. The higher dv levels to evoke a "not acceptable" response from survey participants could be a result of a more-urban scene, with most dominant landscape features being only meters or a kilometer or two distant. It is also possible that the participants in the study were accustomed to and accepting of the lower visibility levels normally found in Beijing. Because the physical characteristics of the scene used in the Beijing study and the distribution of dv levels of the slides used in the study are not known, it is difficult for these results to be integrated with and discussed in the context of the other preference studies discussed previously.

Robustness of Visibility Preference Studies

Smith (2013) argued that preference studies, such as those described previously, are not valid measures of an individual's preferences. Smith argues that preference study results are subject to "framing bias" in that the participants will choose acceptability levels based on the range of visibility conditions to which they are asked to respond. To that end, Smith explored the acceptability ratings of variants of the original Washington, D.C., study described above. The original study, referred to as variant 1, presented 20 different dv levels with five duplicate slides, making a total of 25 slides. The dv levels ranged from a low of 8.8 dv to a high of 38.3 dv. Variant 2 presented only those photographs that corresponded to dv levels of 27.1 and less, resulting in 12 photographs, while in variant 3, two hazier photographs were added that corresponded to dv levels of 42 dv and 45 dv. Three slides corresponding to 24.5 dv, 15.6 dv, and 11.1 dv were removed, as well as five of the duplicate slides used in variant 1.

Preference

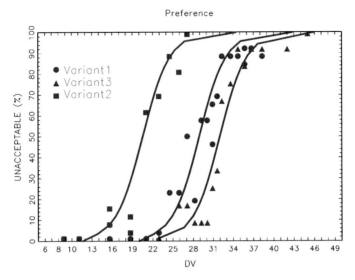

FIGURE 5.20   "Unacceptable" percent levels plotted against deciviews (dv) for each of the images used in the variant studies.

Results of this survey are shown in Fig. 5.20 along with a logit model fit to the datasets.

In Fig. 5.20 the preference data are plotted, as in Smith (2013), as a function of the percent of unacceptability versus deciview, while in Figs. 5.17–5.19, the data are plotted as a function of acceptability. In any case, the 50% acceptability or unacceptability point is the same. Based on the logit fit, the variant 1 acceptability point is 28.9 dv, while for variant 2 it is 20.5 dv, and for variant 3 it is 32.0 dv.

It is worth noting that the instructions first had the participants rate the VAQ and then rate the same set of slides as having acceptable or unacceptable visibility levels. The instructions said, "This time rate the slides according to whether the visibility is acceptable or unacceptable to you." The implicit message was that some of the slides had acceptable visibility while others did not. Therefore, in the variant 2 slide set in which the haziest photo was only at the midpoint, or 50% acceptability point, of the variant 1 slide set, the participants then rated the slide unacceptable, even though in variant 1, 50% of the participants rated it as acceptable. It is interesting to point out that, even though the variant 3 slide set had the haziest slide increase from 38.3 dv to 45 dv and an increase in extinction of almost a factor of 2, or 100%, the 50th percentile acceptability level only increased by about 33%. Furthermore, the contrast of the most-distant feature in variant 1 was about 0.05, while in variant 3 it was about 0.02, consistent with the above hypothesis that, when the more-distant features nearly disappear (contrast levels of about 0.02–0.05), the haze level becomes unacceptable to about 50% of the participants.

There are two points worth making. Study participants should be instructed that it is appropriate to find all of the slides shown to them to have acceptable haze levels, or conversely, that all slides could be rated as having unacceptable visibility levels. Second, the selection of the slide set should have a range of slides that depict the best to the haziest visibility levels that could be expected. This could correspond to slides representing Rayleigh conditions to scenes being entirely "hazed out" in which there are no visible landscape features.

VISIBILITY IN AN URBAN SETTING

There is a dearth of publications on the explicit relationship between visibility and health effects in the urban or nonurban setting. However, as Gesler (2005) pointed out in his editorial titled "Therapeutic landscapes: an evolving theme," there have been conferences organized around the therapeutic landscape concept and also a book addressing the topic (Williams, 1999). In order for a landscape to have therapeutic value, its inherent physical characteristics must be able to be seen, and therein lies the link to visibility.

Velarde et al. (2007) reviewed over 100 articles related to landscape viewing, of which 31 were found to present evidence of health-related effects of landscape views. The reported effects were both positive and negative and included feelings of well-being, anxiety, stress, postoperative recovery, restorative (heart rate, for instance), fatigue, job satisfaction, and number of sick days. Within the context of an urban setting, nature scenes that included visual stimuli such as trees, green grass, rolling hills, the sea, or water tended to evoke positive responses, while impaired landscapes or environments lacking these stimuli had a negative effect.

Other studies have investigated the link between perceptions of air pollution and related health effects, of which perceived visual air quality is only a component of overall air quality awareness. The research that has been conducted in this area has examined awareness of, and attitudes toward, visual air quality and investigated relationships between visual air quality, stress, and human behavior. The author summarized these findings in Malm (1990).

Public Perception of Visual Air Quality

Survey research of public awareness of visual air quality using direct questioning typically reveals that 80% or more of the respondents are aware of poor visual air quality and that poor visibility and media publicity are the primary factors that precipitate the awareness (Cohen et al., 1986). These surveys have also shown that awareness is not uniform across the general population of a given area. Persons with higher incomes and educational levels tend to be more aware of poor visual air quality than those with lower incomes and educational levels.

People are also less aware of pollution in their home area compared to awareness of pollution in areas adjacent to their home (Evans and Jacobs, 1982; Bickerstaff and Walker, 2001). A suggested explanation for this finding is that people cognitively adjust their awareness levels to reduce the dissonance of living in a polluted area with which they are otherwise satisfied or might not be able to leave. The author suggests an alternative explanation: viewing nearby landscape features through a view path with lower optical depths than more-distant features suggests to the uninformed layperson that there is less pollution nearby and more pollution in the more-distant communities.

Attitudes toward poor visual air quality vary with socioeconomic status, health, and length of time an individual has lived in the area (Barker, 1976). Affluent and well-educated people consider poor visual air quality to be a more serious problem than others. People who are not economically tied to sources of air pollution, have respiratory ailments, or are new to an area also show the strongest negative reactions to reduced VAQ.

Visual Air Quality and Stress

Reduced visual air quality is an ambient environmental stressor because it is a relatively constant and unchanging situation over which one has little direct control (Campbell, 1983). The associated stress and lack of control is chronic, not salient, and may be manifested in heightened levels of anxiety, tension, anger, fatigue, depression, and feelings of helplessness (Evans et al., 1987; Zeidner and Shechter, 1988). How one deals with this stress is dependent on coping behaviors and the ability to adapt. The relationship between stress due to poor visual air quality and mental health is poorly understood. However, results from a study conducted by Rotton and Frey (1982) showed that as visual air quality decreased, emergency calls for psychiatric disturbances increased.

Visual Air Quality and Behavior

Evans et al. (1982) found that people who recently moved to Los Angeles from areas with good visual air quality consistently reduced outdoor activities during periods of reduced visual air quality compared with longer-term residents. Studies have also reported reduced altruism and increased hostility and aggression during periods of poor air quality

(Cunningham, 1979; Jones and Bogat, 1978; Rotton et al., 1979). The relationship between aggression, hostility, and visual air quality is curvilinear, with feelings of aggression and hostility increasing to a certain point and then dropping off and yielding to a desire to withdraw and escape from the situation. Evans and Cohen (1987) suggested that individuals adjust to poor visual air quality through adaptation and coping behaviors by altering their judgment of air quality based on current and previous exposure.

VISIBILITY EFFECTS ON VISITOR EXPERIENCE IN NATURAL SETTINGS

Psychological Value of Visibility and Visitor Satisfaction

While VAQ and SBE judgments established a link between haze and these perceptual indicators, they did not address whether good visual air quality was important to park visitors. To that end, National Park Service (NPS) research was directed toward determining if and how VAQ might affect park visitors. Specifically, the NPS has conducted studies that examined

- the importance of clean air compared to other park features,
- how aware visitors are of visibility conditions,
- relationships between visibility conditions and visitor satisfaction and park enjoyment, and
- whether visitor behavior could be linked to haze and visual air quality (Jones and Bogat, 1978; Rotton et al., 1979; Evans and Jacobs, 1982; Evans et al., 1982, 1987; Ross et al., 1985, 1987; Zeidner and Shechter, 1988; Smith et al., 2005).

In general, findings from these investigations suggest that visitors are aware of visibility conditions and that clean, clear air is a basic part of the enjoyment associated with visiting parks. The most important conclusions based on these findings are summarized in the following sections. Again, much of the following material on visitor experience in a natural setting was summarized by the author in Malm (1990).

Finding: An Environment Undisturbed by Man, Including Clean, Clear Air, Is Very Important to Park Visitors

Visitors who participated in the NPS studies conducted at Grand Canyon, Mesa Verde, Mount Rainier, Great Smoky Mountains, and Everglades national parks were given a mixed list of park features (managerial, resource, and social) and asked, on a five-point scale, how important each one was to their recreational experience. Some of the features were the same at all of the parks and some were specific to each park. For example, while "clean, clear air" and "interpretive signs" were listed for all the parks, "viewing canyon rims" was listed only for Grand Canyon. Twenty-four features were included in the list for Grand Canyon and Mesa Verde, 23 for Great Smoky Mountains, 8 for Mount Rainier, and 31 for Everglades national parks.

Results of visitor ratings showed that "clean, clear air" ranked among the top four most important features at all the parks. In fact, 82% of the 638 respondents at Grand Canyon rated it as "very important" or "extremely important" to their recreational experience. Only "cleanliness of the park" and "deep gorges" were rated higher, and that was only by a small amount. Several of the scenery-related features were also rated as "very important," especially at Grand Canyon and Mount Rainier. Flora- and fauna-related features were "very important" at all five parks.

Cluster analysis was used to statistically determine if clusters, or groups of park features, could be identified based on visitors' importance ratings. The analysis identified five clusters of features at Grand Canyon and Mesa Verde, three at Mount Rainier, seven at Great Smoky Mountains, and nine at Everglades. Fig. 5.21 shows the clusters of features and their relative importance at each park, based on the average ratings of all respondents.

The most significant finding reflected in Fig. 5.21 is that a cleanliness cluster, which included clean, clear air, was rated the most important at each park. The features in this cluster, when considered as a group, could be described as a natural environment that is undisturbed by humans. Typically, the second most important group consisted of features associated with the underlying theme of the park and information for which the park is famous. It appears that people visit these parks first to experience a natural setting and second to enjoy specific features associated with the park. Survey results show it is very important to visitors that parks are natural, free of pollution, and undisturbed by humans. These findings suggest that if park resources are allowed to deteriorate, visitor enjoyment of the parks would decline.

Finding: Visitors Are Aware of Visibility Conditions

Visitors were interviewed as they left viewing points in the parks and asked if they had noticed haze, and if so, whether they thought it was slightly, moderately, very, or extremely hazy. A comparison of haze-level responses with measurements of atmospheric extinction/visual range revealed a direct correlation. When visual range was lower, visitors were more aware of haze and were more likely to say it was "very" or

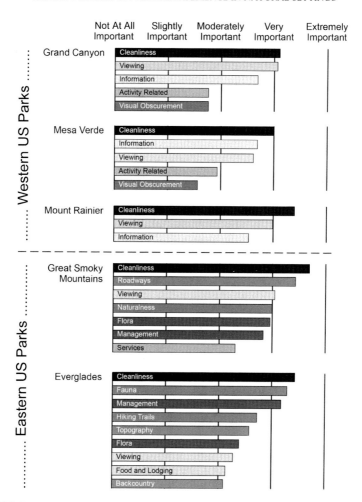

FIGURE 5.21   Mean relative importance of park feature clusters at five national parks.

"extremely" hazy. The varying demographic backgrounds of those interviewed (size of home community, income, age, and sex) did not correlate with awareness of haze.

Finding: Visibility Influences Visitor Enjoyment and Satisfaction

Results of the 1983 NPS visibility and visitor experience research at Grand Canyon and Mesa Verde show that those visitors who said they thought the view was "very" to "extremely" hazy also said they enjoyed the view less, enjoyed the park less, and were less satisfied with visibility conditions than those who said they were not, slightly, or moderately aware of haze. In other words, not only do most visitors notice haze, but also when they consider the view to be relatively hazy, it detracts from their enjoyment of the park.

Visitors at Grand Canyon and Mesa Verde were asked to rate the importance of 33 recreational experiences at the park that they might find satisfying. Examples of these experiences are things such as "learning about the park's history," "knowing that the park's resources and values are being protected," and "doing something with one's family." Those experiences pertaining to the existence for the park, including "knowing that the park resources are being protected," were rated as the most important by visitors at both parks. When combined with the findings discussed earlier, these findings indicate that

- visitors derive the most satisfaction from knowing that the park is there and the park's resources are being protected, and
- visitors feel that the natural environment of the park (which includes clean, clear air) is the most important of all the park resources.

Finding: Visibility Can Affect How Visitors Distribute Their Time in National Parks

Three NPS studies have been conducted to evaluate whether visitors would be willing to spend more time traveling to view vistas to obtain better visual air quality during their park visit. Those who were questioned generally responded that they would be willing to spend more time if visibility conditions were better and less if visibility conditions were worse. These results provide evidence that changes in VAQ can be expected to affect visitors' enjoyment and satisfaction with park visits.

A 1983 NPS study at the Grand Canyon included a smaller study in which 244 visitors were asked to rank possible alternative combinations of travel time to vistas and visibility conditions (illustrated with photographs). Rankings revealed that the average change in the amount of time the visitor was willing to spend traveling to a vista for every unit change in visibility (0.01 km^{-1} extinction coefficient) was between 15 min and 4 h.

The differences in the amount of travel time visitors were willing to spend to obtain better visibility were related to the vista and a hypothesized initial better visibility level. Vistas with more texture and color, dominant distant features, and better initial visibility levels were the most sensitive. For example, the average increase in the amount of time visitors were willing to spend traveling to the San Francisco Peaks vista in exchange for a unit increase in visibility was between 1.5 and 4 h when the initial visual range was more than 200 km and between 1.0 and 1.5 h when the initial visual range was less than 200 km. The average increase in travel time for the Desert View vista, where distant features are not dominant and for which the photographs were taken at a time of day when the vista was heavily shadowed and therefore showed little color or textural detail, was between 15 min and 1 h.

The findings of this study show that visitors are more disturbed by deterioration in visibility when the air is relatively clean and when the vista is highly colored and textured are consistent with those of the human judgment studies discussed earlier in this section. This is also in agreement with the findings of the visitor experience studies that suggest an environment undisturbed by humans is one of the features most valued by park visitors.

THRESHOLDS AND JUST NOTICEABLE DIFFERENCES/CHANGES

There are literally hundreds of journal articles addressing the workings of the human visual system (HVS) and its ability to detect a just noticeable difference (JND) between two images, especially as related to electronic image display systems. The interested reader is directed to an excellent review by Geisler (1989) and more recently by Zhenzhong and Hongyi (2014) and Wang et al. (2004).

The following discussion is not meant to be a review of all the literature as it pertains to the HVS but is presented in the context of edge thresholds and is directed toward research that was designed specifically to address haze within the context of visual impairment of scenic resources. Smoke stack plumes can be viewed perpendicular to or along their axis or at any intermediate angle. Depending on the viewing angle, smoke stack plumes can appear as nearly circular or oval or as a streak across the horizon sky. As such, there was a need to establish contrast thresholds as a function of viewing angle and plume size and shape. What level of contrast between a plume and its background is required to be detectable as a function of plume shape? This problem could just as easily be expressed in the context of JNDs or JNCs between a radiance field with and without a plume.

From a historical perspective, one of the first threshold problems discussed and addressed was that of viewing large landscape features against a uniform background such as the sky. It was from these studies that the concept of visual range, the farthest distance a large, dark object could be seen, evolved. Then, studies of contrast thresholds of a variety of targets, primarily circular in nature, were undertaken during World War II, with an eye toward understanding how much contrast was required to recognize and follow military targets. More recent studies have focused on the perceptibility of sine wave grating stimuli with varying contrast and spatial frequencies in an effort to understand how the human visual system works. In the next sections, some of these studies will be reviewed in the context of how they are applicable to the detectability of plumes, which typically have a Gaussian radiance profile. One basic question was how the contrast sensitivity of Gaussian plumes compared to sine wave gratings of comparable contrast and size.

Definition of Contrast

For the purpose of the following discussion, contrast is defined simply as contrast (Eq. 5.6) or as modulation contrast (Eq. 5.7):

$$C = \frac{N_{max} - N_{min}}{N_{max}} \qquad (5.6)$$

$$C = \frac{N_{max} - N_{min}}{N_{max} + N_{min}}. \qquad (5.7)$$

N_{max} and N_{min} refer to the maximum and minimum radiance levels, respectively, of the presented stimuli. The contrasts defined in Eqs. 5.6 and 5.7 are also known as Weber's contrast and Michelson's contrast, respectively. While Eq. 5.6 is generally used for expressing contrast between an object and its background, modulation contrast is used for grating patterns. Either contrast is appropriate, or one can be derived from the other. Note that when the values for N_{max} and N_{min} differ by a small amount, as they typically do in contrast sensitivity studies, contrast, as defined by Eq. 5.6, is approximately a factor of 2 greater than modulation contrast.

Detection of Large Landscape Features

Over the course of the last 100 or so years, there have been numerous studies and published articles on the human eye–brain system contrast thresholds. As early as the 1700s, Bouguer (1729, 1760) stated that if one looks carefully, a tree-covered mountain is visible if the difference between the brightness of the mountain and background sky is about one part in 51 or a contrast of about 0.02! He also pointed out that it made little difference if the mountain were snow covered (positive contrast) or tree covered (negative contrast). His estimate of the contrast threshold of the human eye–brain system under daylight conditions was, for all practical purposes, about right. Middleton and Knowles (1935) and Middleton (1952) report on threshold studies in natural lighting conditions in Ottawa, Canada, and Mount Washington, Vermont, where median detection thresholds of 1000 and 285 observations were −0.031 and −0.025, respectively. The reported thresholds ranged from 0.01 to 0.10 and were approximately lognormally distributed. Other early threshold experiments were reported by Houghton (1939), Bricard (1944), and Douglas and Young (1945). Johnson (1981) gave an excellent review of these studies and how they relate to instrumental measures of atmospheric scattering and extinction.

Detection of Laboratory Stimuli

Stimuli Characteristics

Typically, stimuli used to measure visual sensitivity consisted of grating patterns of alternating light and dark bars. Sensitivity was derived by varying the contrast of the bars and by changing the size of the bars along one dimension. Generally, the size variation was along the smallest dimension, regardless of whether it was in a vertical or horizontal orientation.

Fig. 5.22 shows how the angular extent (width) of three distributions compares for stimuli that contain one complete cycle per degree (cpd). Differentiating between background and surround luminance is important when comparing luminance distributions such as these. Background luminance is defined as the base luminance upon which the stimulus is presented, and surround luminance is the luminance of the area surrounding the stimulus. Usually, when a sine or square wave stimulus is generated, it is presented on a background luminance level that is midway between the lightest and the darkest portions of one cycle of the waveform. Because the sine and square wave stimuli used to examine

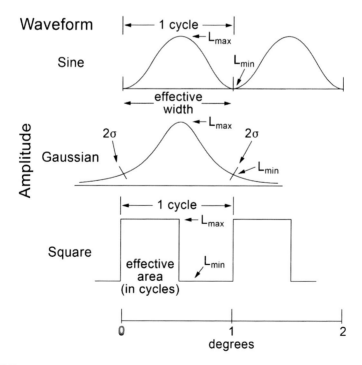

FIGURE 5.22   Schematic showing how spatial frequency is defined for sine, Gaussian, and square wave stimulus profiles.

visual sensitivity have usually been gratings covering the entire area of presentation, the surround has been either dark or set to some level similar to the background luminance.

Unlike a sine wave luminance profile, a Gaussian distribution does not reach a minimum amplitude. Therefore, the specification of the width of a Gaussian stimulus is somewhat arbitrary. One could use the full width at half maximum or a width based on one or more standard deviations. Bijl et al. (1989) defined the width of a Gaussian stimulus as $\pm\sigma$ (σ being one standard deviation), which includes 65% of the area under a Gaussian curve. A width of $\pm2\,\sigma$ incorporates 95% of the area and is the width used in the Ross et al. (1990, 1997) studies.

In the case of the Gaussian plume distribution, the background and the surround are the same. The intent is to show how cycles of sine and square waves would compare with a cycle of a Gaussian distribution if they were presented on the same background luminance level. The width of a stimulus with any of the above waveforms is defined to be the degrees of visual angle subtended between the maximum and minimum luminance of the stimulus and is one over the number of cpd subtended by the stimuli.

Duration of Stimuli

When considering the duration of stimulus presentation, distinguishing between stimulus detection and what will be called stimulus identification is relevant. Stimulus detection is simply defined as the ability of the HVS to detect the presence of a stimulus. Stimulus identification is defined as the task of visually locating and detecting the presence of a stimulus of known or unknown size, shape, and luminance characteristics when it is presented in an unknown area of the visual field. This difference between identification and detection is important when one begins to consider how results of the laboratory experiments can be utilized in a real-world setting.

From a laboratory procedural perspective, the duration of stimulus presentation, location of the stimulus in the visual field, and the subject's knowledge of specific stimulus characteristics must be addressed. Bloch's Law (Rovamo et al., 1984) describes the relationship between the duration of stimulus presentation and detection threshold. According to the law, when a stimulus is presented for a period shorter than some critical duration, threshold is dependent on the exposure time because of the temporal summation characteristics of the photoreceptors. As exposure time increases, the intensity of the stimulus required for threshold detection decreases until a critical time is reached. Once this value is reached, exposure time and threshold are independent of each other. Bloch's Law has been examined for circular targets that range from 0.0059° to 17°, for sinusoidal gratings with spatial frequencies ranging from 0.3 to 10 cpd, and for illuminance levels ranging from 0 to 6500 Trolands (a Troland is luminance in candle/m^2 multiplied by the area of the pupil in mm).

The critical duration required for stimulus detection is dependent on the illuminance level and which type of photoreceptor is active. At higher levels (such as 6500 Trolands, which is similar to normal daylight conditions), the cone system of the eye is active and acuity is best. At lower levels, the rod system dominates and acuity is poorer. The critical duration period is approximately 30 ms for cone photoreceptors and 100 ms for rods. In other words, targets need to be presented only for a short duration for the visual system to reach the "critical time." For more information on this topic, the reader is directed to papers by Deubel and Bridgeman (1995) and Gomez et al. (1994). Presentation time for the Ross et al. (1990, 1997) Gaussian plume studies discussed later was of a long enough duration that it should not be a factor in the reported detection thresholds.

The preponderance of research conducted on the contrast sensitivity characteristics of the HVS has been directed at understanding contrast sensitivity as a function of the spatial frequency characteristics of different stimuli. Most often, the stimuli were gratings that varied in luminance distribution and number of cpd of visual angle. The stimuli were sine wave, square wave, and edge to surround step function luminance distributions. Table 5.6 overviews the most pertinent studies. The stimuli used

TABLE 5.6 Studies on Visual Threshold Contrast Sensitivity for Different Size and Shape Stimuli

Author	Shape	Edge Type	Method	Threshold Measure
Campbell and Robson, 1968	Full length gratings	Square wave	Adjustment	Averaged
Howell and Hess, 1978	Full length gratings	Square wave Sine wave	Adjustment	Averaged
Cannon, 1979	Full length gratings	Sine wave	Magnitude estimation	Averaged
Van der Wildt and Waarts, 1983	Full length gratings	Sine wave	Adjustment	Averaged
Lamar et al., 1947	Rectangles	Square wave	Yes/no	63%
Shapley, 1974	Rectangles	Gaussian	Adjustment	Averaged
Thomas, 1978	Rectangles	Square wave	Rating	Averaged
Thomas et al., 1969	Rectangles	Square wave	Forced choice	50%
Blackwell, 1946	Circles	Square wave	Yes/no	50%
Pointer and Hess, 1989	Circles/ovals	Gaussian	Forced choice	Averaged
Bijl et al., 1989	Circles	Gaussian	Forced choice	50%

to examine contrast sensitivity can be categorized into four basic types: grating, rectangular, oval, and circular.

Grating Stimulus

Campbell and Robson (1968) conducted one of the earliest studies to assess visual sensitivity to gratings with different spatial frequencies. They used sine and square wave gratings of different spatial frequencies and an adjustment psychophysical technique, which has subjects adjust the contrast of the stimulus until it can just barely be detected. Howell and Hess (1978) and Cannon (1979) also conducted studies to establish threshold levels of contrast sensitivity as a function of grating spatial frequency but added an emphasis on visual spatial summation and contrast sensation, respectively. Spatial summation refers to an apparent additive effect between photoreceptors in the same receptive field. This summation may occur between individual photoreceptors within a visual field or between closely located adjacent fields with the same spatial frequency characteristics.

Fig. 5.23 displays results of the Campbell and Robson (1968), Howell and Hess (1978), and Cannon (1979) studies for sine wave gratings. The

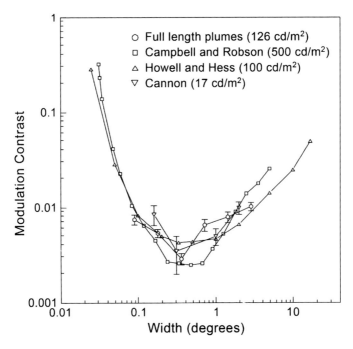

FIGURE 5.23   Comparison of threshold modulation contrast as a function of stimulus width for the full-length plume (Ross et al., 1990, 1997), Campbell and Robson (1968), Howell and Hess (1978), and Cannon (1979) grating studies. Luminance levels used in each of the studies are also included.

data are based on a variable number of cycles for Cannon and Campbell and Robson and on five cycles for Howell and Hess. Howell and Hess also examined sensitivity as a function of the number of cycles, but the only complete curve of contrast sensitivity as a function of spatial frequency is from their five-cycle data. Data from the Ross et al. (1990, 1997) plume detection threshold studies, which are discussed later, are also shown.

Campbell and Robson (1968) conjectured that if less than four cycles of the waveform were present, the contrast required for detection would be greater because of incomplete spatial summation across the visual field. Howell and Hess (1978) surmised that summation was complete when 10 cycles of the grating were present and that there was only a slight change between 5 and 10 cycles. Furthermore, their experiments showed that the detection threshold for 1 cpd stimuli was about a factor of 2 greater than for 5 cpd.

All of these studies show that the contrast required for a threshold level of detection decreases as the size of the stimulus increases, until it approaches approximately 0.33° (3 cpd). Then it increases as size continues to increase. However, the magnitude of the contrast required for threshold detection varied from experiment to experiment. Van der Wildt and Waarts (1983) designed a number of studies to investigate this issue. They found that if the size of the surround (area surrounding the grating stimuli) is limited in relation to the size of the stimulus, the contrast required for threshold detection is increased. Specifically, they found that if the size of the surround is such that it could not contain 10 uninterrupted cycles of the waveform, then the edges of the surround are detected, with a net result of reducing contrast sensitivity.

Rectangular Stimulus

The term rectangular stimulus refers to a stimulus that is limited along two dimensions. Grating stimuli typically subtend a large enough portion of the visual field so that the size of the second dimension has little or no effect on sensitivity. Most of the research conducted on sensitivity to rectangular stimuli used a step function luminance pattern. The studies were designed to investigate how the ratio of the longer to shorter dimensions of a rectangle, the background luminance, and the stimulus area affected visual detection. Lamar et al. (1947) examined the effect of luminance level and stimulus size and area on contrast threshold, using five rectangular sizes with length-to-width ratios of 2, 7, 20, 70, and 200; areas that ranged from 0.50 to 800 square minutes; and luminance levels of 60 and 10,000 cd/m². A luminance level of 10,000 cd/m² is approximately what one would experience in full daylight, in which photopic (cone) vision would dominate. A level of 60 cd/m² corresponds to luminance conditions similar to low-level indoor light, where the scotopic and photopic systems are functional.

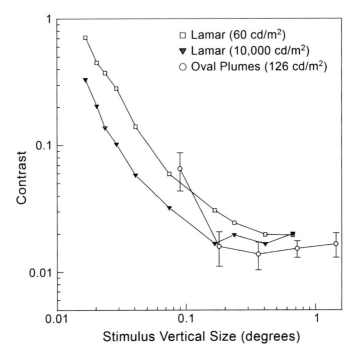

FIGURE 5.24 Comparison of threshold modulation contrast as a function of stimulus vertical size for the oval Gaussian plume (Ross et al., 1997) and rectangular step function study (Lamar et al., 1947). Luminance levels used in each of the studies are also included.

Fig. 5.24 shows results of the Lamar et al. (1947) study that are pertinent to the Ross et al. (1997) oval Gaussian plume study. Also shown in Fig. 5.24 is the Ross et al. (1997) oval plume data. Lamar et al. concluded that for 2×1 rectangular stimuli whose shorter dimension is less than approximately 0.03°, threshold contrast is an inverse function of the area of the rectangle. As the dimensions of a rectangle increase beyond 0.03°, threshold contrast becomes a function of rectangle asymmetry, with threshold contrast decreasing as the stimulus becomes longer and narrower. Finally, when the width of a 2×1 rectangle is larger than approximately 0.15 inches, threshold contrast tends to be invariant of size or area. Shapley (1974) supported this invariance hypothesis. He found that for a rectangular stimulus, sensitivity seemed independent of size when the width of the stimulus exceeded a critical size of approximately 0.2–0.3°. The Lamar et al. (1947) comparisons of threshold contrast values for background luminance values of 60 and 10,000 cd/m² showed that as background luminance decreases, threshold contrast increases. These results were verified and explained further in a study reported by De Valois et al. (1974). They found that as the luminance level decreases, sensitivity shifts to lower spatial frequencies, and this shift results in lower contrast sensitivity. As the luminance level decreases to the point where photopic vision subsides, the loss in sensitivity to the higher spatial frequencies becomes even more pronounced.

There was some question as to whether spatial summation might be dependent on the orientation of the stimuli. Thomas et al. (1969) and Thomas (1978) investigated whether spatial summation was symmetrical for square-wave rectangles by testing whether the contribution of length and width to sensitivity varied with rectangle orientation and if the length-to-width ratio affected sensitivity. No sensitivity difference was found between rectangles oriented vertically and those oriented horizontally. Pointer and Hess (1989) reported similar orientation findings for circular- and oval-shaped stimuli. Thomas et al. and Thomas also concluded that sensitivity is not uniquely determined by rectangle area or perimeter. They also found greater sensitivity to changes in the longer dimension.

Circular Stimulus

Studies conducted by Blackwell (1946) and Bijl et al. (1989) can be used for comparison to the plume study results. During World War II, substantial effort was put into establishing the contrast thresholds for a variety of targets for use in military applications. The overriding goals of the research were to investigate the various factors controlling the apparent contrast of objects, to ascertain the contrast thresholds of the human observer from laboratory experiments, and to evolve algorithms to predict a visual range or detectability of various objects. The early studies on contrast thresholds were carried out by Blackwell (1946) on circular targets and became known as the Tiffany experiment.

The visibility of a uniformly luminous object depends upon the apparent contrast between the object and its background, the angular size of the object, its shape, the contrast threshold of the observer at the level of luminance to which his eyes are adapted, knowledge of target location, search time, and other factors, such as the state of mind of the observer. The Tiffany experiment included a wide range of background luminance and stimulus diameters and over two million observations. Background luminance ranged from 0.001 to 102 ft-L, and stimulus diameters varied from 36° to 0.6°. These data have been supplemented to include cases of larger targets (Taylor, 1960a,b). The general conclusion was that contrast threshold increases with decreasing background luminance and decreasing stimulus size. As luminance falls from daylight to typical interior light levels and down through twilight, the increase in thresholds becomes rapidly more marked. At the limit, when the background luminance is taken down to starlight levels, the increase is very rapid. At these low light levels, luminance versus contrast threshold curves exhibit a discontinuity at around 3×10^{-3} cd/m^2. For high scene luminances, cone vision is used and employs foveal vision. As the luminance reduces, rod vision becomes more efficient and scotopic vision takes over.

The other study was conducted by Bijl et al. (1989), who used circular stimuli with a Gaussian luminance distribution. Bijl et al. examined contrast sensitivities as a function of presentation time and size. The longest presentation time was 1 s, and because all other studies had presentation times of one or more seconds, it is that data that are examined here. The widths of the stimuli ranged from approximately 0.03° to 10.0°. Width was defined at a Gaussian distribution cutoff point of one standard deviation, and threshold was set at a 76% detection level. Bijl et al. found that threshold contrast decreased as stimulus width increased from approximately 0.04° to 0.151°, then was mostly constant as width continued to increase up to 10.0°. The authors concluded that the contrast required for detection of a circle with a Gaussian luminance distribution was a constant when the diameter of the stimulus was between about 0.15° and 10° and when the circle was presented on a large uniform background. Results of the Blackwell (1946) and Bijl et al. (1989) studies are shown in Fig. 5.25.

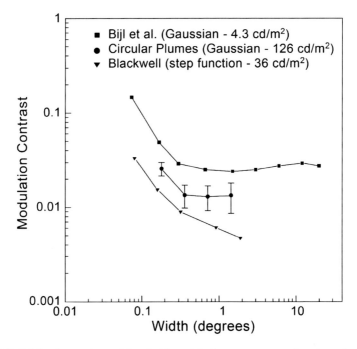

FIGURE 5.25 Comparison of threshold modulation contrast as a function of stimulus width for Gaussian circular plumes (Ross et al., 1997), Gaussian circles (Bijl et al., 1989), and step function circular stimuli (Blackwell, 1946). Luminance levels used in each of the studies are also included.

Gaussian Plume Studies

A primary limitation of the studies summarized in Table 5.5, as they relate to the goals of understanding contrast thresholds of plumes or layered hazes, is that the typical stimulus used to measure visual sensitivity consisted of grating patterns of alternating light and dark bars. Sensitivity was derived by varying the contrast of the bars and by changing the size of the bars along one dimension. Typically, the size variation was along the smallest dimension, regardless of whether it was in a vertical or horizontal orientation. Results from these studies are not sufficient to gauge sensitivity to plumes because a sine wave stimulus does not have a brightness profile that resembles a Gaussian plume. Even the single-cycle sine wave stimuli used in previous threshold studies consisted of bright and dark bands with variations in surround brightness. It is difficult to argue that these stimuli represent what an observer would see when viewing a plume. Furthermore, as orientation changes, the visual angle between the observer and the plume also changes. This can have a significant effect on sensitivity because the observed size of the plume can change along two dimensions. Another limitation is that when the stimulus was varied along two dimensions, the stimulus to surround luminance distribution most often used was a step function. Since the luminance distribution of a plume is more typical of a Gaussian nature, additional knowledge was required to understand visual contrast sensitivity in this context.

In conventional studies on perception thresholds of grating stimuli, width is not defined in terms of angular subtense but rather is presented in terms of the number of cpd subtended by the stimuli. For a single-cycle sine wave stimuli, the width of the stimuli is still one over the number of cpd. The stimuli, however, usually begin and end at the average as opposed to the minimum luminance, as shown in Fig. 5.21. One degree sine wave grating stimuli start at some average value, increase to a maximum, and then decrease to a minimum and return to the average luminance level. Therefore, the sine wave stimuli are not directly comparable to Gaussian profile data, nor are they representative of thresholds for real-world plumes. However, it will be shown that as a function of plume width for Gaussian plume thresholds, the gross features of contrast sensitivity are similar to sine wave grating threshold data.

Three laboratory experiments were conducted (Ross et al., 1990, 1997). The stimuli consisted of a plume plus background sky, which had plumes of varying sizes and contrasts with a Gaussian luminance distribution superimposed on the Rayleigh sky. In a real-world setting, the vertical and horizontal visual angular subtense of a plume depends on the observer-to-plume geometry. For example, if an observer is close to a plume, it will appear larger than it would if that same observer

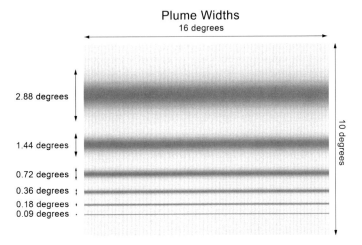

FIGURE 5.26   Relative sizes of the plumes used in the full-length-plume studies. Only one plume at a time was shown during the studies.

was farther away. Similarly, if an observer was down range and looking straight into the plume, it would appear to be more circular than if the observer was at another orientation. Because the purpose of the studies was to develop an empirical algorithm to predict the probability of visual detection of plumes, the vertical and horizontal angular subtenses of the plumes varied between studies to represent a realistic range of observer–plume geometries.

Study 1 used plumes with a horizontal visual extent of 16° and vertical extents of 0.09°, 0.18°, 0.36°, 0.72°, 1.44°, and 2.88°. The plumes used for Study 2 had a 1 × 2 vertical to horizontal extent ratio, with vertical angular dimensions of 0.09°, 0.18°, 0.36°, 0.72°, and 1.44°. Study 3 used circular plumes with angular extents of 0.09°, 0.18°, 0.36°, 0.72°, and 1.44°. Based on the shape of the stimuli, Study 1 will be referred to as "full length," Study 2 as "oval," and Study 3 as "circular." Fig. 5.26 illustrates the relative sizes of the plumes used for Studies 1–3.

A yes/no, high-threshold psychophysical procedure (Green and Swets, 1966) was used as the methodology to conduct the laboratory experiments. The yes/no method consisted of randomly presenting subjects with stimuli that show either a blue-sky background or a background with a plume. The subjects' task was to indicate whether a plume could be seen in each of the presentations. The experiments were conducted using a Telex Caramate 3200 slide projector equipped with a Buhl f5.7-mm lens and a high-intensity ELH 300 watt bulb. All ambient light sources were removed from the room during testing, and the room was painted a dark, flat gray to minimize reflection. The luminance level from the high-intensity bulb was such that the photopic portion of the visual system would be dominant and was similar to levels found in medium to high indoor lighting.

Subjects

Colorado State University students served as subjects. All subjects had normal or corrected-to-normal vision and were free of any color vision deficiencies. Subjects were tested in pairs for each experiment during 12 sessions, each of which lasted approximately 90 min. Short breaks were given at 30-min intervals. Responses were recorded on a PC, using subject-controlled response boxes with "yes" and "no" buttons. A partition between the subjects prevented them from observing each other's responses.

Four sets of test slides were used, with 50% of the slides in each set being background sky and 50% being plume plus background sky. Slides were arranged in random order in slide carrousels, and the carrousels were presented in random order across subjects and sessions. Slides were presented to the subjects at a rate of one every 2.5 s, with subjects indicating for each whether or not they could detect a plume. Only one plume size was presented during a given session. Subjects underwent training prior to testing to familiarize them with study procedures and stimuli. Subjects, seated in the positions they would be occupying during testing, were shown each of the test stimuli and were allowed to view each stimulus until they were certain they either could or could not see the plume. After they were fully familiar with the stimuli, a trial test session was conducted. This allowed them to become familiar with the use of the response box and the timing sequence for image display. It also provided an opportunity to furnish some pre-experiment feedback about how liberal or conservative they were being in their responses. Finally, the trial session allowed subjects to ask questions and review slides.

Probability of Detection and Sensitivity

Possible responses using a yes/no methodology include a "hit" (correctly detecting a plume), a "miss" (failing to identify the presence of a plume), a "false alarm" (falsely detecting a plume), and a "correct rejection" (correctly not detecting a plume). The raw-hit rate, or percentage of times correct, for a given stimulus is not usually a good measure of a subject's probability of detection, since the latter may partially depend on the subject's decision criteria. In an exaggerated example, an extremely liberal subject might respond "yes" to every stimulus. The raw-hit rate would be 100%, which would suggest perfect sensitivity at all contrast levels. However, since only 50% of the stimuli contained plumes, the false-alarm rate would also be 100% and would inform the experimenter that the subject was merely saying "yes" to every stimulus and not really recognizing the signal. Therefore, raw-hit rates are often adjusted using one of several parametric or nonparametric methods to compensate for variation in subject decision criteria (De Valois and De Valois, 1990).

Results

Study 1: Full-length Plumes

Sixteen subjects participated in Study 1. The stimuli used for this experiment consisted of plumes with vertical angular sizes of 0.09°, 0.18°, 0.361°, 0.720°, 1.44°, and 2.88° and a horizontal angular extent of 16.0°. Contrast values of 0.050, 0.040, 0.030, 0.020, 0.017, 0.015, 0.013, 0.011, and 0.005 were used for all sizes. Fig. 5.22 shows the predicted probability of detection curves for five of the plume sizes for one of the subjects. The Campbell and Robson (1968), Howell and Hess (1978), Cannon (1979), and Gaussian plume curves show a similar shape, with a rapid increase in probability of detection for a small change in contrast. Also, notice the pattern of how the curves shift along the abscissa as the vertical angular size of the plumes changes.

Results of Study 1 are compared with sine wave grating studies in Fig. 5.23. Threshold contrast for all of the studies was defined as a 50% probability of detection, and modulation contrast was used in Study 1 to make it comparable to the other studies. The results shown in Fig. 5.23 for the full-length plumes indicate that the contrast required for a 50% probability of detection decreases as the angular subtense of the plume increases from 0.09° up to approximately one-third of a degree, and then increases as the angular extent of the plume continues to increase to 2.88°.

The results for the full-length-plume study are surprisingly similar to results from the other studies in that they all show a maximum sensitivity when the stimulus is about 0.33° in width (3 cpd). It should be pointed out, however, that the positioning of the data points for the full-length-plume data is somewhat arbitrary when comparing it to sine wave stimuli. If the width was defined to be $\pm\sigma$ instead of $\pm2\sigma$, the width of each plume would be decreased by a factor of 2. Furthermore, the stimulus profile for a one-cycle sine wave is different from the Gaussian data in that sine wave stimuli begin and end at the average instead of at minimum luminance.

The magnitude of the threshold contrast varies between studies for a variety of reasons. Most significantly, the surround was different for each experiment. The Van der Wildt and Waarts (1983) data show that a surround that is less than 10 cycles from the edge of the stimuli serves to reduce contrast sensitivity. For instance, the Howell and Hess (1978) data suffered from "edge effects" in that each stimulus was shown with a surround that either caused a dark line or a dark background that was less than 10 cycles from the edge of each stimulus. Notice that at 3 cpd their data show sensitivities significantly less than the Campbell and Robson (1968) data.

Study 2: Oval Plumes

Four subjects participated in Study 2 because of the extensive amount of time required to test a large number of subjects, such as in Study 1. It is quite common for visual sensitivity studies to employ only one or two subjects because, for visually healthy subjects tested under well-controlled viewing conditions, there is only slight variation in visual sensitivity between subjects.

The plumes used for Study 2 were oval shaped with a 1×2 vertical to horizontal angular extent ratio. The vertical angular sizes of the plumes used were 0.09°, 0.18°, 0.36°, 0.72°, and 1.44°. Plume contrast values were 0.007, 0.014, 0.018, 0.020, 0.023, 0.036, 0.037, 0.041, and 0.045 for the 1.44° and 0.72° plumes; 0.004, 0.008, 0.010, 0.012, 0.015, 0.017, 0.022, 0.031, and 0.050 for the 0.36° and 0.18° plumes; and 0.019, 0.030, 0.033, 0.046., 0.062, 0.075, 0.080, 0.099, and 0.116 for the 0.09° plumes. Contrast values were altered somewhat from those used in Study 1 to provide a full range of values to represent detection levels along the section of most-rapid change in the probability of detection curves. Fig. 5.24 shows the threshold level of detection (defined as a 63% probability of detection) as a function of stimulus width to be 60 and 10,000 cd/m^2 for the Lamar et al. (1947) study and 126 cd/m^2 for the oval-plume study. The error bars around the oval-plume study are 95% confidence intervals. The Lamar et al. data were based on a step function brightness profile.

The two datasets agree quite well in that the threshold contrast decreases as the plume size increases to about 0.2° and then stays fairly constant for larger sizes. At the larger sizes, the threshold contrast for the oval-plume study is somewhat less than that for the step function data. One possible explanation for this discrepancy is that in the Lamar et al. study, the limited surround to stimulus size ratio may have depressed sensitivity and increased threshold contrast for larger-sized stimuli because of incomplete spatial summation of the visual system.

Study 3: Circular Plumes

Five subjects participated in Study 3. The plumes were circular with a Gaussian intensity distribution and subtended visual viewing angles of 0.09°, 0.18°, 0.36°, 0.72°, and 1.44°. Plume contrast values of 0.015, 0.020, 0.025, 0.030, 0.040, 0.050, 0.070, 0.090, and 0.100 were used with each size. All five subjects had a difficult time detecting the smallest plume (0.09°) and typically had a hit rate of 0% at the halfway point in testing. During debriefing, subjects indicated that they had been vigilant and knew what they were looking for but could not detect the plume in any of the trials, although it occurred at a 50% rate. Because of this, data for the 0.09

circular plume were of no practical value and were not included in subsequent analysis. The same analytical procedures as those used to evaluate the data for the full-length- and oval-plume studies were used for this study. The 76% detection level is used for purposes of comparison to other studies.

As in the oval-plume study, the contrast required for detection decreased until the diameter of the plume reached 0.36°, and then remained fairly constant as size continued to increase. A comparison of the Blackwell (1946), Bijl et al. (1989), and circular-plume data is shown in Fig. 5.25. Since Bijl et al. set the threshold level at a 76° probability of detection, a similar level was calculated for the Blackwell data. The magnitude of contrast required for threshold detection at any given width is smaller for the Blackwell study, which used step function stimuli, than for the other two studies. The difference between the circular-plume and Blackwell data is similar to the differences found among contrast sensitivity studies that used grating stimuli having square or sine wave luminance distributions (De Valois and De Valois, 1990). The Blackwell data also show that the threshold contrast continues to decrease up to the largest stimuli tested instead of becoming a constant. The contrast detection thresholds are higher (show lower sensitivity) for the Bijl et al. study because of the low background luminance used in that study.

Summary

Contrast thresholds for single-cycle, full-length-plume stimuli with Gaussian luminance distributions are similar to those developed using sine wave grating distributions in terms of sensitivity as a function of plume width. Observers are most sensitive to plumes that subtend an angle of about 0.33°, with sensitivities falling off for smaller or larger subtended angles. However, the magnitude of the threshold contrast varies from experiment to experiment because of surround and edge effects. The single-cycle, full-length-plume thresholds are greater than the lowest threshold published for sine wave gratings.

The contrast sensitivities for the oval plumes with a length-to-width ratio of 2:1 as a function of vertical size are similar to full-length plumes in the size ranges less than 0.33°. However, for larger sizes, the contrast sensitivity decreases but at a much smaller rate than for the full-length-plume stimuli. Furthermore, the magnitude of the threshold contrasts is greater for oval than for full-length plumes. Finally, the circular-plume sensitivities mimic those of the oval plume in terms of the functional dependence of size. However, circular plumes require greater contrasts to be detected.

The full-length, oval, and circular-plume contrast threshold data have been incorporated into a linear interpolation algorithm that allows plumes of any size to be estimated. The algorithm is referred to as PROBDET or probability of detection algorithm.

PROBABILITY OF DETECTION ALGORITHM (PROBDET)

The full-length, oval, and circular Gaussian plume data form a self-consistent dataset from which contrast detection thresholds can be estimated for any size and shape of Gaussian stimulus. The PROBDET algorithm (Ross et al., 1997) is a linear interpolation scheme that allows any probability of detection level to be estimated for plumes that fall within the bounds defined by the full-length, oval, and circular-plume stimuli. Three runs of the PROBDET algorithm demonstrate how estimated probability of detection values varies as a function of plume contrast and plume horizontal angular size when the vertical visual size is constant. Vertical angular plume size was fixed at 0.36°, and the horizontal extent was set at 0.36°, 0.50°, and 1.00°. Contrast ranged from 0.006 to 0.095 in 0.001 increments. A vertical extent of 0.36° was chosen because the HVS is most sensitive to this size.

Results of the runs are shown in Fig. 5.27. The confidence intervals, which are normally output by PROBDET, were not included to minimize

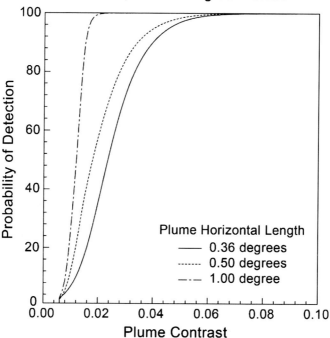

FIGURE 5.27 Estimated probability of detection curves for stimuli with a vertical size of 0.36° and horizontal sizes of 0.36°, 0.5°, and 1.0°.

the number of curves on the graph. Fig. 5.27 shows that as the horizontal extent of the plume increases to 0.5° and 1.0°, the slope of the probability of detection curve increases significantly. If one were to define a detection threshold at a probability of detection value of 75%, for example, the threshold for the 0.36°, 0.5°, and 1.0° horizontal plumes would occur at contrast values of 0.034, 0.029, and 0.017, respectively.

EXTINCTION CHANGE REQUIRED TO EVOKE A SUPRATHRESHOLD JNC

Whereas the preceding discussion pertained to the contrast threshold associated with just being able to detect a plume or layered haze or large, dark landscape features as seen against a background sky, the following analysis addresses the ability of an observer to see a change in an already visible landscape feature, resulting from an increase or decrease in atmospheric extinction.

As discussed previously, typical laboratory studies are done with side-by-side comparisons of a baseline scene or stimulus and some altered or degraded stimulus. Side-by-side comparisons of two scenes, one more degraded than the other, are unrealistic in a real-world setting. An observer will necessarily view a scene at two different time periods, requiring the person to remember what the scene looked like at some previous time. Illumination and meteorological conditions may have changed, and so judging whether the scene has changed because of an increase or decrease in haze is much more difficult than if images are viewed side by side at the same time. However, the side-by-side comparison stimulus sets a lower bound of suprathreshold sensitivity. It was shown in Chapter 4 that, for a scene with sharp contrast edges, a theoretical lower limit of about 4% to 10% change in extinction would evoke one JNC.

JNC Based on VAQ Ratings of Scenes with Complex Scenic Elements

In the Malm et al. (1981) VAQ judgment studies, participants were asked to rate the VAQ of vistas impaired by varying levels of haze on a 1–10 scale. The study and its conclusions were discussed in some detail previously. The average standard deviation was approximately $\sigma = 1.5$ for 30 ratings of a slide depicting a level of VAQ. Given these statistics, it is possible to define a JNC associated with a judged VAQ change (VAQ JNC) in any number of ways. One could calculate a confidence interval and define a VAQ JNC as that interval that gives a 90% confidence (or any other percentile) that the rating is some number, VAQ = y. If a VAQ judgment fell just outside that interval, it would be said to correspond to one JNC.

A second possibility is to form a distribution of differences between paired VAQ judgments and to test the hypothesis that the mean of the distribution of differences is zero for a single observation or assessment. This approach is problematic because it is quite possible that, on a one-time-judgment basis, an observer is more sensitive to a change in haze than a difference of only one VAQ increment implies. For instance, the equation for the VAQ ratings of Mount Trumbull as a function of Mount Trumbull's contrast shown in Fig. 5.5 can be written as a function of optical depth τ:

$$VAQ = 6.2e^{-\tau} + 1 \qquad (5.8)$$

where $\tau = (b_{ext})96$. A change in VAQ of one corresponds to at least a 55% change in extinction. Observers may well be more sensitive than that.

The paradigm adopted here corresponds to two sets of observations by a group of observers n at some appropriate vista overlook point. The VAQ changes over some increment of time, say on the order of an hour, as a result of a frontal passage or a local source of haze such as a forest fire. The question posed is what change in haze conditions would be required for this group of people, on average, to notice a difference in VAQ? This paradigm is essentially equivalent to performing a standard t-test on the mean of two sets of normally distributed VAQ judgments.

The equation for ΔVAQ that corresponds to one VAQ JNC as a function of distribution standard deviation σ, number of observers n, and t value for a level of confidence is

$$\Delta VAQ = \frac{\sqrt{2}\sigma}{\sqrt{n}}t. \qquad (5.9)$$

For confidence levels of 80% and 90%, the t values or $F(x)$ values are 0.85 and 1.3, respectively. $F(x)$ is defined to be

$$F(x) = \frac{1}{\sigma\sqrt{2\pi}}\int_{-\infty}^{x} e^{\frac{-1}{2}\left(\frac{v-u}{\sigma}\right)^2} dv \qquad (5.10)$$

where u is the mean of the distribution and x is the value that corresponds to the desired level of significance. Calculated using Eq. 5.8 for $n = 30$ observers, ΔVAQ values are 0.33 and 0.5 for the 80% and 90% confidence levels, respectively. Using these ΔVAQ numbers and Eq. 5.8, it is possible to calculate the percent change in extinction required to evoke one VAQ JNC for the Grand Canyon Mount Trumbull vista.

Fig. 5.28 is a plot of the percent change in extinction required for one VAQ JNC as a function of optical depth. The percent change in extinction curve versus optical depth reaches a minimum at $\tau \approx 1.8$ and corresponds to percent changes in extinction of 15% and 25% for 80% and

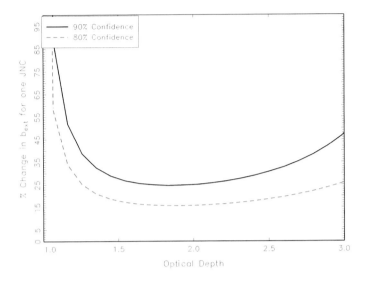

FIGURE 5.28   The percent change in extinction to evoke a noticeable change in a landscape feature plotted as a function of optical depth for 80% and 90% confidence levels.

90% confidence levels, respectively. These levels of extinction change correspond to approximately 1.5 and 2.5 dv, respectively.

Fig. 5.28 is similar to Fig. 4.6 where the percent change in extinction required to evoke a one JNC when viewing a large, dark landscape feature against a sky background as a function of optical depth τ is plotted. That calculation was based on laboratory studies in which visual stimuli were compared on a side-by-side basis and is expected to yield the highest, but unrealistic in the real world, contrast threshold sensitivity as a function of changes in extinction. The laboratory studies suggest that, at an optical depth of about 2, a 4% change in extinction could evoke a one JNC, but Fig. 5.28 suggests that it would take a 15% to 20% change in extinction to be noticed. These results are consistent with the Henry (2002, 2005) studies discussed in the following section.

JNC Based on Colorimetric Measurements

As discussed in Chapter 4, colorimetric measurements of perceived lightness and colorfulness could both be represented as functions of optical depth, using identical equations of $PL = PL_o e^{-\tau}$ and $M = M_o e^{-\tau}$, where PL and PL_o and M and M_o are perceived lightness and colorfulness at some impaired and reference optical depth τ and τ_o, respectively (Henry, 2002, 2005). This relationship between perceived lightness and colorfulness and optical depth of individual landscape features is the same as VAQ judgments of single landscape features and optical depth. Note this is also the approximate relationship between universal apparent contrast C_r and inherent universal contrast C_o when a landscape feature is viewed against the horizon sky.

Henry went on to define a JNC in the context of these measurements as the change in optical depth required to have 95% confidence that two measurements of colorfulness or perceived lightness were different. This definition, of course, is dependent on how accurately an observer can make a measurement using the colorimeter and is therefore to some degree instrument dependent and independent of inherent human-observer thresholds. Given that PL and M are exponential functions of optical depth, Henry derived curves similar to Fig. 5.28, showing how one JNC, expressed in dv, varied as a function of optical depth. For the more sensitive conditions, it would take about a 15% change in extinction or about a dv change of 1.5 to evoke a JNC. For a confidence level of 80%, a percent change in extinction of less than a dv would evoke a JNC.

In general, it seems that under somewhat ideal real-world conditions, a 10–15% change in atmospheric extinction (1.0–1.5 dv) will, on average, be noticed and that a landscape feature viewed against some uniform background with a universal contrast of around −0.02 will be seen. The contrast levels of plumes as viewed against some uniform background are dependent on both size and shape. For full-length plumes subtending about one-third of a degree and with a Gaussian profile, universal contrast levels of as low as 0.01 can be noticed, while circular plumes, again subtending about one-third of a degree and with a Gaussian profile, require universal contrast levels of 0.04 or more.

References

Abt Associates, 2001. Assessing Public Opinions on Visibility Impairment due to Air Pollution: Summary Report. Abt Associates Inc, Bethesda, MD.

AZ DEQ (Arizona Department of Environmental Quality), 2003. Recommendation for a Phoenix Area Visibility Index, available at http://phoenixvis.net/pdf/vis_031403final.pdf.

Barker, M., 1976. Planning for Environmental Indices: Observer Appraisals of Air Quality. Perceiving Environmental Quality. Plenum Press, New York.

Bell, P., 1985. Impact of impaired visibility on visitor enjoyment of the Grand Canyon: a test of an ordered logit utility model. Environ. Behav. 17, 459–474.

Bickerstaff, K., Walker, G., 2001. Public understandings of air pollution; the localization of environmental risk. Global Environ. Change 11, 133–145.

Bijl, P., Koenderink, J.J., Toet, A., 1989. Visibility of blobs with a Gaussian luminance profile. Vision Res. 29 (4), 447–456.

Blackwell, H.R., 1946. Contrast thresholds of the human eye. J. Opt. Soc. Am. 36 (11), 624–643.

Bouguer, P., 1729. Essai d'Optique sur la gradation de la lumiere, Paris.

Bouguer, P., 1760. Traite d'Optique dur la lumiere, de la Caille, Ed., Paris.

Bricard, J., 1944. La visibilite des objets eloignes a travers le brouillard. Ann. Geophys. 1, 101–112.

Buhyoff, G.J., Leuschner, W.A., 1978. Estimating psychological disutility from damaged forest stands. For. Sci. 24 (3), 424–431.

Campbell, F.W., 1983. Ambient stressors. Environ. Behav. 15, 355–380.

Campbell, F.W., Robson, J.G., 1968. Application of Fourier analysis to visibility of gratings. J. Physiol. 197, 551–566.

Cannon, Jr., M.W., 1979. Contrast sensation: linear function of stimulus contrast. Vision Res. 19 (9), 1045–1052.

Chestnut, L.G., Dennis, R.L., 1997. Economic benefits of improvements in visibility: acid rain provisions of the 1990 Clean Air Act amendments. J. Air Waste Manage. Assoc. 47, 395–402.

Cohen, S., Evans, G.W., Stokols, D., Krantz, D.S., 1986. Behavior, Health, and Environmental Stress. Plenum Press, New York.

Craik, K.H., Zube, E.H., 1976. Perceiving Environmental Quality: Research and Applications. Plenum Press, New York.

Cunningham, M., 1979. Weather, mood, and helping behavior: quasi-experiments with the sunshine Samaritan. J. Per. Soc. Psych. 37, 1947–1956.

Daniel, T.C., Boster, R.S. Measuring landscape esthetics: The scenic beauty estimation method, USDA Forest Service Research Paper RM-167, Fort Collins, CO, 1976.

Daniel, T.C., Wheeler, L., Boster, R.S., Best, P.R., 1973. Quantitative evaluation of landscapes: an application of signal detection analysis to forest management alternatives. Man-Environ. Syst. 3 (5), 330–344.

De Valois, R.L., De Valois, K.K., 1990. Spatial Vision (Oxford Psychology Series 14). Oxford University Press, New York.

De Valois, R.L., Morgan, H., Snodderly, D.M., 1974. Psychophysical studies of monkey vision. 3. Spatial luminance contrast sensitivity tests of macaque and human observers. Vision Res. 14, 75–81.

Deubel, H., Bridgeman, B., 1995. Perceptual consequences of ocular lens overshoot during saccadic eye-movements. Vision Res. 35 (20), 2897–2902.

Douglas, C.A., Young, L.L. Development of a Transmissometer for Determining Visual Range, U.S. Dept. Commerce, C.A.A. Tech. Dev. Rep. No. 47, 1945.

Ely, D.W., Leary, J.T., Stewart, T.R., Ross, D.M. The establishment of the Denver Visibility Standard, 84th Annual Meeting of the Air and Waste Management Association, June 16–21, 1991.

Evans, G.W., Cohen, S., 1987. Environmental stress. In: Stokols, D., Altman, I. (Eds.), Handbook of Environmental Psychology. Wiley, New York.

Evans, G.W., Jacobs, S.V., 1982. Air pollution and human behavior. In: Evans, G.W. (Ed.), Environmental Stress. Cambridge University Press, New York.

Evans, G.W., Jacobs, S.V., Frager, N.B., 1982. Behavioral responses to air pollution. In: Baum, A., Singer, J. (Eds.), Advances in Environmental Psychology, vol. 4. Erlbaum, New York.

Evans, G.W., Jacobs, S.V., Dooley, D., Catalano, R., 1987. The interaction of stressful life events and chronic strains on community mental-health. Am. J. Commun. Psych. 15, 23–24.

Fajardo, O.A., Jiang, J., Hao, J., 2013. Assessing young people's preferences in urban visibility in Beijing. Aerosol Air Quality Res. 13, 1536–1543.

Geisler, W.S., 1989. Sequential ideal-observer analysis of visual discriminations. Psychol. Rev. 96 (2), 267–314.

Gesler, W., 2005. Therapeutic landscapes: an evolving theme. Health Place 11 (4), 295–297.

Gomez, C., Atienza, M., Vazquez, M., Cantero, J.L., 1994. Saccadic reaction times to fully predictive and random visual targets during gap and non-gap paradigms. In: Delgado Garcia, J.M., Godaux, E., Vidal, P.-P. (Eds.), Information Processing Underlying Gaze Control, Pergamon Studies in Neuroscience No. 12. Elsevier Science, Ltd, Oxford, UK.

Green, D.M., Swets, J.A., 1966. Signal Detection Theory and Psychophysics. Wiley, New York.

Henry, R.C., 2002. Just noticeable differences in atmospheric haze. J. Air Waste Manage. Assoc. 52, 1238–1243.

Henry, R.C., 2005. Estimating the probability of the public perceiving a decrease in atmospheric haze. J. Air Waste Manage. Assoc. 55, 1760–1766.

Houghton, H.G., 1939. On the relation between visibility and the constitution of clouds and fog. J. Aer. Sci. 6, 408–411.

Howell, E.R., Hess, R.F., 1978. Functional area for summation to threshold for sinusoidal gratings. Vision Res. 18, 369–374.

Johnson, R.W., 1981. Daytime visibility and nephelometer measurements related to its determination. Atmos. Environ. 15, 1835–1845.

Jones, J.W., Bogat, G.A., 1978. Air-pollution and human aggression. Psych. Rep. 43, 721–722.

Lamar, E.S., Hecht, S., Schlaer, S., Hendely, C.D., 1947. Size, shape, and contrast in detection of targets by daylight vision. 1. Data and analytical description. J. Opt. Soc. Am. 37 (7), 531.

Latimer, D.A., Hogo, H., Daniel, T.C., 1981. The effects of atmospheric optical conditions on perceived scenic beauty. Atmos. Environ. 15, 1865–1874.

Loehman, E.T., Park, S., Boldt, D., 1994. Willingness-to-pay for gains and losses in visibility and health. Land Econ. 4, 478–498.

Malm, W.C. Visibility: Existing and Historical Conditions – Causes and Effects, National Acid Precipitation Assessment Program, State of Science and Technology Report 24, Section 6. Government Printing Office, Washington, DC, October 1990.

Malm, W.C., Pitchford, M.L. The use of an atmospheric quadratic detection model to assess change in aerosol concentrations to visibility, Air & Waste Management Association 82nd Annual Meeting, paper 89-67.3, June 25–30, 1989.

Malm, W.C., Kelley, K., Molenar, J., Daniel, T., 1981. Human perception of visual air-quality (uniform haze). Atmos. Environ. 15 (10/11), 1875–1890.

Malm, W.C., MacFarland, K.K., Molenar, J.V., Daniel, T., 1983. Human perception of visual air quality (layered haze). In: Rowe, R.D., Chestnut, L.G. (Eds.), Managing Air Quality and Scenic Resources at National Parks and Wilderness Areas. Westview Press Inc., Boulder, CO, pp. 27–40.

Malm, W.C., Molenar, J.V., Pitchford, M.L., Deck, L.B. Which visibility indicators best represent a population's preference for a level of visual air quality?, Paper 2011-A-596-AWMA, Air & Waste Management Association 104th Annual Conference, Orlando, June 21–24, 2011.

Middleton, W.E.K., 1952. Vision through the Atmosphere. University of Toronto Press, Toronto, Canada, pp. 219–220.

Middleton, W.E.K., Knowles, W.E., 1935. How far can I see? Sci. Mon. 41, 343–346.

Middleton, P., Stewart, T.R., Dennis, R.L., 1983. Modeling human judgments of urban visual air-quality. Atmos. Environ. 17 (5), 1015–1021.

Middleton, P., Stewart, T.R., Ely, D., 1984. Physical and chemical indicators of urban visual air-quality judgments. Atmos. Environ. 18 (4), 861.

Mumpower, J., Middleton, P., Dennis, R.L., Stewart, T.R., Veirs, V., 1981. Visual air quality assessment: Denver case study. Atmos. Environ. 15, 2433–2441.

Pointer, J.F., Hess, R.F., 1989. The contrast sensitivity gradient across the human visual-field: with emphasis on the low spatial-frequency range. Vision Res. 29 (9), 1133.

Pryor, S.C., 1996. Assessing public perception of visibility for standard setting exercises. Atmos. Environ. 30 (15), 2705–2716.

Ross, D.M., Malm, W.C., Loomis, R.J. The psychological valuation of good visual air quality by national park visitors, 78th Annual Meeting of the Air Pollution Control Association, Pittsburgh, 1985.

Ross, D.M., Malm, W.C., Loomis, R.J., 1987. An examination of the relative importance of park attributes at several national parks. In: Bhardwaja, P.S. (Ed.), Transactions: Visibility Protection: Research and Policy Aspects. APCA, Pittsburgh, PA.

Ross, D.M., Malm, W.C., Loomis, R.J., Iyer, H., 1990. Human visual sensitivity to layered haze using computer generated images. In: Mathai, C.V. (Ed.), Visibility and Fine Particles. Air & Waste Management Association, Pittsburgh, PA.

Ross, D.M., Malm, W.C., Iyer, H., 1997. Human visual sensitivity to plumes with a Gaussian luminance distribution: experiments to develop an empirical probability of detection model. J. Air Waste Manage. Assoc. 47, 370–382.

Rotton, J., Frey, J., 1982. Atmospheric conditions, seasonal trends, Psychiatric Emergencies. Replication and Extensions. American Psychological Assoc, Washington, DC.

Rotton, J., Frey, J., Barry, T., Milligan, M., Fitzpatrick, M., 1979. Air-pollution experience and physical aggression. J. Appl. Psych. 9, 397–412.

Rovamo, J., Leinonen, L., Laurinen, P., Virsu, V., 1984. Temporal integration and contrast sensitivity in foveal and peripheral vision. Perception 13 (6), 665–674.

Schulze, W.D., Brookshire, D.S., Walther, E.G., Macfarland, K.K., Thayer, M.A., Whitworth, R.L., Bendavid, S., Malm, W.C., Molenar, J., 1983. The economic-benefits of preserving visibility in the national parklands of the Southwest. Natural Resources J. 23 (1), 149–173.

Shapley, R., 1974. Gaussian bars and rectangular bars: influence of width and gradient on visibility. Vision Res. 14, 1457–1462.

Smith, A.E., 2013. An evaluation of the robustness of the visual air quality "preference study" method. J. Air Waste Manage. Assoc. 63, 405–417.

Smith, A.E., Kemp, M.A., Savage, T.H., Taylor, C.L., 2005. Methods and results from a new survey of values for eastern regional haze improvements. J. Air Waste Manage. Assoc. 55, 1767–1779.

Stewart, T.R., Middleton, P., Downton, M., Ely, D., 1984. Judgements of photographs versus field observations in studies of perception and judgement of the visual environment. J. Environ. Psychol. 4 (4), 283–302.

Taylor, J.H., 1960a. Visual Contrast Thresholds for Large Targets, Part I: The Case of Low Adapting Luminances, SIO 60-25, Scripps Inst. of Oceanography, University of California, LaJolla.

Taylor, J.H., 1960b. Visual Contrast Thresholds for Large Targets, Part II: The Case of High Adapting Luminances, SIO 60-31, Scripps Inst. of Oceanography, University of California, LaJolla.

Thomas, J.P., 1978. Spatial summation in fovea: asymmetrical effects of longer and shorter dimensions. Vision Res. 18, 1023–1029.

Thomas, J.P., Padilla, G.J., Rourke, D.L., 1969. Spatial interactions in identification and detection of compound visual stimuli. Vision Res. 9, 1373–1380.

Van der Wildt, G.J., Waarts, R.G., 1983. Contrast detection and its dependence on the presence of edges and lines in the stimulus field. Vision Res. 23 (8), 821–830.

Velarde, M.D., Fry, G., Tveit, M., 2007. Health effects of viewing landscapes: landscape types in environmental psychology. Urban Forestry Urban Greening 6, 199–212.

Wang, Z., Bovik, A.C., Sheikh, H.R., Simoncelli, E.P., 2004. Image quality assessment: from error visibility to structural similarity. IEEE Trans. Image Process. 13 (4), 600–612.

Williams, A. (Ed.), 1999. Therapeutic Landscapes: The Dynamic between Wellness and Place. University Press of America, Lanham, MD.

Zeidner, M., Shechter, M., 1988. Psychological responses to air pollution: some personality and demographic correlates. J. Environ. Psych. 8, 191–208.

Zhenzhong, C., Hongyi L. JND Modeling: Approaches and Applications. In: Proceedings of the 19th International Conference on Digital Signal Processing, Hong Kong, August 2014.

Zube, E.H., 1974. Cross-disciplinary and intermode agreement on description and evaluation of landscape resources. Environ Behav 6 (1), 69–89.

Image Processing Techniques for Displaying Haze Effects on Landscape Features

The most effective way to communicate the visibility effects of air quality on any specific scene is with an image that displays those effects. In Chapter 1, a number of photographs are presented to visually depict how various spatial distributions of aerosol concentrations contribute to visibility impairment.

It is possible to measure various aerosol chemical and physical properties that reflect quantifiable aerosol optical properties such as atmospheric extinction, atmospheric transmittance, contrast transmittance, and many other related variables. However, communicating the effects that various types of spatially varying aerosol concentrations in combination with varying lighting and meteorological conditions have on visibility to the general public or the informed scientist presents a difficult problem. Color photography or digital images that demonstrate the visual impacts of current or predicted air quality conditions have a number of important uses, including providing stimuli for visibility perception and valuation surveys that hold all of the scene and lighting conditions constant and aiding policy makers and the public faced with decisions concerning development or implementation of air emission regulations and permitting for new emission sources.

If aerosol and optical variables were measured throughout the view shed at the same time color images were collected, it would be possible to establish a database that would show the correspondence between measured values and the appearance of the scenic resource. While such an approach has the potential for being most accurate in terms of characterizing the relationship between the visibility impacts and atmospheric properties, it requires extensive measurements of atmospheric aerosol and optical properties and the collection of many photographic or electronic images. An attempt to use the tens of thousands of scenic photographs

Visibility: The Seeing of Near and Distant Landscape Features. http://dx.doi.org/10.1016/B978-0-12-804450-6.00006-1

with collocated aerosol data collected as part of the IMPROVE program to display the visual impacts of various air quality conditions for over 40 locations is available at http://vista.cira.colostate.edu/improve/Data/IMPROVE/Data_IMPRPhot.htm.

An alternative, and in many ways a more satisfactory approach, for obtaining images to depict visual conditions of any scene is to use image processing techniques to synthetically create the appearance of scenic or urban landscape features. This, of course, requires an assumed aerosol mass concentration spatial distribution; a theoretical estimation of aerosol optical properties, such as the atmospheric mass scattering and absorption coefficients and the volume scattering function; and knowledge of overall illumination characteristics in relation to the observer–landscape geometry. This information must then be integrated into algorithms to solve the general radiative transfer Eq. 3.2.

THE RADIATION TRANSFER PROBLEM

Discussions of the volume scattering function and aerosol attenuation coefficients, along with a basic discussion of the transfer of radiation through the atmosphere, were presented in Chapters 2 and 3. However, calculation of contrast transmittance for arbitrary observer–sun angle geometries, general landscape features, and spatially inhomogeneous distributions of aerosol concentrations requires sophisticated computer-based models. There are many radiative transfer models that are used in a fairly routine manner, but this chapter presents a discussion of the few models that have been used in and specifically developed for visibility applications and are currently in use to create synthetic images of landscape features impaired by uniform and layered hazes.

For the interested reader, Thomas and Stanmes (2002), Clough et al. (2005), and Kokhanovsky (2013) provided a rather complete review of the many radiative transfer models currently in use and discussed the many varied applications of these models.

Following the notation of Chapter 3, the general radiation transfer equation is given by

$$\frac{dN_r(\theta,\phi,\vec{r})}{dr} = \underbrace{-b_{ext}N_r(\theta,\phi,\vec{r})}_{(loss)} + \underbrace{N_*(\theta,\phi,\vec{r})}_{(gain)} \qquad (6.1)$$

where $N_r(\theta,\phi,\vec{r})$ is the apparent radiance at some vector distance r from a landscape feature, $N_*(\theta,\phi,\vec{r})$ (referred to as the path function) is the radiant energy gain within an incremental path segment, and $b_{ext}N_r(\theta,\phi,\vec{r})$ is radiant energy lost within that same path segment. Although not explicitly stated, it is assumed that each variable in, and each variable derived from, Eq. 6.1 is wavelength dependent. The parenthetical variables $(\theta,\phi,\vec{r})$ indicate that N_r

and $N_*(\theta,\phi,\bar{r})$ are dependent on both the direction of image transmission and the position within the path segment. For the sake of brevity, the parenthetical variables will be dropped in the following equations.

Now N_* can be expressed as

$$N_* = h_s\sigma' + \int_{4\pi} N\sigma'd\Omega \qquad (6.2)$$

where h_s is the sun irradiance at the scattering volume, and N is the apparent radiance of the sky, moon, ground reflectance, etc. σ' is the scattering function as defined in Eq. 2.5. The first term in Eq. 6.2 is first-order scattering, while the second term is generally referred to as second-order or a multiple scattering term.

The path radiance (air light), the radiance reaching the eye (or detector) from some direction $\bar{r}$, is then

$$N_r^* = \int_0^r N_* T_r dr. \qquad (6.3)$$

If the limit is from $0\to\infty$, then N_r^* is just the sky radiance N_s.

In Fig. 6.1, the equation for the radiance when viewing a plume against the background sky is

$$_pN_r = N_{r_1}^* + \left(N_p^* + N_sT_p\right)T_{r_1} \qquad (6.4)$$

where $N_{r_1}^*$ is the path radiance between the plume and the observer, N_p^* is the inherent radiance of the plume, N_sT_p is the sky radiance attenuated by the plume, and T_{r_1} is the transmittance between the plume and the observer. If the plume is viewed against some landscape feature, then Eq. 6.4 becomes

$$_pN_r = N_{r_1}^* + \left[N_p^* + \left(_bN_oT_{r_2} + N_{r_2}^*\right)T_p\right]T_{r_1} \qquad (6.5)$$

where $_bN_o$ is the inherent radiance of the background landscape feature, and T_{r_2} and $N_{r_2}^*$ are the transmittance and path radiance due to the atmosphere between the plume and landscape feature.

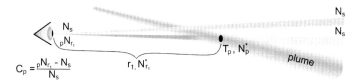

FIGURE 6.1   Schematic for viewing a plume with the sky as a background.

The first use of image processing techniques to show the pictorial appearance of haze focused on the visual appearance of coal-fired power plant plumes such as the plume shown in Fig. 6.1. Williams et al. (1978, 1980, 1981) used Brasslau and Dave's (1975) iterative technique to solve Eqs. 6.2 and 6.3 for the background sky. It assumes that the atmosphere is made up of a number of plane parallel layers with uniformly distributed aerosols. Implicit to this assumption is that the earth is flat, and therefore earth curvature does not play a role in sky radiance values. This means that the sky at the horizon is in equilibrium and as such would lack color or appear white or gray.

The plume radiance was estimated in a similar way, assuming infinite planes oriented normally to the observer's line of sight and using the background radiance calculation as boundary conditions to the plume calculation (Williams et al., 1980).

Malm et al. (1983) developed an approach to simulate the effect of uniform haze on scenic landscape features. Their method relied on extracting sky radiance values from a base photograph of some selected scene with near-zero haze conditions and then using approximations to estimate the inherent radiance of each image pixel and the path radiance between that pixel and the observer resulting from added aerosol content. Inherent to their calculation is that path radiance can be approximated by Eq. 3.8:

$$N_r^* = N_s\left(1 - T_r\right). \tag{6.6}$$

It is then assumed that the radiance associated with each pixel is approximated by

$$_lN_r^i = {}_lN_o^i\, T_{r_i} + N_s^i\left(1 - T_r\right) \tag{6.7}$$

where $_lN_r^i$ and $_lN_o^i$ are the inherent and apparent radiance of each pixel i, N_s^i is the sky radiance behind each pixel, and T_{r_i} is the transmittance between the landscape feature associated with pixel i and the observer.

These first attempts to image the effects of haze on complex scenes with a variety of landscape features were successful, but they required several limiting assumptions. The sky radiance was not reproduced as a function of aerosol concentrations for all viewing angles and for all extinction levels. Under clear conditions, flat-earth plane parallel geometries seriously misrepresented sky radiance and resulting color near the horizon, and therefore, the assumptions leading to Eq. 6.6 are violated. Spatial variations in aerosol concentrations leading to layered or suspended haze layers could not be handled nor could the presence of clouds.

To address these limitations, Davis et al. (1985) and Weissbluth et al. (1987) invoked the use of Monte Carlo methods for line of sight solutions to the radiation transfer equation without the limiting assumptions described previously. Discussions of the Monte Carlo method can be found in Cashwell and Everett (1959), Collins and Wells (1965), McKee and Cox (1974), and Marchuc et al. (1980). The goal of a Monte Carlo calculation is to simulate the transfer of photons from the top of the atmosphere, primarily the sun, through various scattering mediums and reflections from terrestrial or aquatic surfaces and then count the photons that arrive at an observer with iris aperture of dA along some line of sight $d\Omega$. This brute-force approach is not viable even with modern-day computing technology.

Davis et al. (1985) presented an outline for a forward Monte Carlo approach, referred to as the local estimate approximation, that allows for inhomogeneous distributions of aerosols. As outlined in Davis et al., the integral form of the equation of transfer can be written as

$$\Phi(\vec{r}) = \int_R l(\vec{r}',\vec{r}) f(\vec{r}') d\vec{r}' \tag{6.8}$$

where

$$l(\vec{r}',\vec{r}) = \frac{b_s}{b_{ext}} \frac{\exp(-\tau)}{d^2} \frac{\sigma^{\cdot}(\mu)}{4\pi} \Delta\Omega. \tag{6.9}$$

$\Phi(\vec{r})dSd\omega$ is the number of photons that cross an area element dS from within a cone of directions $d\omega$. $f(\vec{r}')$ is the collision density and $l(\vec{r}',\vec{r})$ represents a transition density, which when multiplied by $d\vec{r}$ gives the probability that a photon that had an interaction at $\vec{r}'$ will have its next interaction in $\vec{r}' + d\vec{r}'$. $\frac{b_s}{b_{ext}}$ is the scattering albedo, τ and d are the optical depth and the distance between $\vec{r}'$ and $\vec{r}$, respectively, $\sigma^{\cdot}(\mu)$ is the scattering function, and μ is the cosine of the scattering angle. This form of the transition density is referred to as the local estimate.

Fig. 6.2 shows the kind of geographic geometry that can be used with the Monte Carlo approach. The geometry was developed in a rectangular geometry, and as such, did not account for earth curvature. The terrain and haze are represented by a series of rectangular volumes and surfaces with varying physical and aerosol characteristics. The terrestrial surfaces are assumed to be Lambertian reflectors, while the rectangular volumes are assumed to have varying optical characteristics such as volume scattering functions and scattering albedos. This approach allows for introducing plumes, suspended or ground-based haze layers, or uniform haze (Molenar et al., 1994).

Initial investigations by Molenar et al. (1994) highlighted the effect of multiple scattering on estimated vista radiance values as well as the effect of aerosols on vista radiance as a function of sun-angle–observer geometry.

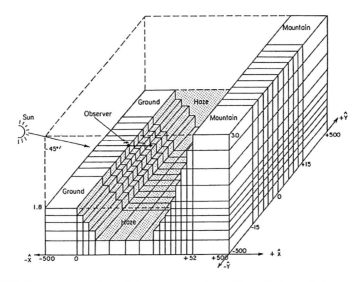

FIGURE 6.2   Example of a model geometry used in a Monte Carlo radiation transfer calculation. It represents an observer viewing a layered haze trapped in a canyon that is in front
of a mountain range. Coordinates are in kilometers. The atmosphere extends from the top of
the mountain to 50 km above the surface.

For a view of a landscape feature through a plume, vista radiance was decreased with the addition of an aerosol when the scattering angle was at
90° but was increased in the forward-scattering geometry with concurrent
changes in vista–sky apparent contrast. Because the atmosphere was not
in optical equilibrium, the vista–sky or plume–sky apparent contrast was
not well predicted by the Koschmieder relationship (Koschmieder, 1924).
They did demonstrate that apparent spectral contrast was nearly invariant
as a function of single-particle scattering albedo, again as predicted by the
Koschmieder equation.

Weissbluth et al. (1987) extended this work to include the effect of earth
curvature and to increase the computational efficiency over the conventional Monte Carlo approach. Instead of tracking all photons from the
radiation source (sun) to the observer (forward approach), they employed
the backward Monte Carlo model (BackMC) as described by Collins et al.
(1972). In the backward Monte Carlo approach, photons are released from
the observer for some line of sight and tracked "backward" through the
atmosphere, allowing for a statistical representation of all possible photon,
aerosol, and terrestrial interactions.

For many, if not most, scenes that contain multiple landscape features, it is the contrast between adjacent scenic elements that determines

the overall scenic quality or beauty of the view. To that end, Weissbluth et al. (1987) introduced the concept of contiguous contrast or the contrast between adjacent scenic elements. The change or attenuation of contiguous contrast resulting from increased aerosol concentration is not described by the Koschmieder relationship nor is there a simple algorithm that captures this relationship. It is only through the use of Monte-Carlo-type calculations that changes in apparent radiance of individual scenic elements, and therefore contiguous contrast, can be evaluated as a function of distance from the scenic elements or changes in atmospheric aerosol content.

Weissbluth et al. applied BackMC to a complex Grand Canyon National Park scene that precluded the use of any simplifying symmetry assumptions and that had many adjacent landscape features that allowed for assessing contiguous contrast as a function of changing aerosol concentration and light conditions. One significant finding was that without the inclusion of multiple scattering terms, contiguous contrast estimations were in error by about 40%, while apparent radiance estimations were in error by over 100%. Also it was evident that under Rayleigh-type atmospheric scattering conditions, accounting for earth curvature is essential. The background sky, even at zenith viewing angles of 90° (horizontal view), the sky never reaches equilibrium, so that the basic assumptions required for the use of the Koschmieder contrast equation are violated, and the assumed atmospheric contrast transmittance is in significant error.

Sky radiance generated with BackMC compares favorably with analytical solutions for Rayleigh atmospheres, published results from other existing radiation transfer models, and physical simulations in which the scattering physics can be exactly defined and measured (Johnson and Molenar, 1990). However, it is desirable to compare modeled radiance estimates to measured values under complex visibility conditions. The Dallas–Fort Worth Winter Haze Project (DFWWHP) presented an opportunity for such a comparison (McDade et al., 1997a).

Extensive, high-quality surface and airborne aerosol, gaseous, optical, and meteorological measurements were taken during the winter of 1994–1995 for the DFWWHP. These data were used to characterize the spatial, temporal, chemical, and optical properties of the urban and background haze (McDade et al., 1997b). On four days, February 18–21, 1995, a Photoresearch Model 703-PC SpectraScan telespectrophotometer was used to measure the radiance spectra of the horizon sky and various terrain features at 2-nm intervals throughout the visible spectrum, 400–700 nm (Mahadev and Henry, 1997). Measurements were taken approximately every 30 min from 10:30 am to 4:00 pm. The instrument was calibrated both before and after the study period for radiance as a function of wavelength using NIST (National Institute of Standards and Technology) traceable calibration standards, following the manufacturer's recommended procedures.

BackMC was run using measured aerosol and gaseous data to simulate high-resolution sky spectral radiance measurements made during the study. The reader is referred to Molenar et al. (2002) for details of the calculation. Fig. 6.3 is a plot of the measured versus modeled sky radiances for February 20, 1995. Although the measurements were made at 2-nm intervals, the model was exercised at 10-nm intervals from 400 to 700 nm. The plots indicate a remarkable agreement between the measured and modeled sky radiances. The fluctuations in the measured sky spectrum (especially above 600 nm) are due to water vapor absorption bands, which are not currently accounted for in the Monte Carlo model. Fig. 6.4 is a scatter plot of all measured versus modeled sky radiance pairs. An ordinary least squares regression with modeled sky radiance as the independent and measured sky radiance as the dependent variable yields an $R^2 = 0.86$ and a slope of 1.01, indicating rather remarkable agreement between measured and modeled radiance values.

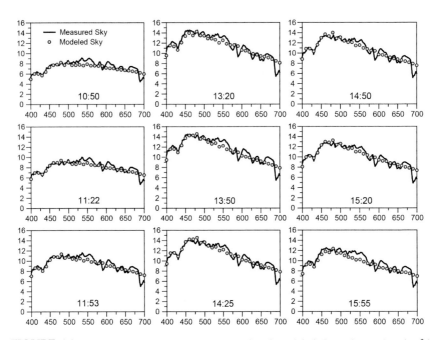

FIGURE 6.3   Dallas, February 20, 1995, measured and modeled sky radiance ($\mu w/cm^2/sr/nm$) plotted as a function of wavelength (400–700 nm).

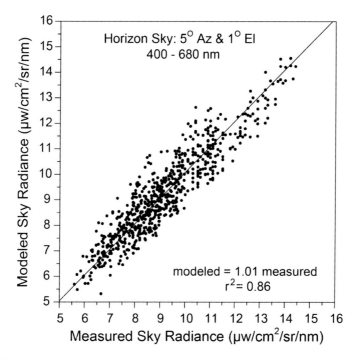

FIGURE 6.4 Scatter plot of measured and modeled horizon sky radiance with an ordinary least squares regression line forced through zero.

THE USE OF IMAGE PROCESSING TECHNIQUES TO VISUALLY REPRESENT EFFECTS OF HAZE ON SCENIC LANDSCAPE FEATURES

A publicly available image processing system, incorporating many of the radiative transfer concepts discussed previously and referred to as WinHaze, has been developed to demonstrate the effects of haze on a number of scenic landscapes found in federally protected lands managed by the National Park Service, U.S. Forest Service, and U.S. Fish and Wildlife Service. Also included in the program are a number of urban images. The WinHaze program currently only addresses uniform haze under clear sky conditions. The WinHaze program and the BackMC code can be downloaded at http://www.air-resource.com/resources/downloads.html.

Fig. 6.5 is a flow diagram of the various steps incorporated into the WinHaze program. The first step requires a photograph or electronic camera image that was taken under near-Rayleigh conditions and under cloud-free, uniform illumination conditions. Because this sort of atmospheric condition can be quite rare, even in pristine environments typically found

---

ORIGINAL IMAGE
As a starting point, an image, either in digital or film format, representing a scene under near-Rayleigh conditions is required. The physio-chemical characteristics of the aerosol at the time the image was acquired must be known or estimated. If the image was taken using a film format, the slide must be digitized to yield red, green, and blue density (DN) values between 0 and 255 for each pixel value acquired. Typically, the pixel size is on the order of 25 μm.

DISTANCE MASK
All terrain features in the image are separated into distance categories based on the geography of the scene and camera position.

COLOR CALIBRATION
The system must be color calibrated not only for exposure vs. density characteristics of the system used to collect the image but also for the display system, whether it is a color slide or a computer display screen. The camera system is calibrated to a series of known exposures that include a variety of colors and brightness levels and measuring detector or film response. The same is done for the film writing or display system.

CALCULATION OF OPTICAL PROPERTIES OF AEROSOLS
A variety of models are available to calculate the optical characteristics of aerosols. Particles are usually assumed to be spherical and externally mixed; however, multicomponent, internally mixed aerosols with varying hygroscopicity characteristics have also been used. Alternatively, it is also possible to use measured optical parameters such as the volume scattering function and extinction and scattering coefficients.

RADIATION TRANSFER MODELING
For the assumption of uniform haze the BackMC model is used to calculate sky radiance values as a function of aerosol optical properties, and when the particulate concentrations are great enough to assume that the atmosphere is in optical equilibrium the path radiance between each pixel is estimated using $N_r = N_s(1 - T_r)$. When nonuniform haze is modeled, path radiance is calculated using the BackMC model.

OUTPUTTING MODIFIED IMAGE
The calibrated DN values and modeled sky and path radiance are used to generate a new landscape radiance field that can be displayed on a computer screen or used to generate pictures or slides of the modified image.

FIGURE 6.5   Flow diagram of image processing steps required to model the effects of haze on a view of landscape features.

in many national parks, it usually is necessary to select the image from a database of images that have been collected over some extended timeframe, usually by some automated image collection system. The federal land management community operated automated 35-mm camera systems that were used to collect images on Kodachrome 64 slide film for a period of years, and these scenes have been, for the most part, incorporated into WinHaze. Currently, there are many webcam systems operated by various individuals and agencies, and in some cases the images are being archived. Images collected by electronic camera systems, as well as images collected using a film format, can be used in a WinHaze-type application.

Image Digitization and Exposure Estimation

Color film or an electronic camera may be regarded as a measurement tool that creates a map of an incident image spectral radiance field. Film employs red, green, and blue (RGB) emulsion layers to collect radiation and chemically convert it to exhibit varying density values related to the initial scene radiance. The density (DN) of each picture element or pixel is determined by illuminating it with a given light intensity through a specific filtering system, measuring the light transmitted by the film and converting it from transmittance to a perceptually uniform DN scale. Usually, the slide is digitized through three wide-band filters selected such that, when combined with the spectral response of the detector, the spectral distribution of the illumination employed by the scanning device and the spectral characteristics of the film yield nearly Gaussian "color" response curves with peaks centered near 650 nm (R), 550 nm (G), and 450 nm (B), with little overlap between the curves. The digitizing device is calibrated by scanning slides of gray and color scales immediately preceding and following the scan of the input image. The RGB densities of each element of these calibration slides are measured with a traceable densitometer with specific filter responses and are correlated to the DN values in the resulting digital image of each slide. The RGB DN values for each pixel in the input image file are then calibrated to these corresponding density data.

Electronic imaging devices such as webcams can be programmed to output RGB DN values directly that can be calibrated to gray and color scales, which have known reflectance characteristics. The image is captured by an array of charge coupled device (CCD) detectors covered with a matrix of RGB filters. When exposed to light, different sections of the CCD build up electrical charges proportional to the light's intensity. Measurement of a charge is precisely proportional to how bright that section of the image is. Each image pixel corresponds to a single section of the CCD. The time interval for which the film or CCD detector is exposed to the scene, multiplied by the radiance of the scene, is referred to as the exposure. Because every pixel is exposed for the same time interval, the varying densities are directly related to the initial scene radiance N_r. The measured pixel RGB densities are converted to RGB exposures by using density versus log-exposure curves published by the manufacturer of the film or the detector. The RGB exposures for each pixel are determined from these curves and used as the observed image radiances.

The upper left panel of Fig. 6.6 shows a color photo of the La Sal Mountains taken from Canyonlands National Park, Utah. The remaining three photos show black and white images of the DN values derived from measuring the film density in the RGB wavelengths.

FIGURE 6.6   Color image (top left) of the La Sal Mountains as viewed from Canyonlands National Park. The images marked as red, green, and blue are the black and white re-
productions of the measured densities in the red, green, and blue portions of the spectrum.

Distance Masking

Each terrain pixel in an image is assigned a specific distance, elevation angle, and azimuth angle with respect to the observer position by
using detailed topographic maps of the area and an interactive image
processing display system. The image is masked by tracing outlines of
various contiguous features on the displayed digital image. The surface of each masked feature is then approximated by a separate suitable
geometric model that adequately describes the distance of each pixel
in the feature from the observer. The distance models can be as simple
as vertical planes, semicomplex representations as intersecting incline
planes, or highly detailed, multiple-ordered, solid geometrical surfaces.
The more detailed the approximation, the more realistic the simulation.
The elevation and azimuth angles of each pixel are a simple function of
the row and column position, the horizon, and the lens used to take the
image. The distance mask for the image shown in Fig. 6.6 is presented
in Fig. 6.7.

FIGURE 6.7   Black and white image of the distance mask for the scene shown in Fig. 6.6. Each feature is assumed to be at a unique but same distance from the observer.

Atmospheric Aerosol Model

The aerosol model assumes spherical particles and incorporates stable algorithms to obtain Mie solutions for externally mixed, homogeneous, or internally mixed, coated aerosols. The resulting bulk optical properties such as extinction, scattering, and absorption coefficients, single scattering albedo, and the volume scattering function are incorporated into radiative transfer models. As an alternative, particle mass concentrations and their respective extinction efficiencies can be input directly into WinHaze, or the extinction and scattering coefficients along with an assumed relative humidity value can be input.

Sky Radiance

The BackMC model discussed previously is used to calculate sky radiances. The wavelength-dependent, scattered radiation, intensity, and polarization parameters are computed as a function of the observer's elevation and azimuthal viewing angles. The model domain is determined by a set of three-dimensional boxes extending beyond the horizon in all directions and to the top of the modeled atmosphere. Each box is defined as a horizontal surface, a vertical surface, or free atmosphere. The free atmosphere boxes can be further defined to have any molecular or aerosol optical properties desired, including specific wavelength-dependent extinction, scattering, and absorption coefficients and phase functions for every gas or aerosol species in each individual box. Lambertian reflection by the ground and elevated terrain are also included. Any general distribution of terrain, solar position, cloud distribution, or extinction can be modeled, provided the specific optical properties can be associated with each feature of every box. This includes complex terrain, uniform haze, layered haze, elevated plumes, clouds, or any combination thereof. This model has been thoroughly tested and compared to analytical solutions for Rayleigh atmospheres, published results from various model calculations of aerosol atmospheres, and sky and plume radiance measurements, with excellent results (Molenar et al., 2002).

Terrain Radiance

The inherent radiance of each terrain pixel is estimated by solving Eq. 6.7 for $_lN_o^i$, with the transmittance calculated for each line of sight, assuming the atmospheric extinction level corresponding to the time the picture was recorded.

New Image Radiances

In the default WinHaze model, the modeled image radiance field is calculated by first using the new extinction value and distance to each terrain pixel to calculate a new path transmittance. The new path radiance is calculated using this transmittance and the modeled sky radiance in Eq. 6.6. The assumption is made that the inherent terrain radiance field is not modified due to the change in extinction. Then, the new apparent image radiance field is calculated using these values in Eq. 6.7. These new image radiance fields are then used in the image processing modules to generate new DN values that are used to display the image on a computer screen or output to a hardcopy printing system. Alternatively, path radiance and transmittance can be estimated for every line of sight (pixel position) using the BackMC model and any aerosol spatial distribution desired. This approach can be quite computer intensive depending on the assumed detail of the aerosol spatial distribution and the number of line of sights used.

EXAMPLES OF IMAGE PROCESSING TO DISPLAY EFFECTS OF HAZE ON LANDSCAPE FEATURES

Fig. 6.8 is a photo of the La Sal Mountains and their foreground features as seen from Canyonlands National Park. The mountains are approximately 50 km distant from the observation point, while the nearby rock

FIGURE 6.8 Photos representing (a) a particle-free atmosphere and (b) 20% lowest, (c) average, and (d) 20% highest fine mass concentration days.

face is only 4 km distant. Fig. 6.8 shows the appearance of the scene under Rayleigh conditions (a) and on the average lowest 20% (b), average (c), and average 20% highest (d) fine mass concentration days. Photos (a), (b), (c), and (d) correspond to deciview (dv) values of 0, 5.65 (b_{ext} = 0.018 km^{-1}), 8.8 (b_{ext} = 0.024 km^{-1}), and 11.5 (b_{ext} = 0.032 km^{-1}), respectively. These photos effectively communicate the range of visibility conditions of one view found in Canyonlands National Park.

Fig. 6.9 demonstrates how relative humidity changes the haze level when hygroscopic particles contribute to the ambient fine particle concentration. In each of the photos, the assumed atmospheric concentrations of hygroscopic and nonhygroscopic particles were 4.0 μg/m^3 and 3.6 μg/m^3, respectively. In photos (a), (b), (c), and (d), the assumed relative humidity was 0.0%, 40%, 90%, and 95%, respectively. As relative humidity increases, hygroscopic particles such as ammonium sulfate and nitrate absorb water and grow in size. The effect is nonlinear in that the growth in particle size from 0% to 40% is substantially less than from 50% to 90% and less than from 90% to 95% relative humidity. This is demonstrated in Fig. 6.9 in which the change in haziness between photos (a) and (b) is substantially less than between (b) and (c) or even (c) and (d), which corresponds to only a 5% change in relative humidity.

The photos in Fig. 6.10 show the difference in haze levels that are caused by 20 μg/m^3 of fine or coarse mass concentrations. Photo (a)

FIGURE 6.9   Photos showing the effect of relative humidity. Each picture corresponds to an aerosol concentration of 4.0 $\mu g/m^3$ and 3.6 $\mu g/m^3$ of hygroscopic and nonhygroscopic particles. Photos (a), (b), (c), and (d) correspond to 0%, 40%, 90%, and 95% relative humidity, respectively.

FIGURE 6.10   Photos showing the relative visibility impairment caused by the same amount of fine and coarse mass concentrations. In photo (a), the fine and coarse mass concentrations are 0.0 and 20.0 $\mu g/m^3$, respectively, while in photo (b) the fine mass concentration is 20.0 $\mu g/m^3$ and the coarse mass concentration is 0.0 $\mu g/m^3$.

corresponds to a haze level associated with 20 $\mu g/m^3$ of coarse mass, while photo (b) was created assuming a 20 $\mu g/m^3$ level of fine mass. The physical particle cross section, which is responsible for scattering light, of fine over coarse mass concentrations increases in proportion to the ratio of the radius of the particles making up the fine and coarse mass distributions. Also, fine mass particles are scattered with greater efficiency

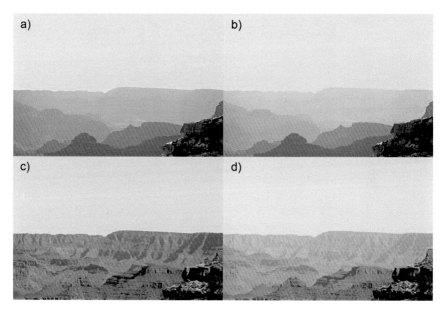

FIGURE 6.11 Photos showing the effect of sun angle on how haze modifies the appearance of a scene. Photos (a) and (c) correspond to a view of Desert View, Grand Canyon, Arizona, in the morning and afternoon with a nearly particle-free atmosphere, while photos (b) and (d) show the same scene with a dv level of 13.9 (0.04 km^{-1}).

than are coarse particles. Hence, as demonstrated by the difference in haze levels between photos (a) and (b), the same amount of fine particle mass causes substantially more haze than the same mass concentration of coarse particles.

Fig. 6.11 shows the effect of haze in combination with the effect of sun angle on a scenic resource. Photos (a) and (c) show a Grand Canyon scene, under near Rayleigh conditions, where the most distant cliff feature is 37 km distant and the middle cliff feature, known as Desert View, is 27.5 km away from the observation point. Photo (a) was taken in the morning when nearly the entire scene is in shadow, while photo (b) was taken at 3:00 pm when the scene was directly illuminated. In the shadowed scene, the cliff faces lack detail and color and the contiguous contrast of adjacent features is near zero, while photo (c) shows high, contiguous contrast edges and the inherent color of cliff faces is clearly visible. Visibility degradation due to haze is generally caused by the addition of path radiance or air light between the observer and the scenic element, and second, the inherent radiance of a landscape feature is attenuated as it travels from the landscape feature to the observer. In photo (a) the inherent landscape radiance is near zero, so the change in the scene resulting from adding haze is primarily due to the addition of path radiance, while in photo (d)

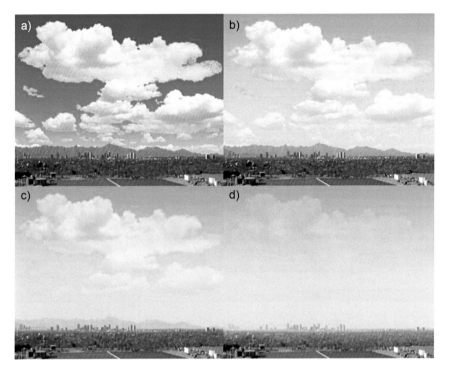

FIGURE 6.12 Photos showing a Phoenix, Arizona, scene with overhead cumulus clouds. The distant mountains features are 27 km away. Photos (a), (b), (c), and (d) correspond to dv levels of 0 (0.01 km^{-1}), 10 (0.027 km^{-1}), 20 (0.074 km^{-1}), and 30 (0.2 km^{-1}), respectively.

the change in the appearance of landscape features is a result of image attenuation as well as added path radiance.

The next series of photographs (Fig. 6.12) shows a scene of the down-town area of Phoenix, Arizona, where the distant mountain range is 27 km away (Molenar and Pitchford, 2008). The base photograph, (a), was taken under near-Rayleigh conditions and under direct sunlight and cloud-free conditions. The cloud field was added synthetically for purposes of demonstrating how the overall scenic quality or beauty of a scene is enhanced with the addition of certain types of meteorological conditions and then to show how these scenic features are degraded.

The final demonstration of the use of image processing techniques to assess the effects of haze on scenic landscape features is demonstrated in Fig. 6.13 (Molenar et al., 1994). The scene again is of the Grand Canyon; however, in this case the photograph is taken from Desert View at the east end of the Grand Canyon, and the view is down canyon, viewing toward the west. Under certain meteorological conditions, haze can be transported into the canyon and trapped below the canyon rim as demon-strated in Fig. 1.19. Here, these conditions were simulated on a background

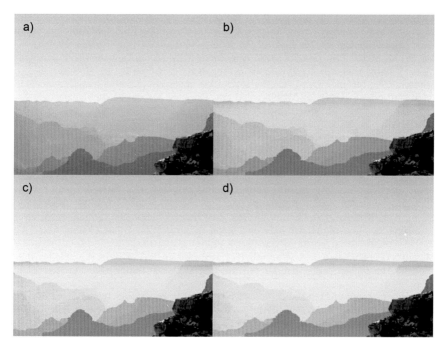

FIGURE 6.13 Simulated layered sulfate haze in Grand Canyon National Park: (a) 0.02 km^{-1} background haze; (b) 0.035 km^{-1} layered haze; (c) 0.067 km^{-1} layered haze; and (d) 0.09 km^{-1} layered haze.

haze of 0.022 km^{-1} (Rayleigh + aerosol), which is shown in Fig. 6.13. The images in Fig. 6.13b–d correspond to an incremental increase in haze, as represented by atmospheric scattering, of 0.035 km^{-1}, 0.067 km^{-1}, and 0.090 km^{-1}, respectively. These increases in haze scattering correspond to about 2.6 μg/m^3, 5.1 μg/m^3, and 6.8 μg/m^3, respectively, at 90% relative humidity. The path radiances for the various viewing directions were directly simulated using the BackMC radiation transfer model.

References

Brasslau, N., Dave, J.V., 1975. Direct solution of spherical harmonics approximation to radiative-transfer equation for an arbitrary solar elevation: 1. Theory. J. Atmos. Sci. 32 (4), 790–798.

Cashwell, E.D., Everett, C.J., 1959. A Practical Manual on the Monte Carlo Method for Random Walk Problems. Pergamon, New York, p. 153.

Clough, S.A., Shephard, M.W., Mlawer, E.J., Delamere, J.S., Iacono, M.J., Cady-Pereira, K., et al., 2005. Atmospheric radiative transfer modeling: a summary of the AER codes. J. Quant. Spectrosc. Ra. 91, 233–244.

Collins, D.G., Wells, M.B. Monte Carlo Codes for Study of Light Transport in the Atmospheres: Vol. 1, Descriptions of Codes, Report RRA-T54, Radiation Research Associates, Inc, Fort Worth, 100 pp, 1965.

Collins, D.G., Wells, M.B., Blattner, W.G., Horak, H.G., 1972. Backward Monte Carlo calculations of polarization characteristics of radiation emerging from spherical-shell atmospheres. App. Opt. 11 (11), 2684.

Davis, J.M., McKee, T.B., Cox, S.K., 1985. Application of the Monte Carlo method to problems in visibility using a local estimate: an investigation. Appl. Opt. 24 (19), 3193–3205.

Johnson, C.E., Molenar, J.V., 1990. The application of several radiative transfer models to aircraft plume visibilities. Visibility and Fine Particles. Air and Waste Management Association, Pittsburgh, PA.

Koschmieder, H., 1924. Theorie der horizontalen Sichtweite. Beitr. Phys. Freien. Atmos. 12, 33–55.

Kokhanovsky, A. (Ed.), 2013. Light Scattering Reviews 8. Springer Praxis Books, New York, p. 632.

Mahadev, S., Henry, R.C., 1997. Application of color appearance models in visual air quality research. Visual Air Quality: Aerosols and Global Radiation Balance. Air and Waste Management Association, Pittsburgh, PA.

Malm, W.C., Molenar, J.V., Chan, L.Y., 1983. Photographic simulation techniques for visualizing the effect of uniform haze on a scenic resource. J. Air Pollut. Control Assoc. 33, 126–129.

Marchuc, G.I., Mikhailov, G.A., Nazaraliev, M.A., Darbinjan, R.A., Kargin, B.A., Elepov, B.S., 1980. The Monte Carlo Methods in Atmospheric Optics. Springer-Verlag, New York, p. 739.

McDade, C., Tombach, I., Seigner, C., 1997a. The Dallas-Fort Worth Winter Haze Project: description and findings. Visual Air Quality: Aerosols and Global Radiation Balance. Air and Waste Management Association, Pittsburgh, PA.

McDade, C., Tombach, I., Hering, S., 1997b. Optical properties of urban aerosol in Dallas-Fort Worth. Visual Air Quality: Aerosols and Global Radiation Balance. Air and Waste Management Association, Pittsburgh, PA.

McKee, T.B., Cox, S.K., 1974. Scattering of visible radiation by finite clouds. J. Atmos. Sci. 28, 1187.

Molenar, J.V., Pitchford, M.L. Sky color and clouds appearance: sensitive indicators of urban visual air quality, Extended Abstract 18, presented at the Air & Waste Management Association Visibility Specialty Conference, Moab, April 29 to May 2, 2008.

Molenar, J.V., Malm, W.C., Johnson, C.E., 1994. Visual air quality simulation techniques. Atmos. Environ. 28, 1055–1063.

Molenar, J.V., Henry, R., Mahadev, S., 2002. Comparison of measured and modeled high resolution sky spectral radiance data. Visual Air Quality: Aerosol and Global Radiation Balance. Air and Waste Management Association, Pittsburgh, PA, pp. 407–418.

Thomas, G.E., Stanmes, K., 2002. Radiative Transfer in the Atmosphere and Ocean (Cambridge Atmospheric and Space Science Series). Cambridge University Press, Cambridge.

Weissbluth, M.J., Davis, J.M., Cox, S.K., 1987. A modeling study of visibility in the Grand Canyon. Atmos. Environ. 21 (3), 703–713.

Williams, M.D., Wecksung, M.J., Leonard, E.M., 1978. Computer simulation of the visual effects of smoke plumes. In: Proceedings SPIE 0142: Optical Properties of the Atmosphere. SPIE, Bellingham, WA, pp. 135–141.

Williams, M.D., Treiman, E., Wecksung, M.J., 1980. Plume blight visibility modeling with a simulated photographic technique. J. Air Pollut. Control Assoc. 30, 131–134.

Williams, M.D., Chan, L.Y., Lewis, R., 1981. Validation and sensitivity of a simulated-photographic technique for visibility modeling. Atmos. Environ. 15, 2151–2169.

Monitoring Visibility

This chapter starts with easy-to-understand descriptions of how a number of measurement techniques can be used to extract physical and optical characteristics of atmospheric particulate matter that relate to visibility. Concepts that are not covered in other books or published literature are explored in more depth, while literature references are provided for more readily characterized topics.

As discussed in earlier chapters, visibility is not defined by a single parameter, which makes it more difficult to design a single monitoring methodology or strategy. However, the currently used monitoring methods can be divided into three categories: view, optical, and aerosol monitoring (Malm, 1992). Visibility, in the most general sense, reduces to understanding the effects that various types of aerosol and lighting conditions have on the appearance of landscape features. Many visibility indexes quantify the appearance of a scene; however, a photograph relating the effects particles have on the appearance of landscape features is the most simple and direct form of communicating visibility impairment. Therefore, a systematic photography or digital imaging program (view monitoring) that records the appearance of the scene under a variety of lighting conditions and aerosol concentrations is a key part of any visibility monitoring program. However, because it is difficult to routinely extract quantitative optical data from photographs or digital images that can be related to atmospheric aerosols, some direct measure of a fundamental optical property of the atmosphere, such as atmospheric scattering or extinction, is desirable. Finally, if the goal is to relate visibility impairment to emissions sources, particle measurements must be made in conjunction with optical measurements. Size and composition are the two dimensions of particle characterization that are most useful in attributing source contribution to visibility impairment.

Developing a comprehensive visibility monitoring program covering view, atmospheric optical variables, and aerosols contributing to haze is a complex task. Addressing all possible methodologies for measuring

Visibility: The Seeing of Near and Distant Landscape Features. http://dx.doi.org/10.1016/B978-0-12-804450-6.00007-3

visibility-related variables is beyond the scope of this book. The U.S. EPA published a visibility monitoring guidance document (U.S. EPA, 1999), Malm and Walther (1980) reviewed instruments measuring visibility-related optical variables, and Hinds (1999) and Baron and Willeke (2001) authored books covering many aerosol measurement techniques.

Table 8.1 lists 14 regional-scale monitoring programs, many of which contained aerosol, optical, and photographic monitoring features. The first monitoring program in the United States designed specifically for assessing visibility concerns was the Visibility Investigative Experiment in the West (VIEW), started in 1978. In 1987, VIEW was incorporated into the Interagency Monitoring of Protected Visual Environments (IMPROVE) program, which is ongoing.

BASIC PRINCIPLES OF MEASUREMENT OF SCATTERING AND EXTINCTION

The scattering coefficient is a measure of the ability of particles to scatter photons out of a beam of light, while the absorption coefficient is a measure of how many photons are absorbed. Each parameter is expressed as a number proportional to the amount of photons scattered or absorbed per distance. The sum of scattering and absorption is referred to as extinction or attenuation.

Fig. 7.1 is a diagram showing a beam of light made up of photons with varying wavelengths that are incident on a concentration of particles and absorbing gas. By knowing the number of photons incident on a concentration of particles and measuring the number of photons

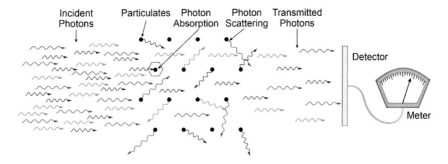

FIGURE 7.1   Placement of a light source and detector as shown here is known as a transmissometer. As photons pass through a concentration of particles and gases, they are either scattered out of the light path or they are absorbed. Thus, a detector placed as indicated measures only those photons that are transmitted the length of the light path. Because this instrument is sensitive to both scattering and absorption, it can be calibrated to measure the extinction coefficient.

successfully passing through the particulate concentration, one can cal-culate the number of photons scattered and absorbed. The instrument that measures extinction (sum of scattering and absorption) is known as a transmissometer.

The light source is usually an incandescent lamp, and the receiver is a telescope fitted with an appropriate detector. The light source and detec-tor can be placed 1–10 km apart, and the measurement is usually referred to as a long-path measurement.

A similar light source-detector configuration can be used to mea-sure just the scattering ability of particles and gases. If the detector is placed parallel to the incident photons, only those photons that are scat-tered will be detected. This type of instrument is called a nephelometer (Fig. 7.2). If the detector is so aligned as to measure scattering in only one direction, it is referred to as a polar nephelometer. On the other hand, if all photons scattered in forward, sideways, and backward directions are allowed to hit the detector, the instrument is referred to as an integrating (summing) nephelometer. The instrument is usually constructed in such a way so as to have the sampling chamber and light source confined to a small volume so that the instrument makes a "point" or localized mea-surement of scattering.

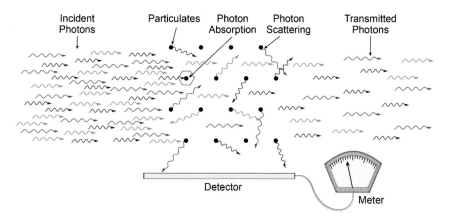

FIGURE 7.2   Placement of a detector for the measurement of the number of photons scattered by a concentration of particles and gas.

BASIC CONCEPTS OF MEASUREMENT OF PARTICLES IN THE ATMOSPHERE

Particle measurements are generally made in conjunction with optical measurements to help infer the cause of visibility impairment and to es-timate the sources of visibility-reducing aerosols. Size and composition are the two dimensions of particle characterization most often used in visibility monitoring programs. Particle sizes between 0.1 and 1.0 μm are most effective on a per mass basis in reducing visibility and tend to be associated with manmade emissions. Fig. 7.3 shows a diagram of a cyclone-type particle monitor that separates out all those particles less than a specified size (usually 2.5 μm) and collects them on a filter substrate for additional analysis. The air is caused to spin in much the same way as a merry-go-round or carousel. The heavier particles, those larger than 2.5 μm, fall off the merry-go-round, impact on the side of the sampler, and are discarded to the bottom of the sampler. Those particles staying in the air stream pass through a filter where they are extracted for further analysis. Particles are speciated into sulfates, nitrates, organic material, elemental carbon (soot), and soil. The speciation of particles helps determine the chemical-optical characteristics and the ability of the particle to absorb water (RH effects) and is important to separate out the origin of the aerosol.

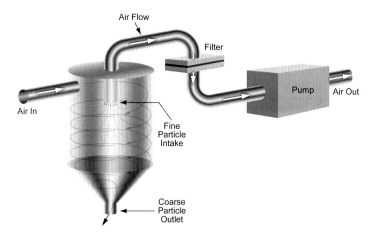

FIGURE 7.3   Diagram of a cyclone-type particle monitor. The air is spun around inside the cyclone in such a way so as to cause larger, more massive particles to impinge on the inside of the monitor and deposit to the bottom, while smaller particles continue in the air stream and are collected on a filter substrate for analysis.

A MORE IN-DEPTH LOOK AT OPTICAL MONITORING PRINCIPLES RELATED TO VISIBILITY

Optical measurements that have been routinely used to assess visibility impairment are outlined in Table 7.1. Measurements are grouped into four categories: scene monitoring, scattering, absorption, and extinction,

TABLE 7.1 An Outline of the Types of Instruments Used to Measure Scene/Landscape Characteristics, Atmospheric Scattering, Absorption, and Extinction

	Instrument	What Is Measured	Most Closely Related Optical Variable	Other Derived Optical Variables	Routine or Special Study Monitoring and Limitations
Scene and radiance	Teleradiometer or telephotometer	Radiance (N) Luminance (B)	Universal and modulation contrast (C_r and m_r) of a landscape feature, plume contrast (C_p)	Extinction coefficient (b_{ext}), plume extinction, or opacity	Has been used for routine monitoring programs but currently not in the United States. Accurate measurement of radiance but high degree of uncertainty interpreting as b_{ext} or V_r.
	Digital camera	N' or B' proportional to radiance or luminance	C_r and m_r, equivalent contrast (C_{eq}), scene average contrast, mean square radiance fluctuation ($\overline{N_r^2}$), C_p	b_{ext}, plume extinction or opacity	Routine visibility monitoring; many webcams currently in use but few used to routinely extract visibility metrics.
	Photographic camera	N' or B' proportional to radiance or luminance	C_r and m_r, C_{eq}, scene average contrast, mean square radiance fluctuation ($\overline{N_r^2}$), C_p	b_{ext}, plume extinction or opacity	Same as above but few used routinely. Replaced by webcams.
	Human eye	Distance	Visual range (V_r)	b_{ext}	Historically used for routine monitoring of visual range. Not in use today.
Scattering	Integrating nephelometer	$N = const\ I_o \int_{\beta_1}^{\beta_2} \sigma\, d\Omega$	Atmospheric scattering coefficient (b_{scat})	b_{scat}	Routinely used. When scattering chamber is enclosed, there is a risk of aerosol heating, so measurement is not ambient.
	Polar nephelometer	$N(\beta) = const\ I_o \sigma(\beta)$ for any β	Volume scattering at a number of scattering angles β.	Atmospheric scattering coefficient $b_{scat} = \int_{4\pi} \sigma(\overline{\tau}, \beta) d\Omega$	Primarily used in special studies.

(Continued)

TABLE 7.1 An Outline of the Types of Instruments Used to Measure Scene/Landscape Characteristics, Atmospheric Scattering, Absorption, and Extinction (*cont.*)

	Instrument	What Is Measured	Most Closely Related Optical Variable	Other Derived Optical Variables	Routine or Special Study Monitoring and Limitations
	Forward scatter meter	$N(\beta = ?) = const\ I_o \sigma(\beta = ?)$ $0° < \beta < 90°$	Volume scattering at a scattering angle $\beta < 90°$	Atmospheric scattering coefficient $b_{scat} = const\ \sigma(\beta = 45°)$	Routinely used to estimate b_{scat} and $PM_{2.5}$ but high degree of uncertainty because measurement is dependent on particle size.
	Backward scatter meter	$N(\beta = ?) = const\ I_o \sigma(\beta = ?)$ $160° < \beta < 180°$	Volume scattering at a scattering angle $160° < \beta < 180°$	Atmospheric scattering coefficient $b_{scat} = const\ \sigma(\beta \approx 180°)$	Routinely used to estimate b_{scat} and $PM_{2.5}$ but high degree of uncertainty because measurement is dependent on particle size.
Absorption	Filter absorption	$P = P_o e^{-b_{abs} r}$	Atmospheric absorption coefficient (b_{abs})	None	Routinely used but biases when a filter-based measurement is interpreted as ambient atmospheric b_{abs}.
	Photo acoustic	$P =$ acoustic pressure	b_{abs}	None	Currently only used in special studies but could be used for routine measurements of b_{abs}.
Absorption + scattering	Transmissometer	$H(r) = \dfrac{I_o}{r^2} e^{-b_{ext} r}$	Atmospheric extinction coefficient (b_{ext})	None	Has been used for routine monitoring. Problems of routine maintenance associated with window contamination.
	Cavity ring-down	$V(t) = V_{offset}$ $+ V_o \exp\{-(b_{scat} + b_M)ct\}$	b_{ext}	None	Special study.

which is the sum of scattering and absorption. In most cases, the measurement is associated with some type of detector from which a voltage or current is extracted; however, for each instrument there is an optical variable that it is designed to measure directly. For instance, a teleradiometer is designed to measure radiance, while an integrating nephelometer is designed to measure atmospheric scattering. The table also summarizes the type of optical and perceptibility variables that can be derived from the more fundamental measurement. The subsequent discussion will follow the sequence of monitoring instrumentation as listed in Table 7.1. The discussion will outline the basic principles of operation and indicate how various optical variables can be calculated either by reference to the equations in Chapters 2 and 3 or explicitly derived in this section. However, specific instrument manufacturers and inherent instrument uncertainties will not be addressed here.

SCENE AND RADIANCE MEASUREMENTS

Human-Observed Visual Range

Observations of visual range have been made at airports since 1919 and computer archived since 1947 (NOAA, 1982). Trained observers would attempt to identify preselected landscape features at known distances, ideally dark landscape features viewed against a background sky (WMO-No.8, 1996). Identifying a dark feature against a background sky that is just visible establishes the visual range, and within the context of the limiting assumptions of the Koschmieder relationship (see Chapter 3), $V_r = 3.912/b_{ext}$, the atmospheric extinction can be estimated.

A drawback of using this type of visibility observation is a general lack of landscape features corresponding to the current visibility conditions. However, identifying the most distant feature observable, say at a distance r, establishes that the visibility is r or greater, or conversely, the closest feature that cannot be seen, say at distance r', establishes that the visibility is less than r'. Making these types of observations over time allows for the development of a cumulative frequency distribution of visual range values that in turn allow for establishing a visual range percentile level, such as that 80% of the time the visibility is 120 km or less even though there may not be a landscape feature at exactly 120 km. This type of analysis allows for tracking some percentile level of visual range over time. Trijonis (1982a,b) used this approach to link smelter emissions to visibility reduction in the southwestern United States.

Teleradiometer Measurements

Teleradiometers or telephotometers are essentially telescopes fitted with some sort of electronic detector with a light filter (filters out all light except for the band of wavelengths desired). Teleradiometers have narrow-band filters to make radiance (N) measurements, while telephotometers use broadband filters that mimic the sensitivity of the human eye to make brightness (B) measurements. The following discussion presents various ways in which teleradiometric measurements can be used to derive a number of visibility-related metrics.

Apparent Contrast

Measuring the radiance of two adjacent landscape features or a landscape feature and adjacent sky radiance allows for a direct and simple calculation of apparent universal contrast C_r or modulation contrast m_r, as defined by Eq. 3.15 or 3.16. Furthermore, teleradiometer measurements can be used to measure the radiance of a plume or layered haze from which the contrast of that plume of layered haze can be calculated. Apparent contrast in and of itself has been shown to relate to various judgments of visual air quality as outlined in Chapter 5.

Atmospheric Extinction

Apparent contrast with some limiting assumptions can be used to estimate atmospheric extinction either of the ambient atmosphere or of the transmittance (opacity = $1/T$) of a plume. Fig. 7.4 outlines the measurement of a sky–landscape-feature's C_r. Eq. 4.5 relates apparent contrast to atmospheric extinction:

$$b_{ext} = -\frac{1}{r}\ln C_r/\gamma C_o.$$
(7.1)

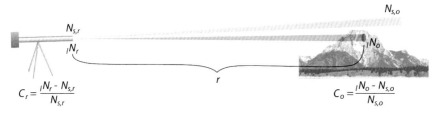

FIGURE 7.4   Schematic of a teleradiometer measurement of sky–landscape apparent contrast.

Now, if the sky at the landscape feature and observation point are equal ($\gamma = 1$) and if C_o is known, b_{ext} can be easily calculated. However, $\gamma = 1$ requires assumptions that are not usually met. The first criterion that must be met is uniform illumination between the observer and basically infinity. In the presence of cloud fields this rarely happens. Second, in a clean background atmosphere, the earth's curvature will result in an atmosphere where $\gamma \neq 1$, as will an observation zenith angle other than 90°. And then there is the variability of C_o as the sun's illumination of the landscape feature changes throughout the day, cloud cover shadows the landscape, and in the case of natural landscape features, vegetation changes from one season to another, as do things like snowcover.

If the natural landscape feature shown in Fig. 7.4 is replaced by a near-black artificial target, $C_o \approx -1$, as shown in Fig. 7.5, many of the issues and errors associated with not knowing C_o are addressed. However, uniform illumination is still required, so the atmosphere must be free of clouds or uniformly overcast.

If two adjacent landscape features, as outlined in Fig. 7.6, have different inherent radiance values, $_1N_o$ and $_2N_o$, Eq. 3.13 can be used to estimate extinction:

$$b_{ext} = -\frac{1}{r_1}\ln\left\{ \frac{(_1N_{r_1} - {}_2N_{r_1})}{(_1N_o - {}_2N_o)} \right\} \tag{7.2}$$

FIGURE 7.5 Schematic of a teleradiometer measurement of sky–artificial-black-target apparent contrast.

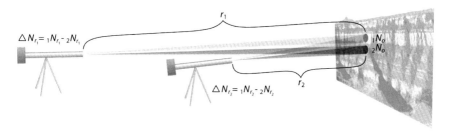

FIGURE 7.6 Schematic of a teleradiometer measurement of the radiance difference of two adjacent landscape features at two different distances r_1 and r_2.

where the variables are defined in Fig. 7.6 but with $r_2 = 0$. This equation takes advantage of the fact that radiance difference transmission through the atmosphere does not depend on illumination of the sight path, and as such, clouds do not affect the estimation. However, the inherent radiance difference must be estimated, and it can change as a function of time of day and cloud cover. It can be estimated when the atmosphere is near Rayleigh conditions in that $b_{ext} = b_{sg}$.

If the geometry of observation and landscape features allow teleradiometers to be placed at two distances from two adjacent landscape features at the same azimuth and zenith viewing angles, Eq. 3.13 applied for each observation distance allows for the cancellation of the inherent radiance difference to yield

$$b_{ext} = -\frac{1}{(r_1 - r_2)} \ln \left\{ \left({_1}N_{r_1} - {_2}N_{r_1} \right) \Big/ \left({_1}N_{r_2} - {_2}N_{r_2} \right) \right\} = \Delta N_{r_1} \Big/ \Delta N_{r_2} . \quad (7.3)$$

Either Eq. 7.2 or 7.3 can be used with artificial light and dark targets as outlined in Fig. 7.7. One innovative experiment (Richards, 1988; Richards et al., 1989) used a spinning disk of alternating white and black wedges painted onto its surface. The disk was spun and a teleradiometer at some distance r_1 was used to measure the alteration or radiance difference between the light and dark areas. Eq. 7.2 was used to estimate atmospheric extinction.

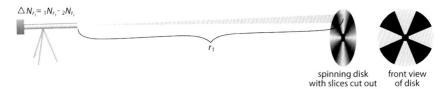

$\Delta N_{r_1} = {_1}N_{r_1} - {_2}N_{r_1}$

r_1

spinning disk   front view
with slices cut out   of disk

FIGURE 7.7   Schematic of a teleradiometer radiance measurement of alternating light and dark targets at some distance r_1.

Estimates of perceptibility parameters such as unimpaired and impaired visual range and deciview can be calculated from a measurement of extinction coefficient.

Limitations of Using Teleradiometers or Telephotometers as Measures of Atmospheric Extinction

A number of routine monitoring programs listed in Table 8.1 employed teleradiometers to measure landscape feature apparent contrast, which was in turn used to develop an estimate of atmospheric extinction. While operating these instruments in combination with integrating nephelometers in a number of special studies, it was determined that in relatively clear environments such as the western United States, the uncertainty in estimated extinction was nearly a factor of 2 because of variability of inherent contrast and changing illumination conditions. However, averaging the measurements over periods of weeks brought the uncertainty down to a more manageable level of about 20–30% (Allard and Tombach, 1981; Malm et al., 1981, 1982). As an initial estimate of atmospheric extinction in environments lacking any visibility measurements, teleradiometers are an effective "first look" at developing a quantitative estimate of existing visibility levels. However, if the goal is to link atmospheric optical to aerosol characteristics, it is desirable to use instruments that have an inherently higher degree of precision and accuracy.

Plume Transmittance

Plume opacity, defined as $1/T_p$, where T_p is the plume transmittance, has been estimated by human observers, and recently it has been suggested that a measurement of plume contrast can be used to make this estimate. The universal contrast equation, Eq. 3.14, can be used to calculate plume contrast as

$$_pC_r = (_pN_r - _bN)/_bN \tag{7.4}$$

where $_pN_r$ and $_bN$ are the measured plume radiance and background radiance values, respectively, as shown in Fig. 7.8, where the background radiance has been set equal to the background sky $_bN = N_s$.

Consider the term $_pN_r$:

$$_pN_r = N^*_{r_1} + \left(N^*_p + N_sT_p\right)T_{r_1}. \tag{7.5}$$

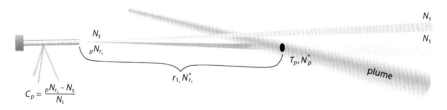

FIGURE 7.8 Schematic of a teleradiometer measurement of plume contrast.

The apparent plume radiance is the sum of the path radiance between the plume and teleradiometer $N^*_{r_1}$, the attenuated plume radiance $N^*_pT_{r_1}$, and the attenuated sky radiance $N_sT_pT_{r_1}$ as it passes the plume and the atmosphere between the plume and the radiometer.

If the path radiance $N^*_{r_1}$ can be approximated as $N_s(1 - T_{r_1})$,

$$_pC_r = (_pN_r - N_s)/N_s = \left(T_p - 1 + \frac{N^*_p}{N_s}\right)T_{r_1}. \qquad (7.6)$$

Now, if it is further assumed that T_{r_1} is approximately one or negligible compared to T_p, Eq. 7.6 yields

$$_pC_r = \left(T_p - 1 + \frac{N^*_p}{N_s}\right). \qquad (7.7)$$

In the general sense, this equation cannot be solved for T_p, the variable of interest. Under the very limiting condition of $N^*_p / N_s \ll 1$, Eq. 7.7 becomes

$$_pC_r = \left(T_p - 1\right). \qquad (7.8)$$

$N^*_p / N_s \ll 1$ implies that the plume path radiance is much less than the sky radiance, meaning the plume is very dark. This could occur either for a plume primarily made up of absorbing particles such as diesel soot or for a plume of larger particles with the sun–plume–radiometer geometry such that the scattered radiance from the plume to the radiometer is a result of backscattering. In such a case, the plume scattering is negligible, and the plume contrast is primarily a result of the attenuation of sky radiance having passed through the plume.

If a plume is viewed against some background instead of a sky background, Eq. 7.5 becomes

$$_pN_r = N^*_{r_1} + \left[N^*_p + \left(_bN_oT_{r_2} + N^*_{r_2}\right)T_p\right]T_{r_1}, \qquad (7.9)$$

which is the sum of the path radiances, plume radiance, and background radiance as seen through the plume. Using Eq. 7.9 and assuming $N^*_{r_1}$ and $N^*_{r_2}$ are small compared to N^*_p and that $T_{r_1}T_{r_2} \approx 1$, Eq. 7.7 becomes

$$_pC_r = \left(T_p - 1 + \frac{N^*_p}{_bN_o}\right), \qquad (7.10)$$

which reduces to Eq. 7.8 only if $N^*_p / _bN_o \ll 1$, which means a light or white background and a dark plume where the plume contrast is primarily dependent on attenuation of background radiance.

The only reliable and general method of using radiometric measurements to measure plume transmittance is to place a light and dark background behind the plume and measure the radiance of the backgrounds both through and outside the plume at the same distances, as schematically indicated in Fig. 7.9.

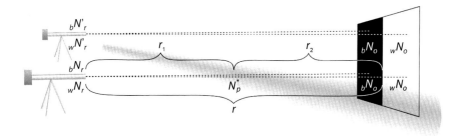

FIGURE 7.9 Schematic of teleradiometer measurements of black and white targets through and outside a plume.

Summing the radiances of white and black backgrounds as seen through the plume yields

$$_b N_r = N^*_{r_1} + N^*_p T_{r_1} + T_p T_{r_1} \left(_b N_o T_{r_2} + N^*_{r_2} \right) \tag{7.11}$$

and

$$_w N_r = N^*_{r_1} + N^*_p T_{r_1} + T_p T_{r_1} \left(_w N_o T_{r_2} + N^*_{r_2} \right). \tag{7.12}$$

Viewing the black and white backgrounds outside the plume, the radiance summation gives

$$_b N'_r = N^*_{r_1} + T_{r_1} \left(_b N_o T_{r_2} + N^*_{r_2} \right) \tag{7.13}$$

and

$$_w N'_r = N^*_{r_1} + T_{r_1} \left(_w N_o T_{r_2} + N^*_{r_2} \right). \tag{7.14}$$

Subtracting Eqs. 7.12 from 7.11 and 7.14 from 7.13 gives

$$_w N_r - _b N_r = T_{r_1} T_{r_2} T_p \left(_w N_o + _w N_o \right) \tag{7.15}$$

and

$$_w N'_r - _b N'_r = T_{r_1} T_{r_2} \left(_w N_o + _w N_o \right). \tag{7.16}$$

Now, dividing Eq. 7.15 by 7.16 yields

$$T_p = \left. \left(_w N_r - _b N_r \right) \middle/ \left(_w N'_r - _b N'_r \right). \right. \tag{7.17}$$

Eq. 7.17 is general and contains only measured quantities. Assumptions do not have to be made concerning path radiances or atmospheric transmittances between backgrounds and the plume or the plume and radiometer. Most importantly, assumptions concerning uniform illumination are not required in that path radiances cancel out in the above derivation. Given plume transmittance, one can solve Eqs. 7.12 and 7.14 for plume radiance N_p^*:

$$N_p^* = \left[{}_wN_r - {}_wN_r'T_p + N_{r_2}^*\left(T_p - 1\right)\right]/T_{r_1}. \qquad (7.18)$$

If atmospheric extinction is concurrently measured with radiometer measurements, T_{r1} and $N_{r_2}^*$ can be estimated. However, if it is assumed that $T_{r_1} \approx 1$, and $N_{r_2}^*(T_p - 1) \ll {}_wN_r$, then

$$N_p^* = \left({}_wN_r - {}_wN_r'T_p \right). \qquad (7.19)$$

Digital and Photographic Camera Measurements

Digital or photographic cameras can be used in much the same way as a teleradiometer. Photographic images, slides, or negatives can be "digitized" by scanning an image in such a way so as to yield millions of "density" values that essentially represent the quantity of radiant energy falling on any part of the film substrate. The photographic medium is usually digitized in multiple wavelengths such that the density values can be used to recreate the original image. Because the response of film to radiant energy is not linear, the photographic "system" must be calibrated such that a film density reading can be made to represent a quantitative radiance or luminance value. Density is defined as

$$DN = 1/T = \log_{10}(B_o/B_t) \qquad (7.20)$$

where DN and T are film density and transmittance, respectively, and B_o and B_t are the incident and transmitted luminances, respectively. Furthermore, the amount of light incident on a film surface is defined to be exposure, and exposure is given by

$$E = B_t \qquad (7.21)$$

where E is exposure and t is time of exposure. Typically, in a camera, t is controlled by the shutter speed, while the quantity of light is controlled by the aperture of the camera. Fig. 7.10 shows a typical density versus log exposure graph. First, notice the nonlinearity in the relationship between DN and E, and although not immediately evident in Fig. 7.10, the narrow dynamic range of film in that the extremes of the ability of film to respond to variation in exposure is quite limited. In other words, it is easy to

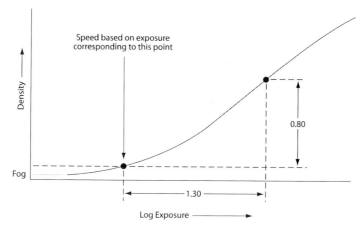

FIGURE 7.10   Hypothetical photographic film density (or readout from a digital camera) versus log of the exposure.

over- or under-expose film; the exposure has to be adjusted to the amount of light in the scene being recorded.

A film camera system can be calibrated for a given exposure by photographing an illuminated grayscale such as the one shown in Fig. 7.11. The actual luminance or radiance entering the camera can be independently

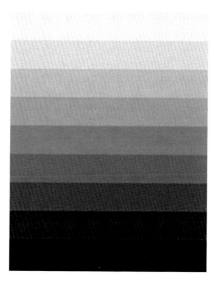

FIGURE 7.11   A grayscale varying from high to low density.

measured with a calibrated radiometer. In this way, luminance can be related to exposure (some combination of f-stop and shutter speed) and thus to density in the film image. In other words, luminance is some function of film density.

A problem of course arises as one adjusts the exposure to match the amount of light present in a given scene. As the exposure changes, so does the calibration curve. If the exposure is kept constant, the relationship between density and luminance is kept constant; however, each new exposure would require a new calibration curve.

Digital cameras can be used to extract the same information from landscape elements as photographic cameras, the difference being the calibration process. Instead of using film to record the image, digital technology allows for the direct "digitizing" of landscape features. Cameras fitted with detector arrays consisting of many megapixels are used to record the image. Typically, each pixel outputs a voltage which is digitized with an 8-, 12-, or 16-bit analog-to-digital converters. An 8-bit converter allows for the voltage, which is proportional to the radiance value, to be digitized into 256 discreet levels, while 12- and 16-bit converters allow 4096 and 65,536 levels of digitization, respectively. The relationship between digital values and absolute or relative radiance can be established in much the same way the density–exposure curve in Fig. 7.11 was determined and the calculations described previously performed.

Some Studies Relating Image Indexes to Visibility

A number of scene-dependent visibility metrics, derived from webcam images, were discussed in Chapter 4 on visibility metrics and many have recently been used in efforts to establish links between atmospheric haze and routine webcam or film-based photography.

Table 7.2 summarizes a number of studies that associate various image-related metrics with independently measured visibility parameters such as visual range or atmospheric extinction or scattering. Broadly speaking, even though all metrics are, in general, related to the effects of haze on the ability to see landscape features, some metrics allow for a direct estimation or calculation of the atmospheric extinction coefficient, while others must be related to physical variables through some appropriate statistical analysis, primarily linear or nonlinear regression analysis. Table 7.2 includes the reference, when and where the study was carried out, the temporal duration of the study, whether the pixel values were calibrated to remove nonlinearity in the exposure response curve, the visibility metric used, the reference measurement to which the metric was compared, whether the derived metric was related to the independent measurement directly or statistically, level of background visibility, and finally, a performance statistic in the form of R or R^2.

TABLE 7.2   Studies Focused on Deriving Quantitative, Visibility-Related Parameters Such as Visual Range and Atmospheric Extinction Coefficient from Routinely Collected Photographic or Electronic Images

Reference	Where/When	Duration	Index	Reference Measurement	Direct/ Statistical	V_r	Performance
Johnson et al., 1985	Eight national parks, 1984	1 year	Contrast single feature	Standard teleradiometer	Direct (calibrated)	100–300 km	$R^2 = 0.8$–0.9
Seigneur et al., 1984	Northern New Mexico, 1981	A few measurements	Contrast of plume	Standard teleradiometer	Direct (calibrated)	100–300 km	$R = 0.9$
Henry, 1967	n/a	Theoretical	Proposed: FFT and average mean square radiance	n/a	n/a	n/a	n/a
Malm et al., 1981	Mesa Verde NP, 1980	Various short time periods	Contrast, extinction coefficient	Standard teleradiometer	Direct (calibrated)	100–300 km	$R = 0.72$
Richards et al., 1989			Radiance difference, extinction		Direct (calibrated)		
Kim and Kim, 2005	Gwangju, Korea, 1999	Extended	Color, saturation, hue, and intensity difference	Transmissometer	Statistical	3–70 km, average ≈20 km	$R^2 = 0.79$
Janeiro et al., 2007	Evora, Portugal, 1996–2005	10 years	Ratio of contrast of two targets, extinction	Human observer	Direct	5–30 km	??
Du et al., 2013	Xiamen, China, Feb–Sept 2011	8 months	Radiance difference using two cameras, extinction	Scatter measurement	Direct (calibrated)	1–20 km	$R = 0.51$–0.86
Pokhrel and Lee, 2011	Incheon and Seoul, Korea, 2006	2 months	Root mean square: RMS	Transmissometer and scattering	Statistical	5–30 km	$R^2 = 0.88$ and 0.71

(Continued)

TABLE 7.2   Studies Focused on Deriving Quantitative, Visibility-Related Parameters Such as Visual Range and Atmospheric Extinction Coefficient from Routinely Collected Photographic or Electronic Images (*cont.*)

Reference	Where/When	Duration	Index	Reference Measurement	Direct/ Statistical	V_r	Performance
Luo et al., 2002	Taichung and Kaoshiung, Taiwan, 1999 and 2000	5 months	Specific brightness: Log(contrast)	Human observer	Statistical	5–10 km	$R = 0.91$
Luo et al., 2005	Kaoshiung, Taiwan, 2003	6 months	Sobel and FFT	Human observer	Statistical	2–16 km	$R^2 = 0.81$ and 0.78
Liaw and Lina, 2010	Taichung, Taiwan	4 months	FFT, FFT+ homomorphic + Gaussian, Harr func + homomorphic	Human observer	Statistical	1–15 km	$R^2 = 0.77, 0.86,$ and 0.91, respectively
Graves and Newsam, 2011	Phoenix, Arizona	Multiple years	Local Sobel, FFT, and dark channel prior (assumes $N_o = 0$)	Transmissometer and nephelometer	Statistical	30–300 km	$R^2 = 0.65, 0.55,$ and 0.24, respectively
Xie et al., 2008	Phoenix, Arizona	Multiple years	Radiance difference and FFT on blocks of image	Transmissometer and nephelometer	Statistical	30–300 km	$R^2 = 0.69,$ and $\approx 0.55,$ respectively
Chen et al., 2013	Beijing, China, 2012	2 days	Dark channel prior (assumes $N_o = 0$)	Forward scatter	Direct	3–4 km	$R = 0.80$
Poduri et al., 2010	Phoenix, Arizona, and Los Angeles, California	Selected days	Sky color gradient	Transmissometer and nephelometer	Statistical	30–300 km	$R = ??$

It is noteworthy to point out that most studies reported on in Table 7.2 were carried out under fairly hazy conditions ($V_r = 3\text{–}30$ km) and in urban environments. For reference, in a clear atmosphere free of particles the visual range would be on the order of 300–400 km, depending on perception thresholds, viewing angles, and earth curvature. Under hazy conditions, the atmospheric optical characteristics are typically closer to satisfying the conditions for the Koschmieder relationship to be valid, primarily uniform lighting conditions. Furthermore, urban scenes tend to be gray, devoid of color associated with vegetation or brightly colored cliffs or terrain faces such as those viewed in many of our national parks and wilderness areas. Bright edges of cloud formations are typically far enough from the observer to be obscured by heavy haze levels.

It is of interest to point out that papers referring to the so-called dark channel prior method are essentially using the Koschmieder relationship, with C_o, the inherent contrast of a landscape feature, set equal to -1.0, an absolutely black feature. Graves and Newsam (2011), Malm et al. (1981), and Chen et al. (2013) all used the Koschmieder contrast relationship and the measurement of the contrast of an isolated landscape feature to estimate the atmospheric extinction coefficient with varying degrees of success. If the assumptions associated with the Koschmieder relationship are met, the method works reasonably well, but under general atmospheric conditions it does not.

A more robust use of the general contrast equation, Eq. 3.17, is to simultaneously measure the contrast of two landscape features or to use two cameras to measure the contrast of the same landscape feature at two different distances. The extinction coefficient is then given by

$$b_{ext} = -\frac{1}{\Delta r} \ln\left(\frac{C_1}{C_2}\right)$$ where b_{ext} is the atmospheric extinction coefficient, Δr

is the distance between the two landscape features or between the two cameras depending on which method is being used, and C_1 and C_2 are the contrasts at distances r_1 and r_2, respectively. If the landscape features have the same reflective characteristics, the effects of changing inherent contrast and lighting conditions to a large degree are canceled out. These two techniques were successfully employed by Janeiro et al. (2007) and Du et al. (2013).

A number of investigators used Eq. 7.3, radiance difference attenuation, to directly estimate atmospheric extinction. Richards et al. (1989), discussed above, proposed using calibrated photographic images to extract radiance differences of the same two landscape features at two different distances and then calculating the ratio of these differences to estimate extinction. Du et al. (2013) proposed the same procedure but with digital cameras, and they were able to achieve correlations between the radiance-difference-estimated extinction and measured scattering as high as 0.86. Xie et al. (2008) implied they used contrast measurements of a single landscape feature to estimate extinction, but in reality they used a measure of radiance difference and called it contrast. The log of their measured radiance difference and measured extinction did correlate at greater than 0.8.

Other investigators have used a variety of image quality indexes using the whole image and developed statistical relationships between these indexes and a measure of atmospheric haze such as extinction or visual range. Pokhrel and Lee (2011) showed a high correlation between the root mean square index and independently measured extinction. Luo et al. (2005) used the Sobel edge detection algorithm over the entire image, while Graves and Newsam (2011) used Sobel edge detection to extract radiance gradients over small subsets of a larger image. In both analyses, average grayscale levels after Sobel filtering compared favorably to human-observer and transmissometer-derived visual range.

Many of the same investigators who employed Sobel filter analysis to extract radiance difference edges also employed fast Fourier transform (FFT) analyses to accomplish the same task. Not surprisingly, because both algorithms extract similar information, the correlations between either the Sobel filter or the FFT index and extinction are about the same. One interesting approach employed by Liaw and Lina (2010) sequentially applied filter techniques or used weighted combinations of two filtering techniques. They reported on the FFT, FFT plus homomorphic filtering, and a combination of Harr function analysis plus homomorphic filtering.

Can Indexes Derived from Routine Webcam Images in Pristine Natural Areas Be Used to Quantify Visibility?

The material presented in the next few pages is unpublished, so it is presented in more depth than other topics in this chapter. The studies summarized in Table 7.2 suggest that in urban areas with significant haze levels, metrics extracted from webcams can be used to track haze over time. The overriding goal or objective of the following analysis is to see if visibility metrics extracted from webcam images collected in relatively clear pristine environments can also be used to quantitatively track changes in haze over time periods of years or decades. The following analysis also demonstrates the use of gradient filters, such as the Sobel operator, and average pixel comparative analysis, such as average contrast. The analysis demonstrates the effects that illumination conditions and meteorology have on visibility metrics and how sensitive these metrics are to the camera color "channels" used in the analysis.

Over the course of the many years, webcam operation technology has changed, and web cameras have been replaced many times as technology improved or older systems wore out. As cameras were upgraded,

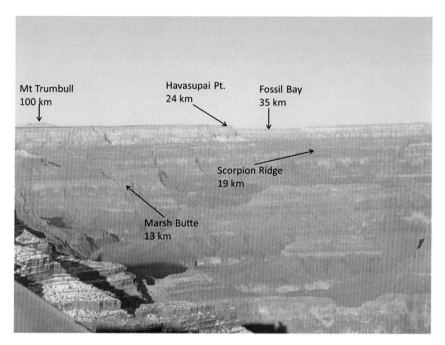

FIGURE 7.12 Grand Canyon National Park looking west at 9:00 am local time. The most distant mountain ridge on the horizon, Mt. Trumbull, is 100 km distant.

image pixel density changed, and in some cases, the focal length of lenses was changed as the information needs of land managers changed. The cameras have been operated in an automatic exposure mode, so as meteorological, illumination, and haze conditions change, exposure is altered for the best "average" picture. So, can an image index be identified that is insensitive to the hardware and operational changes that have occurred, such that these images can be used to quantify changes in haze levels over time?

Images collected in the more pristine national parks and wilderness areas typically have visual range values that are near 300–400 km or near-Rayleigh, particle-free conditions, and the images are of scenic features with widely varying color, meteorological, and lighting conditions. The eastern vistas typically have vegetation that varies in color and density as seasons of the year change. Figs. 7.12 and 7.13 show typical Grand Canyon images. Fig. 7.12 is a westerly view taken at 9:00 am local time under near-Rayleigh conditions from Yavapai Point. The most distant feature, Mount Trumbull, is 100 km distant, while the nearest foreground features are only about 1 km distant. Fig. 7.13 is the same view but at 3:00 pm local time and under cloudy conditions. Notice that Mount Trumbull is no longer visible, and note the significant color shift and additional contrast

FIGURE 7.13   Grand Canyon National Park looking west at 3:00 pm under cloudy conditions.

edges of the clouds, cloud shadows, and shadows caused by sun angle shift. Unlike some of the simpler and significantly hazier conditions discussed previously where contrast edges were primarily modified because of changes in haze levels, images such as those collected at Grand Canyon have contrast edges shifted and modified significantly by natural occurring events such as changing lighting conditions associated with meteorological conditions and sun angle.

Even though it is expected that contrast edges identified in images such as those shown in Figs. 7.12 and 7.13 will be modified by haze, it is not clear whether haze effects on these edges are significant enough to be detected independent of the effects of naturally occurring phenomenon such as changing sun angle and meteorological events.

Malm et al. (2015) analyzed about 180,000 Grand Canyon (GRCA) webcam images that were collected between 2001 and 2014 to try to answer that question. Pixel values associated with the collected images were not calibrated to ambient radiance levels but were used "as is." However, as camera focal length changed, the size of the physical image captured changed as well. Therefore, the webcam images were cropped such that the landscape features depicted in the image were the same across all years of data. Average contrast and equivalent contrast as represented by Eqs. 4.33 and 4.34 and a Sobel index given by Eq. 4.26 were calculated for the red, green, and blue channels of the collected images. Fig. 7.14 shows the green channel of Fig. 7.13 after convolving the green channel pixel values with the Sobel kernels given in Eq. 4.26. The radiance gradients associated with the many scenic landscape features are evident, as are the contrast edges of the clouds and cloud shadows. The Sobel index is just the average value of all the index numbers that make up the pixel array shown in Fig. 7.13. Fig. 7.14 shows the average green contrast as represented by Eq. 4.33. Whereas the Sobel filter picks out the high radiance gradient areas, resulting in an image made up of contrast edges, the average contrast index can be thought of as a grayscale image representing the continuous contrast changes across the entire scene.

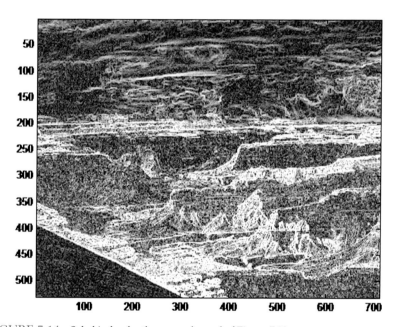

FIGURE 7.14   Sobel index for the green channel of Figure 7.13.

Fig. 7.15 shows the pixel-by-pixel average contrast of Fig. 7.13 as represented by Eq. 4.33. The average or equivalent contrast index is just an average of all the individual pixel contrast values in Fig. 7.15. For a scene that is totally obscured by haze, the average scene contrast index would be zero, as would the Sobel index.

The so called red, green, and blue image "channels" associated with webcam images will be referred to as red, green, or blue indexes, even though the camera spectral response of each of these three channels is

FIGURE 7.15   Average green contrast for the green channel of Figure 7.13.

quite broad, covering more than a single wavelength as implied by referring to them as red, green, and blue.

Fig. 7.16 shows a plot of green average contrast and the Sobel index
for the time period of Julian Day 2007.33 to 2007.47 (April 30, 2007, to
June 20, 2007). First, both indexes show a clear diurnal pattern reflective of sun angle and associated shadowing change over the course of a

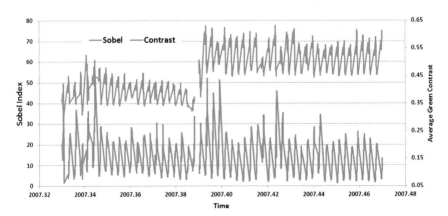

FIGURE 7.16   Temporal plot of green average contrast and the Sobel index for the time
period of Julian Day 2007.33 to 2007.47 (April 30, 2007, to June 20, 2007) derived from the
GRCA webcam images.

day. However, on 2007.388 (May 21, 2007) the focal length of the lenses was changed to wide angle. The subsequent cropping to compensate for the wider angle of view changed the resolution of the webcam image, which is reflected in a discontinuity of the Sobel index. This example shows that the sensitivity of a Sobel-type filter, or any other edge detection technique such as FFT filtering that is sensitive to changes in image resolution, makes it undesirable to use as a quantitative haze index if various cameras with varying image resolution or lenses with varying focal lengths are used. Therefore, the following analysis will focus on the use of contrast-type indexes.

In Tables 7.3 and 7.4, R, G, and B are the indexes in the red, green, and blue channels, respectively, and Cr refers to average contrast. b_{sp} is the atmospheric scattering coefficient as measured by an integrating nephelometer. Because average and equivalent contrasts are nearly perfectly correlated with each other at greater than 0.99, equivalent contrast is not shown in these tables. This high degree of correlation implies that average and equivalent contrast are capturing the same information on a channel-by-channel and image-by-image basis, although equivalent contrast is about a factor of 2 greater than average contrast.

Correlations between color channels are high at greater than 0.8; however, it is clear that the R, G, and B channels are changing over time with respect to each other because of either haze effects or changes in landscape feature characteristics as a function of time and meteorological conditions or both.

TABLE 7.3   Statistical Summaries of b_{sp}, CrR, CrG, and CrB for GRCA

Variable	Mean	Std Dev	Minimum	Maximum
b_{sp}	10.62	12.07	1.00	200
CrR	0.19	0.12	0.02	0.97
CrG	0.19	0.09	0.02	0.98
CrB	0.16	0.06	0.02	0.97

TABLE 7.4   Correlation between b_{sp}, CrR, CrG, and CrB for GRCA

Variable	b_{sp}	CrR	CrG	CrB
b_{sp}	1.00	−0.22	−0.26	−0.27
CrR	−0.22	1.00	0.96	0.81
CrG	−0.26	0.96	1.00	0.92
CrB	−0.27	0.81	0.92	1.00

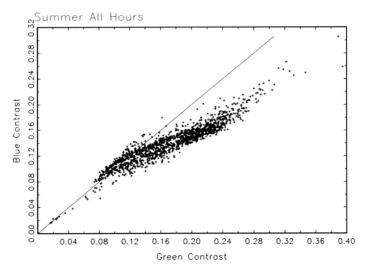

FIGURE 7.17   Scatter plot of *CrG* and *CrB*.

Effects of Time of Day and Lighting Conditions on *Cr*

Fig. 7.17 is a scatter plot of green and blue *Cr* for the year 2005 for all daylight hours, along with a 1:1 line. The lower *Cr* values correspond to morning hours when the east-facing wall of the Grand Canyon is under direct sunlight, while the higher contrast values correspond to afternoon hours, when the sun creates more and more shadows in the scene as the day progresses. Initially, blue and green contrasts follow the 1:1 line; however, in the afternoon, blue is less than green contrast by about 30%, primarily because of enhanced blue over green path radiance scattering.

Fig. 7.18 shows a plot similar to Fig. 7.17 but with green and red contrast values. Notice that during morning hours, red is lower than green contrast, while in the afternoon it is higher. During morning hours, the red cliffs are under direct lighting conditions, and the red contrast in any portion of the scene is low, including the sky. The red radiance levels for all parts of the scene are near the overall average scene red radiance level. As landscape features become shadowed, the red radiance levels of those shadowed areas decrease and the red *Cr* increases. The point is again that landscape features and their changes with illumination conditions contribute significantly to changes in image metrics, and they are different for different metrics.

Fig. 7.19 is a 3-D plot that further highlights the differences in *CrG* as a function of time of day and month, clearly affected by illumination

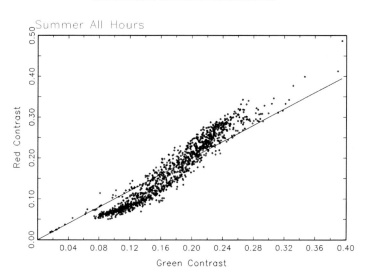

FIGURE 7.18   Scatter plot of *CrG* and *CrR*.

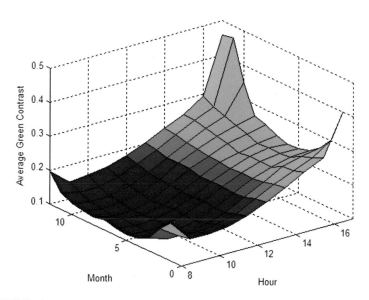

FIGURE 7.19   Average *CrG* plotted as a function of time of day and month for the GRCA dataset.

changes, primarily sun angle. During the winter months when the sun is low, CrG is greater than any other time of the year and is highest in late-afternoon hours. At about noon, the contrast is lowest and is nearly independent of time of year. Not shown is a plot for red and blue contrast, but again, the trends are the same but slightly enhanced in the red and less dramatic in the blue portion of the spectrum. Embedded in these Cr changes as function of lighting conditions are the changes due to increasing and decreasing haze levels.

Table 7.4 shows that the correlation between b_{sp} and Cr on an hour-by-hour basis is low at 0.26. Changing lighting conditions account for much of the hour-by-hour change in contrast, as shown in Fig. 7.19. It seems evident that the relationship between b_{sp} and Cr can be improved by selecting only certain hours for developing Cr values that correlate with b_{sp}. Fig. 7.20 shows the distribution of CrG for 9:00 am, noon, and 3:00 pm local time for the years 2011–2013. The solid line represents a normal distribution fit to a histogram of average CrG values. The normal distribution fit allows for the extraction of the mean and standard deviation (std) for each hour, as well as the R^2 value for the fit of each distribution. The mean and std associated with the normal distribution are somewhat different from the arithmetic mean and std because of the influence of outliers in the latter calculation. Fig. 7.20 shows that all three measured distributions have an increased frequency as compared with the normal distribution at low Cr levels as well as at levels of Cr greater than 0.4 (not shown in Fig. 7.20). These contrasts are due to rain contamination of the window of the camera enclosure and other meteorological conditions.

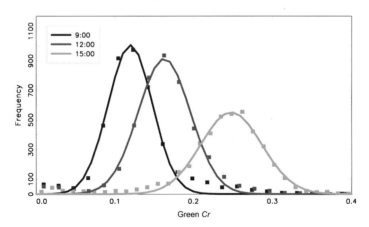

FIGURE 7.20 Frequency distribution of CrG for 9:00 am, noon, and 3:00 pm local time for the GRCA dataset.

The normal distribution derived mean ± std (NDM) for 9:00 am, noon, and 3:00 pm is 0.12 ± 0.028, 0.16 ± 0.033, and 0.24 ± 0.039, respectively. All R^2 values are high at 0.99. Notice, as suggested by Fig. 7.20, that the mean average CrG is lowest in the morning and increases throughout the day, as does the std. The std for the afternoon distribution is 40% greater than in the morning, suggesting that meteorological variables such as cloud shadowing play a bigger role in varying CrG in the afternoon than in the morning, assuming that aerosol scattering tends not to vary over the course of the day.

If meteorological effects on Cr average out over some appropriate time period, the remaining changes in Cr over time periods greater than the averaging time should be due to changes in haze levels. This possibility is explored in the next section.

Grand Canyon Images

Fig. 7.21 is a scatter plot of NDM CrG and the monthly average atmospheric scattering coefficient in inverse megameters (Mm^{-1}) for morning (9:00 am to noon), afternoon (1:00 to 4:00 pm), and all day for 2004–2013. The goal, as stated above, is to average out the effects of lighting and meteorologically generated variability in image contrast and to see if the averaging scheme would be more effective for different times of the day. As indicated in Fig. 7.19, the average contrast values are lowest in

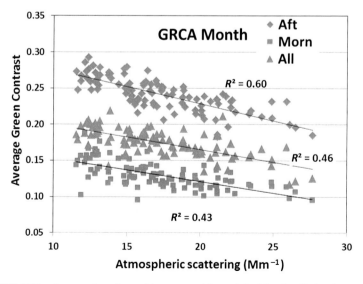

FIGURE 7.21   Scatter plot of monthly averaged b_{sp} and CrG for the GRCA dataset using all data, morning hours (9:00 am to noon local time), and afternoon hours (1:00 to 4:00 pm).

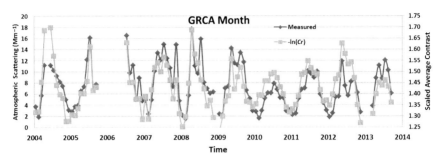

FIGURE 7.22   Temporal plot of b_{sp} and $-\ln(CrG)$ scaled to b_{sp} for the afternoon monthly average GRCA dataset.

the morning, when the scene is illuminated by direct sunlight and the lighting is "flat," and higher in the afternoon, when shadowing occurs because sun angle changes. When all hours of the day are averaged, the contrast values are between the morning and afternoon values. To a large degree, the effects of light and meteorological influences are averaged out independent of time of day. However, the best fit between monthly average CrG and b_{sp} is observed during afternoon hours when the $R^2 = 0.60$. The R^2 value is lowest in the morning at 0.43 and higher at 0.46 for the full daytime dataset.

Fig. 7.22 shows a time series plot of monthly averaged measured b_{sp} and $-\ln(CrG)$ for the morning dataset discussed above. The left-hand axis represents the monthly average atmospheric scattering coefficient, and the right-hand axis is $-\ln(CrG)$ of monthly average CrG values. The seasonal variability of atmospheric scattering is nicely captured in both datasets, and over the course of the 10-year dataset, b_{sp} has stayed about the same during the winter, while there appears to be a slight reduction of b_{sp} during summer months.

The same analysis was carried out on a seasonal basis where 3 months of data were averaged in the hope that the comparison between b_{sp} and CrG could be improved upon. The analysis showed that the R^2 values remained almost the same for the monthly and seasonal comparisons. Similar analyses were also carried out for CrR and CrB, and in both cases, the comparison between b_{sp} and Cr values was degraded; the comparison was not improved by using equations similar to Eq. 4.36, in which all three color Cr values are combined to represent a color shift as opposed to just an overall contrast change.

Can webcam images of scenic landscape features in near-Rayleigh conditions be used to extract quantitative atmospheric optical variables? In urban, hazy conditions, landscape features tend to be monochromatic, and changes in sun angle over the course of the day and changing meteorological conditions do not change apparent lighting conditions. Many investigators (see Table 7.2) have shown that a number of image-derived visibility indexes relate well to variables such as the atmospheric extinction coefficient or visual range. The above analysis suggests that under atmospheric conditions where changing lighting conditions have a significant effect on indexes, one can still extract meaningful relationships between the indexes and atmospheric optical variables if the indexes are averaged over sufficient time to average out the effects of changing lighting, leaving longer-term changes caused by haze to be effectively extracted.

SCATTERING MEASUREMENTS

First, it should be noted that if only scattering is measured, any reduction in visibility due to absorbing aerosols can only be accounted for by assuming some constant b_{abs} to b_{ext} ratio or constant scattering albedo. The scattering albedo can easily vary by a factor of 2 or more. As shown in Table 7.1, optical monitoring is meant to refer to more or less direct measurements of either the volume scattering function σ or the extinction, scattering, or absorption coefficients, b_{ext}, b_{scat}, or b_{abs}, respectively. None of these instruments measures these parameters directly but rather some voltage that is directly proportional to these variables. For instance, the first instrument to be discussed is an integrating nephelometer, which is designed such that voltage from a detector is directly proportional to the atmospheric scattering coefficient.

Integrating Nephelometer

The basic configuration of an integrating nephelometer is shown in Fig. 7.23 (Beuttell and Brewer, 1949). Assume the light source has the property that its intensity in any given direction is represented by $I_o \cos\beta$.

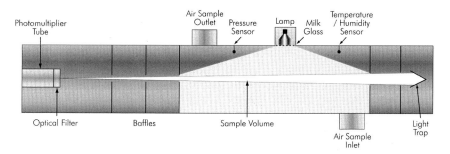

FIGURE 7.23 Schematic of a typical integrating nephelometer.

Given the geometry of the instrument shown in Fig. 7.23, it can be shown that the luminance or radiance at the detector is given by

$$B = (I_o / L) \int_0^\pi \sigma(\beta) \sin(\beta) d\beta \qquad (7.22)$$

where L is the distance between the light source and center of the center line of the detector volume. Comparing Eq. 7.22 to Eq. 2.4 shows that, if the light source is a Lambertian surface, the luminance measured at the detector is directly proportional to atmospheric scattering, whether it is from particles or the atmosphere itself (Rayleigh scattering):

$$b_{sp} = \int_{4\pi} \sigma(r, \beta) d\Omega = 2\pi \int_0^\pi \sigma(r, \beta) \sin(\beta) d\beta. \qquad (7.23)$$

The instrument can be fitted with particle size inlets such that only scattering associated with fine or coarse particles is considered, and filters can be placed in front of the detector so that wavelength-dependent scattering can be measured. Some instruments have been fitted with a shutter such that only backscattering is measured (Charlson et al., 1974; Heintzenberg, 1978). Backscattering is important to the earth's radiation balance calculation.

Limitations of Using Nephelometers to Measure Atmospheric Scattering

An issue with the use of this type of instrument is that the limits of angular integration are dependent on the geometry of the enclosure, light source, and detector. Typically, the integration is from about 10° to 170°, and therefore calibration to a standard gas with one volume scattering function and measuring particles with a different scattering function can yield systematic errors or uncertainties. Perhaps more importantly, because the sampling chamber is enclosed and heated by radiation from the lamp and nearby electronics, the difference between chamber and ambient temperatures can be as high as 10° centigrade or more. Therefore, hygroscopic particles such as sulfates and nitrates are dried out, and measured scattering is substantially less than in the ambient atmosphere. It is essential to either pre-dry the aerosol before it enters the nephelometer chamber, such that "dry" particle scattering is measured, or measure the temperature/relative humidity inside the sampling chamber, such that ambient scattering can be estimated if that is the goal of the measurement. In any case, measuring scattering at some unknown relative humidity is a relatively useless measurement. It cannot even be used to set a meaningful upper or lower bound estimation of scattering.

Polar Nephelometers

If it is of interest to attribute visibility impairment of landscape features, which can be substantially linked to path radiance, it is essential to understand how path radiance varies as a function of sun–landscape feature–observer geometry. This requires understanding the angular scattering dependence of the ambient aerosol, the measurement of which can be obtained with a polar nephelometer.

Fig. 7.24 shows volume scattering functions for four types of particles and highlights why a measure of scattering as a function of scattering angle can be important (see Chapter 2). The solid line represents the scattering function for an ammonium sulfate aerosol lognormal mass size distribution with $D_g = 0.2$ μm and $\sigma_g = 1.7$, while the dotted line also represents a sulfate aerosol but with $D_g = 0.5$ μm and $\sigma_g = 2.0$. The silica aerosol was represented with $D_g = 5.0$ μm and $\sigma_g = 1.7$, while the elemental carbon (EC) mass size distribution corresponded to $D_g = 0.2$ μm and $\sigma_g = 1.7$. The difference between scattering in the forward versus backward direction for the 5.0 μm aerosol is nearly 10,000, while for sulfate it is still substantial at over 100.

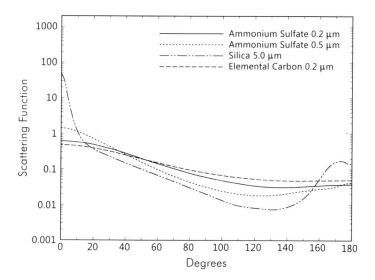

FIGURE 7.24   Scattering function for typical fine and coarse particle mass size distributions.

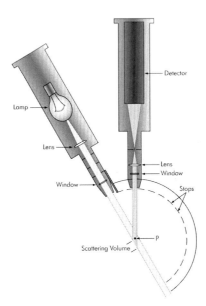

FIGURE 7.25 Schematic of a polar nephelometer.

Polar nephelometers, schematically shown in Fig. 7.25, measure scattered radiant energy from some small volume at specific angles between $0° < \beta < 180°$, either through moving a detector through multiple angles such as shown in Fig. 7.25 or by arranging detectors at multiple angles around the scattering volume V. One of the first polar nephelometers was used for photometric measurements of the upper atmosphere (Waldram, 1945). As shown in Eq. 3.10, the scattered radiant intensity J can be represented by

$$J = H\sigma(\beta)V \tag{7.24}$$

so,

$$\sigma(\beta) = \frac{J(\beta)}{HV} = constJ(\beta) \tag{7.25}$$

where J is the scattered radiant intensity of the incident beam (watt/steradian), H is the irradiance of the incident beam (watt/m²), V is the scattering volume (m³), β is the scattering angle, and $const$ is a constant of calibration. Therefore, the units on the volume scattering function are watt/m · steradian or energy per unit time per unit distance per solid angle.

Integrating the measured volume scattering function $\sigma(\beta)$ over all angles yields

$$b_{scat} = 2\pi \int_0^\pi \sigma(\beta)\sin\beta d\beta. \tag{7.26}$$

Forward- and Backscatter Measurements

Forward- and backscatter meters are similar to polar nephelometers in that they do not measure total scattered radiant energy that is proportional to b_{scat} but only radiant energy scattered into some prespecified solid angle:

$$J = H\sigma(\beta = 45°)V \tag{7.27}$$

and

$$\sigma(\beta = 45°) = \frac{J(\beta = 45°)}{HV} = const J(\beta = 45°) \tag{7.28}$$

where *const* is a calibration constant.

Fig. 7.26 is a schematic of a forward-scatter instrument designed to measure scattering at $45° \pm \theta$ over some solid angle Ω (Richards et al., 1996). θ varies from instrument to instrument. There are a wide variety of forward-scatter meters on the market, and it is worth noting that

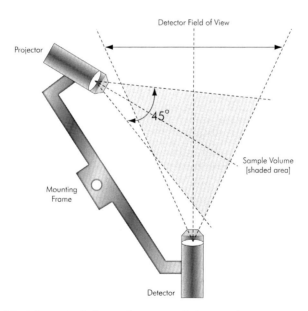

FIGURE 7.26   Schematic of a forward-scatter nephelometer that measures scattering at a scattering angle of approximately $45° - \theta < \beta < 45° + \theta$.

starting in 1994 the U.S. National Weather Service (NWS), in cooperation with the Federal Aviation Administration (FAA), deployed nearly 1000 forward-scatter meters. The system, which involves other weather-related instrumentation, is referred to as the Automated Surface Observing System (NOAA, 1998). Forward-scatter instrument manufacturers purport to measure "visibility" under all types of weather and haze conditions.

Limitations of Using Forward- or Backscatter Measurements to Estimate Atmospheric Scattering

Examination of Fig. 7.24 shows that the volume scattering function varies by orders of magnitude for specific scattering angles, with larger particles scattering more in the forward direction and less in the backward direction. The least amount of variability in the volume or scattering phase function occurs at about 45°, hence the selection of 45° as the angle between the light source and detector for most forward-scatter meters. However, even at 45° the ratio of scattering to total scattering is not constant. The percent difference in light scattered into a solid angle encompassing 40–50° for an ammonium sulfate aerosol with a lognormal mass size distribution with $D_g = 0.2$ μm and $\sigma_g = 1.7$ and a sulfate aerosol with $D_g = 0.5$ μm and $\sigma_g = 2.0$ is 10%. So a small shift in the accumulation-mode particle size distribution can result in uncertainties on the order of 10–15%. Because the index of refraction for organics, light absorbing carbon, and ammonium nitrate are all about the same (see Table 2.2), the overall uncertainty is greater due to particle size shift than due to variation in the relative composition of the accumulation mode. However, the percent difference in light scattered into a solid angle encompassing 40–50° for an aerosol with $D_g = 0.2$ μm and one with $D_g = 5.0$ μm is on the order of a factor of 2, or a 100% uncertainty. Large or coarse particle scattering, if calibrated to a fine particle aerosol, will be underestimated significantly by forward-scatter meters, and vice versa, if the instrument is calibrated with a large particle aerosol, such as dust, fine particle scattering will be overestimated. Furthermore, the difference between reported and true visual range for a pollution haze versus fog or cloud droplets would be substantial.

Forward-scatter meters are also used as atmospheric particle mass concentration meters. Here, the error is even more significant because the density of the particles comes into play. Dust particles have a density near 4 g/cm³, while organic-type particles have densities in the range of 1.2–1.4 g/cm³. So dust and organic particles with the same scattering cross section could have mass uncertainties associated with density differences of 100–200%, depending on which particle is chosen as the reference. That is, the reported mass concentration could be in error by more than a factor of 2, just based on particle density variation!

ABSORPTION MEASUREMENTS

Measurement of absorption of light by atmospheric particulate matter has received a lot of attention with concerns about its role in climate forcing and visibility impairment. Light absorption is typically associated with black carbon (BC), elemental carbon (EC), brown carbon, and light absorbing carbon (LAC), all of which seem to be used somewhat interchangeably in the literature (Bond and Bergstrom, 2006; Petzold et al., 2013). EC typically refers to carbon that is indeed black and absorbs light equally in all wavelengths, such as diesel emissions. BC, which is a result of incomplete combustion, typically absorbs more in the shorter wavelengths of light, and the term LAC is meant to be inclusive of all carbon aerosols that absorb light. It should be also pointed out that soil or dust particles in the atmosphere also contribute to significant amounts of atmospheric absorption. Aerosol absorption measurements are categorized into three general categories: photoacoustic absorption spectrometry, filter transmission, and incandescence measurements (Moosmüller et al., 2009).

Photoacoustic Absorption Spectrometry

The principle of photoacoustic spectrometry is outlined in Fig. 7.27. Pulsed electromagnetic energy is absorbed by a particle and converted to heat that is just higher energy states of various vibrational, translational, and rotational modes of the molecules making up the particle. Because of the relatively high thermal conductivity of the particles, this energy is rapidly transferred to the surrounding air, resulting in an increase in the volume of pressure of the surrounding air. If the air volume is placed in a resonator tuned to the frequency of the excitation light, the resultant pulsed pressure gradient can be detected with a sensitive microphone.

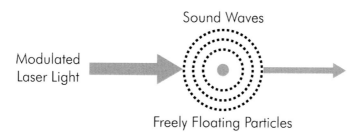

FIGURE 7.27   A modulated light source causes an increase in the internal energy of a particle. This energy is transferred to the surrounding air molecules, resulting in the formation of an acoustic wave.

A schematic of a photoacoustic spectrometer is shown in Fig. 7.28. The instrument typically consists of a chamber designed to be of a length consistent with the wavelength of the induced acoustic standing wave. An acoustic notch filter at either end of the chamber ensures, through destructive interference, a stable standing wave within the chamber. The excitation light source is typically a pulsed laser, and the detector is a microphone that essentially records the amplitude of the induced sound wave (Arnott et al., 1999, 2005).

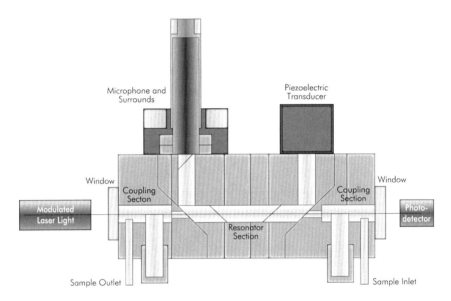

FIGURE 7.28   Schematic of a photoacoustic spectrometer.

The governing equation is

$$b_{abs} = P_m \frac{1}{P_L} \frac{A_{res}}{\gamma - 1} \frac{\pi^2 f_o}{Q} \tag{7.29}$$

where b_{abs} is the absorption coefficient, P_m is the measured acoustic pressure, P_L is the laser beam power, A_{res} is the resonator cross section, Q is the resonator quality factor, and f_o is the resonant frequency. γ is the ratio of isobaric to isochoric specific heats of air. Q is the ratio of the energy stored in the resonator to the energy dissipated per radian. The instrument can be fitted with lasers of different wavelengths, thus offering a measure of the wavelength dependence of atmospheric particle absorption.

One of the advantages of the photoacoustic spectrometer is that it measures the absorptive properties of a free particle or an ambient particle as opposed to particles collected on a filter where interferences associated with particle buildup and filter light-scattering properties come into play.

Filter Absorption

There are a variety of instruments on the market that are based on filter absorption techniques. Fig. 7.29 schematically shows a filter-based measurement of atmospheric absorption, usually referred to as a laser-integrated plate measurement (LIPM). Aerosol is collected on a filter substrate, typically Teflon, over a period of time. Then, either in near real time or after removing the filter and transporting it to a laboratory, the transmission of the filter is measured from which an estimate of b_{abs} can be derived. Taha et al. (2007) reviewed a variety of LIPM techniques and discussed the many problems associated with this type of measurement.

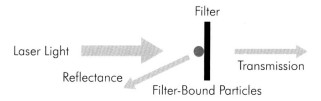

FIGURE 7.29   Diagram showing the absorption of laser light by particles collected on a filter substrate.

The governing equations are

$$P = P_o \exp(-b_{abs}X) \tag{7.30}$$

where P and P_o are in incident and transmitted radiant flux, respectively, and X is the length of the column of sampled air. X can be estimated by $X = V/A$, where V is the volume of air sampled and A is the cross of the filter. Therefore,

$$b_{abs} = (A/V)\ln(1/T) \tag{7.31}$$

where T is the transmittance of the filter (P/P_o) after exposure.

Limitations of Using Filter-Based Measurements to Estimate Atmospheric Absorption

Filter-based measurements of atmospheric absorption are fraught with many potential biases, both negative and positive. For instance, multiple scattering of photons by the sample and sample substrate tends to enhance the estimated absorption, while "shadowing" by particles over-laid on other particles can cause an underestimation of absorption. Consequently, many configurations of light source and detector have been put forth, all in an attempt to correct for possible biases. The many and varied instrumentation configurations will not be discussed here, but a few references pointing to various designs are given for the interested reader to further investigate (Fischer, 1973; Lindberg, 1975; Lindberg et al., 1994; Heintzenberg, 1982a, 1982b; Hitzenberger et al., 1996; Marley et al., 2001; Moosmüller et al., 2009).

TRANSMISSION MEASUREMENTS

Conventional Transmissometer

A measure of atmospheric transmission allows for a direct determination of the average extinction coefficient over a path in which the transmission measurement was made. Historically, transmissometry has been thought of as teleradiometric measurements of the intensity of a light source placed at some distance r from an observation point (Middleton, 1968; Hall et al., 1975; Malm and Walther, 1980). The equation governing the amount of radiant energy received at the observation point is

$$H(r) = (J_o/r^2)\exp(-b_{ext}r) \tag{7.32}$$

where $H(r)$ is irradiance at some distance r from the light source, J_o is radiant intensity, and b_{ext} is the atmospheric extinction coefficient. J_o/r^2 is essentially a calibration term that can be determined by comparison of transmission measurements to other optical measurements, such as teleradiometer measurements made under "standard" lighting conditions or to integrating nephelometer measurements made on days that are near the Rayleigh scattering limit. A second calibration technique, proposed by Hall et al. (1975), utilizes measurements of $H(r)$ at two different distances. Measuring r_1, r_2, $H(r_1)$, and $H(r_2)$ and assuming the atmosphere is homogeneous over distances r_1 and r_2 allows for the calculation of J_o. A third calibration technique involves measuring $H(r)$ when the receiver and light source (transmitter) are placed within a few hundred feet of each other and assuming $\exp(-b_{ext}r) \approx 1$. Then $J_o = H(r_2)r_2$, where r is the distance between the teleradiometer and the transmitter.

To minimize the effect of background or ambient illumination, the transmitted light beam is modulated, while the receiver electronics are designed to measure the difference between background radiance (transmitter off) and background plus transmitter radiance (transmitter on). The difference between the two signals is proportional to the irradiance associated with the radiant energy from the transmitter light source.

Limitations of Using Transmission Measurements to Estimate Atmospheric Extinction

There are a number of interferences associated with long-path transmission measurements. Any stable density variations of the atmosphere between the transmitter and the receiver will cause preferential "bending" or refraction of the radiation beam, resulting in a decrease in measured irradiance that does not have anything to do with attenuation and in an upward bias in estimated extinction. The effect is usually caused by a light beam passing close to a terrestrial surface over which a stable atmospheric inversion layer has built up. Therefore, if the path over which the transmission measurement is made is kilometers in distance, the path should be hundreds of feet above the terrestrial surface.

A second common interference is caused by atmospheric turbulence. If a collimated beam of radiant energy is used, such as laser light or a focused incoherent light source, turbulence will cause the beam to "spread." That is, light is preferentially refracted out from the beam, again resulting in a decrease in measured radiance that is not caused by particle or gas attenuation and therefore a bias in estimated extinction. The transmitted beam should be isotropic over a large enough solid angle such that as much light that is refracted out of the beam due to turbulence is refracted back into the transmission path.

Cavity Ring-Down Transmission Measurement

Application of cavity ring-down spectroscopy was discussed by Wheeler et al. (1998). A diagram of a cavity ring-down atmospheric extinction instrument is shown in Fig. 7.30. Typically, the light source is a pulsed laser at a frequency near the peak response of the human eye–brain system (500–600 nm). The sampling cavity is on the order of 1 m in length with highly reflective mirrors ($R \approx 99.995\%$) at either end of the chamber. Clean, purged air is drawn across the mirror face to prevent contamination over the course of extended sampling periods. A photomultiplier is mounted behind the second mirror (Moosmüller et al., 2005).

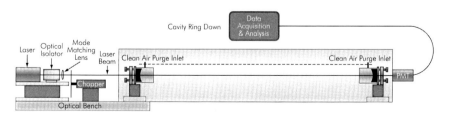

FIGURE 7.30   Schematic of a cavity ring-down spectrometer.

A fraction of the light passes through the mirror and is reflected back and forth across the chamber. On each reflection cycle, it is attenuated by extinction of the atmosphere contained within the chamber and lost due to a small transmittance through each mirror on each reflection cycle. The equation governing the decrease in detector signal as a function of time is

$$V(t) = V_{offset} + V_o \exp\{-(b_{scat} + b_M)ct\} \tag{7.33}$$

where $V(t)$ is the detector output signal or voltage, V_{offset} is the electronics background/dark current signal or offset, V_o is the amplitude or initial PMT voltage before attenuation, and b_{scat} is the atmospheric scattering coefficient including scattering by air molecules. t and c are time and speed of light, respectively. b_M is a pseudo-attenuation coefficient associated with the reflectance of the mirrors. If Eq. 7.33 is rewritten as

$$\ln(V(t) - V_{offset}) = \ln(V_o) - b(ct), \tag{7.34}$$

a plot of $\ln(V(t) - V_{offset})$ versus ct yields a linear plot whose slope is the decay constant b. b_M can be estimated from the known reflectance of the mirrors or determined experimentally. Fast electronics allow Eq. 7.34 to be on a real-time basis such that the instrumentation "output" is in terms of the atmospheric scattering coefficient b_{sp}.

As an alternative to measuring the decay constant of energy stored in the cavity, the energy transmitted through the cavity can also be used to estimate optical extinction. Instruments have been fitted with both types of attenuation measurements and are complementary to each other.

Other instruments are used to estimate the mass concentration of absorbing particles from which absorption can be approximated and will not be discussed in any detail but only mentioned. Thermal optical reflectance and transmittance (TOR/TOT) are thermal evolution techniques in which evolved material is analyzed for carbon (Chow et al., 2004). The single-particle soot photometer (SP2) and the aerosol mass spectrometer (AMS) are two more instruments that are effectively used to determine the mass of absorbing soot particles from which particle absorption can be estimated (Gao et al., 2007). In the SP2, laser-induced incandescence of single particles is empirically related to absorbing carbon mass, while in the traditional AMS, the absorbing carbon is assumed to be the refractory component of the particles and is obtained by mass balance of the measured nonrefractory organics and the total particle mass.

References

Allard, D., Tombach, I., 1981. The effects of non-standard conditions on visibility measurement. Atmos. Environ. 15, 1847–1858.

Arnott, W.P., Moosmüller, H., Rogers, C.F., Jin, T., Bruch, R., 1999. Photoacoustic spectrometer for measuring light absorption by aerosols: instrument description. Atmos. Environ. 33, 2845–2852.

Arnott, W.P., Hamasha, K., Moosmüller, H., Sheridan, P.J., Ogren, J.A., 2005. Towards aerosol light absorption measurements with a seven-wavelength aethalometer: evaluation with a photoacoustic instrument and a three-wavelength nephelometer. Aerosol Sci. Technol. 39, 17–29.

Baron, P.A., Willeke, K. (Eds.), 2001. Aerosol Measurement: Principles, Techniques and Applications, second ed. Wiley, New York.

Beuttell, R.G., Brewer, A.W., 1949. Instruments for the measurement of the visual range. J. Sci. Instrum. Phys. Ind. 26, 357–359.

Bond, T.C., Bergstrom, R.W., 2006. Light absorption by carbonaceous particles: an investigative review. Aerosol Sci. Technol. 40, 27–67.

Charlson, R.J., Porch, W.M., Waggoner, A.P., Ahlquist, N.C., 1974. Background aerosol light-scattering characteristics—Nephelometric observations at Mauna Loa Observatory compared with results at other remote locations. Tellus 26, 345–360.

Chen, X., Lu, C., Liu, W., Zhang, Y., 2013. A new algorithm for calculating the daytime visibility based on the color digital camera. Int. J. Comput. Sci. Issues 10 (1), 694–814.

Chow, J.C., Watson, J.G., Chen, L.-W.A., Arnott, W.P., Moosmüller, H., Fung, K., 2004. Equivalence of elemental carbon by thermal/optical reflectance and transmittance with different temperature protocols. Environ. Sci. Technol. 38 (16), 4414–4422.

Du, K., Wang, K., Shi, P., Wang, Y., 2013. Quantification of atmospheric visibility with dual cameras during daytime and nighttime. Atmos. Meas. Technol. 6, 221–2130.

Fischer, K., 1973. Mass absorption coefficients of natural aerosol particles in the 0.4 to 2.4 μm wavelength interval. Contr. Atmos. Phys. 46, 89–100.

Gao, R.S., Schwarz, J.P., Kelly, K.K., Fahey, D.W., Watts, L.A., Thompson, T.L., et al., 2007. A novel method for estimating light-scattering properties of soot aerosols using a modified single-particle soot photometer. Aerosol Sci. Technol. 41 (2), 125–135.

Graves, G., Newsam, S. Using visibility cameras to estimate atmospheric light extinction, presented at the IEEE Workshop on Applications of Computer Vision (WACV), Kona, Hawaii, 2011.

Hall, J.S., Mikolaj, J., Riley, L., Etitio, K.J., 1975. Extinction measurements of sensitive assessment of air quality. J. Air Pollut. Control Assoc. 25 (10), 1045–1048.

Heintzenberg, J., 1978. The angular calibration of the total scatter/backscatter nephelometer, consequences and application. Staub Reinhalt Luft 38, 62–63.

Heintzenberg, J., 1982a. Measurement of the light absorption and elemental carbon in atmospheric aerosol samples from remote locations. In: Wolff, G.T., Klimisch, R. (Eds.), Particulate Carbon, Atmospheric Life Cycle. Plenum Press, New York, pp. 371–386.

Heintzenberg, J., 1982b. Size-segregated measurements of particulate elemental carbon and light absorption at remote Arctic locations. Atmos. Environ. 16 (10), 2461–2469.

Henry, R.C., 1967. The application of the linear system theory of visual acuity to visibility reduction by aerosols. Atmos. Environ. 11, 697–701.

Hinds, W.C., 1999. Aerosol Technology: Properties, Behavior, and Measurement of Airborne Particles, second ed. Wiley, New York.

Hitzenberger, R., Dusek, U., Berner, A., 1996. Black carbon measurements using an integrating sphere. J. Geophys. Res. 101, 19601–19606.

Janeiro, F.M., Wagner, F., Ramos, M.R., Silva, A.M. Atmospheric visibility measurements based on a low-cost digital camera, in Proceedings of the ConfTele, 6th Conference on Telecommunications, Peniche, Portugal, 2007.

Johnson, C.D., Malm, W.C., Persha, G., Molenar, J.V., Hein, J.R., 1985. Statistical comparisons between teleradiometer-derived and slide-derived visibility parameter. J. Air Pollut. Control Assoc. 35, 1261–1265.

Kim, K.W., Kim, Y.J., 2005. Perceived visibility measurement using the HIS color difference method. J. Korean Phys. Soc. 46 (5), 1243–1250.

Liaw, J.J., Lina, S.B., Huang, Y.F., Chen, R.C., 2010. Using sharpness image with Harr function for urban atmospheric visibility measurement. Aerosol Air Qual. Res. 10, 323–330.

Lindberg, J.D., 1975. Absorption coefficient of atmospheric dust and other strongly absorbing powders: an improvement on the method of measurement. Appl. Opt. 14 (12), 2813–2815.

Lindberg, J.D., Douglass, R.E., Garvey, D.M., 1994. Absorption-coefficient-determination method for particulate materials. Appl. Opt. 33 (19), 4314–4319.

Luo, C.H., Liu, S.H., Yuan, C.S., 2002. Measuring atmospheric visibility by digital image processing. Aerosol Air Qual. Res. 2 (1), 23–29.

Luo, C.H., Wen, C.Y., Yuan, C.S., Liaw, J.J., Lo, C.C., Chiu, S.H., 2005. Investigation of urban atmospheric visibility by high frequency extraction: model development and field test. Atmos. Environ. 39, 2545–2552.

Malm, W.C., 1992. Characteristics and origins of haze in the continental United States. Earth Sci. Rev. 33, 1–36.

Malm, W.C., Walther, E.G. A Review of Instrument-Measuring Visibility-Related Variables, EPA-600/4-80-016, U.S. Environmental Protection Agency, Environmental Monitoring Systems Laboratory, Las Vegas, February 1980.

Malm, W.C., Pitchford, A., Tree, R., Walther, E., Pearson, M., Archer, S., 1981. The visual air quality predicted by conventional and scanning teleradiometers and an integrating nephelometer. Atmos. Environ. 15 (12), 2547–2554.

Malm, W.C., Pitchford, M.L., Pitchford, A., 1982. Site specific factors influencing the visual range calculated from teleradiometer measurements. Atmos. Environ. 16, 2323–2333.

Malm, W.C., Cismoski, S., Schichtel, B.A. Use of webcam images for quantitative characterization of haze, presented at the Air & Waste Management Association annual conference, Raleigh, June 22–25, 2015.

Marley, N., Gaffney, J., Baird, J., Blazer, C., Drayton, P., Frederick, J., 2001. An empirical method for the determination of the complex refractive index of size-fractionated atmospheric aerosols for radiative transfer calculations. Aerosol Sci. Technol. 34, 535–549.

Middleton, W.E.K., 1968. Vision through the Atmosphere (corrected edition). University of Toronto Press, Toronto, ON.

Moosmüller, H., Ravi, V., Arnott, W.P., 2005. Cavity ring-down and cavity-enhanced detection techniques for the measurement of aerosol extinction. Aerosol Sci. Technol. 39, 30–39.

Moosmüller, H.R., Chakrabarty, K., Arnott, W.P., 2009. Aerosol light absorption and its measurement: a review. J. Quant. Spectrosc. Radiat. Transfer 110, 844–878.

NOAA (National Oceanic and Atmospheric Administration), 1982. An Overview of Applied Visibility Fundamentals: Survey and Synthesis of Visibility Literature. FCM-R3-1982. NOAA, Washington, DC.

NOAA (National Oceanic and Atmospheric Administration), 1998. Automated Surface Observing System: ASOS User's Guide. NOAA, Washington, DC.

Petzold, A., Ogren, J.A., Fiebig, M., Laj, P., Li, S.-M., Baltensperger, U., et al., 2013. Recommendations for reporting "black carbon" measurements. Atmos. Chem. Phys. 13, 8365–8379.

Poduri, S., Nimkar, A., Sukhatme, G.S. Visibility monitoring using mobile phones, in Annual Report: Center for Embedded Networked Sensing, 125-127, University of California, Los Angeles, http://research.cens.ucla.edu/about/annual_reports/CENSAnnualReport2010.pdf, 2010.

Pokhrel, R., Lee, H., 2011. Algorithm development of a visibility monitoring technique using digital image analysis. Asian J. Atmos. Environ. 5 (1), 8–20.

Richards, L.W., 1988. Sight path measurements for visibility monitoring and research. J. Air Pollut. Control Assoc. 38 (6), 784–791.

Richards, L.W., Stoelting, M., Hammarstrand, G.M., 1989. Photographic method for visibility monitoring. Environ. Sci. Technol. 23, 182–186.

Richards, L.W., Dye, T.S., Arthur, M., Byars, M.S., 1996. Analysis of ASOS Data for Visibility Purposes, Final Report STI-996231-1610-FR. Systems Applications International, Inc, San Rafael, CA.

Seigneur, C., Hogo, H., Johnson, C.D., 1984. Comparison of teleradiometric and sensitometric techniques for visibility measurements. Atmos. Environ. 18 (1), 223–227.

Taha, G., Box, G.P., Cohen, D.D., Stelcer, E., 2007. Black carbon measurement using laser integrating plate method. Aerosol Sci. Technol. 41, 266–276.

Trijonis, J.C., 1982a. Existing and natural background levels of visibility and fine particles in the rural East. Atmos. Environ. 16, 2431–2445.

Trijonis, J.C., 1982b. Visibility in California. J. Air Pollut. Control Assoc. 32, 165–169.

U.S. EPA (Environmental Protection Agency). Visibility Monitoring Guidance Document, EPA-454/R-99-003, EPA Office of Air Quality Planning and Standards, Research Triangle Park, NC, June 1999.

Waldram, J.M., 1945. Measurement of the photometric properties of the upper atmosphere. Q. J. Roy. Meteor. Soc. 71, 319–336.

Wheeler, M.D., Newman, S.M., Orr-Ewing, A.J., Ashfold, M.N.R., 1998. Cavity ring-down spectroscopy. J. Chem. Soc., Faraday Trans 94, 337–351.

WMO (World Meteorological Organization), 1996. Guide to Meteorological Instruments and Methods of Observation, Sixth Edition, WMO-No. 8. World Meteorological Organization, Geneva.

Xie, L., Chiu, A., Newsam, S., 2008. Estimating atmospheric visibility using general purpose cameras, in *Advances in Visual Computing*. In: Bebis, G., et al., (Ed.), Lecture Notes in Computer Science (LNCS). International Symposium on Visual Computing, Las Vegas, pp. 356–367.

8

History of Visibility as an Esthetic Concern in the United States

The evolution of an explicit appreciation of the relationship between haze and the ability to see and appreciate scenic landscape features is a recent development starting in the late 1960s and 1970s. The development of a regulatory structure to protect prominent scenic vistas in the United States started with the passage of the Clean Air Act Amendments of 1977 (see Appendix 2). The development of regulations to protect visibility was a long and arduous process. Assessments of the Clean Air Act and associated regulations can be found in a review of the visibility protection program (Patton, 2004) and more generally in a review of National Ambient Air Quality Standards by Bachman (2007). The goal of this chapter is to highlight important features of the Clean Air Act and visibility regulations and, where possible, to present some of the background information and science issues that had to be addressed that are not covered in the above-mentioned publications.

The initial drivers of air pollution regulations were not esthetic concerns such as visibility degradation or impairment but the concern for public health. The significant impact of air pollution on human health sparked scientific investigation into the relationship between anthropogenic emissions of various types and haze. In 1952, Haagen-Smit determined that automobile emissions in the form of nitrogen oxides and hydrocarbons and in the presence of ultraviolet radiation from the sun form the precursors to smog (Haagen-Smit, 1970). This initial understanding that human activity was causing haze and the health impacts of air pollution led to legislative action at the federal level. In 1955, the Federal Air Pollution Control Act provided for research and technical assistance to work toward a better understanding of the causes and effects of air pollution (Stern, 1982). In 1963, the first federal Clean Air Act (CAA) empowered the secretary of the Department of Health, Education, and Welfare to define deleterious air

Visibility: The Seeing of Near and Distant Landscape Features. http://dx.doi.org/10.1016/B978-0-12-804450-6.00008-5

quality levels to human health based on scientific studies. It also provided for grants to state and local air pollution control agencies to study air pollution cause and effect relationships. The CAA of 1963 was amended by the Motor Vehicle Air Pollution Control Act of 1965 so that the federal government can regulate and establish auto emission standards. Then in 1967 the Air Quality Act established a framework for addressing interstate transport and defined "air quality control regions" based on meteorological and topographical factors. For the first time, the federal government conducted extensive ambient monitoring studies and stationary source inspections. In 1969, an executive order established the Environmental Protection Agency (EPA) (Bachman, 2007).

The enactment of the CAA of 1970 authorized the development of comprehensive federal and state regulations to limit emissions from both stationary (industrial) sources and mobile sources. Under the 1970 CAA, the National Ambient Air Quality Standards (NAAQS), State Implementation Plans (SIPs), New Source Performance Standards (NSPS), and National Emission Standards for Hazardous Air Pollutants (NESHAPs) were established (Bachman, 2007).

All of this rather extensive regulatory activity was mainly directed at protecting human health. This early regulatory structure did not explicitly address esthetic impacts of air pollution such as visibility impairment. The legislative protection that indirectly had an effect on visual air quality under the 1970 CAA came in the form of secondary NAAQS, which were levels of air pollutants that were set to protect human welfare and environmental conditions.

VISIBILITY AND ENERGY DEVELOPMENT ON THE COLORADO PLATEAU

With the advent of energy development in the 1960s and 1970s on the Colorado Plateau, the ability to see and appreciate the many unique scenic resources of the area became threatened or compromised. The beginning of the industrialization of the Colorado Plateau arguably began with the building of Hoover Dam. As early as the late 1890s, the Colorado River was diverted to Imperial Valley, California, for irrigation purposes. However, it was not until the 1930s that the decision to build an "arch-gravity dam" for purposes of irrigation and power generation was made. The site chosen for the damming of the Colorado River is just east of Las Vegas near Boulder City, Nevada, a town created for the purpose of supplying the housing and infrastructure needs of the dam building. The dam was to be 660 feet thick at its base and 45 feet at the top. Over 3,250,000 cubic yards of concrete were poured from June 6, 1933, to September 30, 1935, forming what became known as Hoover Dam. The reservoir formed behind the dam is referred to as Lake Mead (Stevens, 1990).

While Hoover Dam itself did not contribute to visibility impairment, the Colorado River dams that followed created reservoirs that could be used for the operation of coal-fired electric power generating facilities, whose emissions contribute to visibility impairment. Reservoirs also allowed for the growth of urban areas and increased recreational activity, all of which result in increased emissions that affect visibility. Davis Dam, built in 1951, created Lake Mohave, which enabled the siting of the Mohave Generating Station in Laughlin, Nevada, about 80 miles downriver from the west end of Grand Canyon National Park (NP). The Mohave Generating Station consisted of two coal-fired, 818-megawatt (MW) units supplied by a coal slurry pipeline that was 18 inches in diameter and 275 miles long. Construction started in 1967 and was completed in 1971. It emitted 40,000 tons of sulfur dioxide (SO_2) per year, or about 4.56 tons of SO_2 per hour. If all the SO_2 were converted to ammonium sulfate, 4.56 tons of SO_2 would correspond to about 9 tons of ammonium sulfate per hour. In addition, the electrostatic precipitators installed on the units were only about 70% efficient, so that under a full load there was an additional 24 tons/h of primary particulate matter or fly ash emitted. Both fly ash and sulfate particles are efficient at reducing visibility (Geue, 1971).

Because of the excessive silt carried by the Colorado River, there was a concern that Lake Mead would rapidly suffer from silt buildup. A second dam was proposed up-river of Hoover Dam and above the Grand Canyon to capture silt and extend the life of Lake Mead. The project's stated main purpose was to better utilize the allocation of the Colorado River water within the river basin. After much haggling between environmentalists, the government, and certain industry groups, they settled on a large dam at Glen Canyon, located just above Lee's Ferry. It was first proposed in 1950, with construction starting in 1956. It was not completed until 1966. The reservoir behind the Glen Canyon Dam is known as Lake Powell (Hundley, 2009; Summitt, 2013).

As the diversion tunnels around Glen Canyon were closed to begin the formation of Lake Powell, a renewed effort to secure Arizona's Colorado River water rights and to address concerns with depleting groundwater use in southern Arizona found the U.S. Bureau of Reclamation proposing another dam along the Colorado River at Marble Canyon, immediately east of the then-current boundary of Grand Canyon NP. Environmentalists raised large concerns over building yet another dam along the Colorado River, particularly so close to the park. The proposed dam was finally abandoned in 1968. Marble Canyon was later incorporated within the boundary of Grand Canyon NP to preserve it from future development (Summitt, 2013).

As with the Hoover Dam, Glen Canyon Dam does not directly contribute to visibility impairment, but it did form the basis for population expansion in the area. Originally a construction city, Page, Arizona, has become a hub for recreational activity and a retirement community. There were substantial coal reserves in the central region of the Navajo Nation and on neighboring Hopi Reservation lands, and the impounding of water behind Glen Canyon Dam gave a readily available source of water. Given the availability of coal and water, a cooperative arrangement between the U.S. Department of the Interior, the City of Los Angeles Department of Water and Power, Nevada Power, the Salt River Project, Arizona Public Service, and Tucson Electric Power proposed another coal-fired power plant known as the Navajo Generating Station (NGS) (Zuniga, 2000). To that end, a coal mine on the northern Black Mesa was leased and developed, an electric rail line from the mine area to the power plant site was built, and the plant was constructed.

The NGS consists of three 750 MW units burning about 8,000,000 tons of coal a year, or about 22,000 tons per day. Its construction began in April 1970, with completion of Units 1, 2, and 3 in 1974, 1975, and 1976, respectively. Together they emit on the order of 71,000 tons of SO_2 per year, which translates into about 390 tons of ammonium sulfate per day. If one assumes 99% control for primary particles, the plant would emit another 22 tons of fly ash per day. While the Mohave Generating Station was built at the west end of the Grand Canyon, NGS was located on the east end of the canyon, just a few miles from Marble Canyon and about 60 miles from Desert View, where the great chasms of the Grand Canyon start (Bechtel Corporation, 1970).

Another large coal-fired power plant complex on the Colorado Plateau, the Four Corners Power Plant, had five units, with Units 1 and 2 completed in 1963, Unit 3 in 1964, Unit 4 in 1969, and Unit 5 in 1970. The five units produced a total of 2040 MW. When the Four Corners plant was initially built, it did not have electrostatic precipitators, so with coal that has 10% ash content, the plant would have emitted on the order of 720,000 tons of particulates per year, or about 2000 tons per day. Particulate controls as well as early controls for SO_2 were placed on the plant in the 1980s. SO_2 removal efficiency was improved on Units 4 and 5 during the 1990s.

The nearby San Juan Generating Station has four units and produces about 135 tons of sulfur dioxide per day, or 270 tons of ammonium sulfate. The Hunter coal-fired power plant facility, in central Utah, has a combined capacity of 1472 MW, and the nearby Huntington plant has a capacity of 960 MW (Vanden Berg, 2010). Their emissions can be approximately scaled to that of other coal-fired power plants on a per-MW basis (U.S. EPA, 2009).

Fig. 8.1 is a map showing the locations of the coal-fired power plants discussed above, as well as other coal-fired power plants on and around the Colorado Plateau. The size of the red circles is scaled to the annual mass of CO_2 emitted from each facility, which is proportional to annual SO_2 and NOx emissions. Also shown are the locations of the vast coal fields

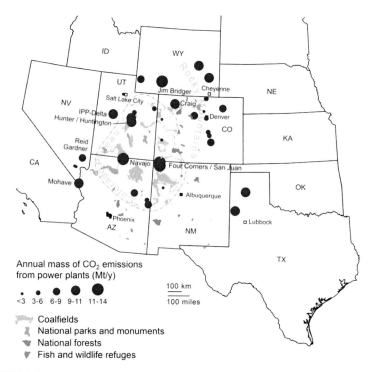

FIGURE 8.1   Map showing the location of coal fields and coal-fired electric power generating facilities on the Colorado Plateau.

used as sources of fuel to fire the power plants. Scenic national parks, national forests, and fish and wildlife refuges are shown in green, black, and blue, respectively. The figure highlights the proximity of coal-fired power plants and their emissions to some of the most scenic federal lands found in the United States.

With all the energy development and associated emissions in the late 1960s and 1970s, there became a general feeling by the residents who lived on the Colorado Plateau that the scenic views in the many national parks in the region were degrading. The author of this book and colleagues many times stood on the rim of the Grand Canyon and wondered what were the relative contributions of near and distant sources to the hazes observed in the canyon. Speculation was that it was local windblown dust, fire, and, of course, nearby coal-fired power plants.

Reflecting citizens' concerns over perceived degradation of scenic quality, a number of scientists attempted to make measurements of air clarity. Hall et al. (1975) reported on what was most likely the first measurements of atmospheric extinction on the Colorado Plateau. They designed and built a transmissometer capable of accurately measuring extinction in a nearly particle-free environment. They made observations over a period of nine nights in October 1973 at a site near Lowell Observatory, Flagstaff, Arizona, a town that is about 80 miles due south of the Grand Canyon. They reported on atmospheric conditions that were not distinguishable from a Rayleigh or a particle-free atmosphere on only one of those nights.

Discussions between the scientists at Lowell observatory and faculty in the physics department at Northern Arizona University (NAU) led to additional measurements of atmospheric extinction and other air-quality-related variables in the Grand Canyon by faculty and students in cooperation with staff at Grand Canyon NP. These measurements included the transmission properties of the atmosphere, yielding a measure of visibility (O'Dell and Layton, 1974), particulate and gas concentrations using high-volume air samplers and gas bubblers, and meteorological variables (Malm, 1974). Another visibility monitoring study, which began in 1976, was carried out near Cedar Mountain, Utah. The monitoring site, which was north of several national parks in southeastern Utah, was initiated by the National Oceanic and Atmospheric Administration (NOAA) and the EPA (Allee et al., 1978).

There was also concern about degraded water quality and the so-called population-boom/bust cycles associated with energy development projects. To that end, a consortium of universities and associated principal investigators developed a project that became known as the Lake Powell Research Project. It was funded by the National Science Foundation (NSF) Research Applied to National Needs (RANN) program and began in 1971 (Anderson, 1971). The project was many faceted in that it addressed the impact of energy development on the social structure of the Navajo Indian Nation, water quantity and quality issues, ecological impacts and, of course, air quality. Air quality was addressed through atmospheric transport and chemistry model development and extensive measurement programs. Air quality measurements started in the winter of 1973, with most measurements ending in 1975. Measurements included particulate and gas concentrations and visibility, including optical depth measurements. Visibility was measured with a camera fitted with a telephoto lens. Contrast of distant landscape features was used to derive visual range. An integrating nephelometer, an instrument designed to measure the atmospheric scattering coefficient (see Chapter 7), and human-observed visual range were also used to estimate visibility (Walther et al., 1977). The study culminated in a report titled "The Excellent but Deteriorating Air Quality in the Lake Powell Region." It was reported that there was a general decrease in visual range over the course of 3 years from 128 km to 115 km, with wintertime visual range being the highest and summer the lowest. As a reference, an atmosphere free of particles would yield a visual range of about 390 km. The conclusion was that the haze was primarily regional in nature, with some local influence associated with the population fluctuation of the local construction town of Page (Walther et al., 1977).

VISIBILITY IS FORMALLY PROTECTED

While scientists were making measurements of various air pollutants and visibility, environmental groups were championing the "visibility cause" in the legislative halls of Washington, D.C. Gordon Anderson (Don't Forget Friends of the Earth, letter to the editor, High Country News, May 16, 1994) writes,

> "As the former Colorado Plateau regional representative over a 10-year period (1974–1984) of Friends of the Earth, … if I am to believe the letter in my files from the congressional staffer most intimately involved with this subject, the issue of visibility impairment caused by coal-fired power plants in the Colorado Plateau initially came to the attention of federal regulators in 1975, when they saw a slide presentation I'd developed, 'Visibility Degradation in the Southwestern Parklands.' It focused on the impacts to regional visibility caused by the Navajo Generating Station.
>
> Seven years later, a TV audience of millions watched a PBS documentary, "The Regulators, Our Invisible Government," which traced the history of the 1977 Clean Air Act Amendments by following the development of the visibility issue through the legislative and federal regulatory processes. While that film made for good television by enhancing my role of photographic documentarian to "former park ranger goes to Washington on a mission from God to save the national parks," the show overlooked the most important aspect of the whole story. It was Friends of the Earth's clean air lobbyist, Rafe Pomerance, who conceived of and wrote a visibility protection amendment. He also arranged for me to speak to key Washington, D.C., audiences on behalf of the canyon country as he steered the amendment through Congress and later led the charge to block efforts to undermine it during the regulatory process.
>
> The Visibility Protection Amendment (Section 169A) of the Clean Air Act of 1977 and its subsequent regulations were specifically written to reduce sulfur dioxide emissions at the Navajo Power Plant. Any later accomplishment toward that end owes an enormous debt of gratitude to the vision and leadership of Rafe Pomerance in providing the legal foundation for that opportunity."

After the enactment of the CAA of 1970, the next milestone in the progression of air quality legislation was the CAA amendments of 1977, which for the first time codified visibility protection, an esthetic value, in certain protected areas referred to as Class I federal areas. The 1977 CAA amendments included visibility-specific protections in a new section labeled "169(A)" (CAA § 169A). This section was closely linked with other new sections of the CAA that codified the "prevention of significant deterioration" regulatory policies developed by the EPA in the mid-1970s (40 CFR 51.166, Prevention of Significant Deterioration of Air Quality). Under that policy and subsequently the CAA, a geographic classification system was established for areas that met the NAAQS. States were given the ability to afford special protection to areas that they considered worthy of maintaining extremely good air quality by designating them Class I. Such areas would only be allowed slight degradation of air quality, among other attributes. All areas of the country with better air quality than NAAQS levels would be established as Class II areas, which allowed for modest amounts of air quality degradation from new point and area air pollution sources. States could also designate Class III areas in which air quality would be allowed to degrade up to the levels of the NAAQS. The CAA amendments of 1977 specified 158 mandatory federal Class I areas in Section 162(B) of the CAA (CAA § 162B). For the purpose of protection of visual air quality within the 158 mandatory Class I areas, it was determined that in 156 of those areas, the inherent scenic qualities were consistent with visibility protection as defined under Section 169(A).

To date, there have been only a handful of redesignations from Class II to Class I under CAA § 164, mainly on tribal lands. Redesignated Class I areas obtain protection under New Source Review programs but are not required to meet the visibility goal of natural visual air quality under Section 169(A) (Reitze, 2001).

Areas designated as mandatory Class I are international parks, national wilderness areas, and national memorial parks that exceed 5000 acres in size, and national parks that exceed 6000 acres in size that were in existence on August 7, 1977. The distinction between 6000 acres for national parks and 5000 acres for other protected federal areas was due to negotiations with Representative James V. Stanton (OH-D, District 20), who would oppose the CAA amendments of 1977 if the emerging Cuyahoga Valley National Park would be designated as a mandatory federal Class I area (Patton, 1998). Fig. 8.2 is a map showing the location of all Class I areas.

All national parks on the Colorado Plateau were designated as Class I; however, many more areas were also recognized to have scenic value worth protecting. Notice that most of the Class I areas are in the western United States; there are very few in the Midwest, and 14 states are devoid of any Class I areas.

CAA § 169(A), part 1, states that "Congress hereby declares as a national goal the prevention of any future, and the remedying of any existing, impairment of visibility in mandatory Class I Federal areas in which impairment results from manmade air pollution" (see Appendix 2 for the visibility protection provisions of the 1977 CAA amendments). Parts 3A, 3B, and 3C state that

"Not later than eighteen months after August 7, 1977, the Administrator shall complete a study and report to Congress on available methods for implementing the national goal set forth in paragraph (1). Such report shall include recommendations for:

(A) methods for identifying, characterizing, determining, quantifying, and measuring visibility impairment in Federal areas referred to in paragraph (1), and
(B) modeling techniques (or other methods) for determining the extent to which manmade air pollution may reasonably be anticipated to cause or contribute to such impairment, and
(C) methods for preventing and remedying such manmade air pollution and resulting visibility impairment.

Federal Mandatory Class I Areas

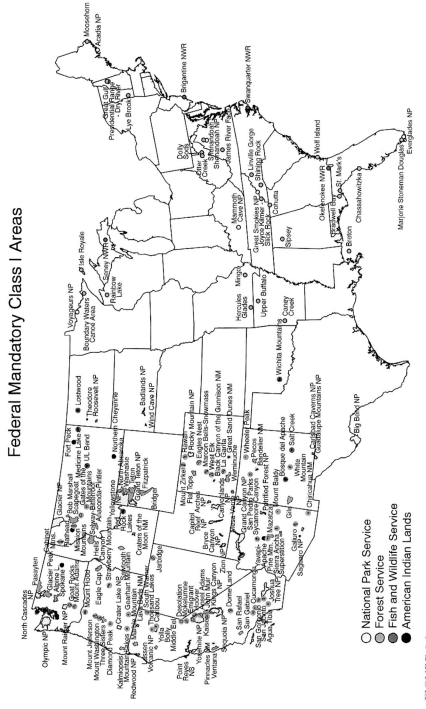

○ National Park Service
◐ Forest Service
◕ Fish and Wildlife Service
● American Indian Lands

FIGURE 8.2 Map of federal Class I areas. Most of the mandatory Class I areas are found in the western United States.

Such report shall also identify the classes or categories of sources and the types of air pollutants which, alone or in conjunction with other sources or pollutants, may reasonably be anticipated to cause or contribute significantly to impairment of visibility."

Some guidance as to what constitutes visibility impairment is in CAA § 169(A), part (g), Definitions (6), which states "the terms 'visibility impairment' and 'impairment of visibility' shall include reduction in visual range and atmospheric discoloration."

Prior to the CAA amendments of 1977, only a few interested parties were investigating possible visibility impairment associated with energy development and population growth on the Colorado Plateau. The amendments gave a legislative mandate to identify those Class I areas where visibility was an important value and to understand the relationship between emissions, transport, and transformation of those emissions into haze and the extent and severity of that haze as it relates to all Class I areas.

As a consequence of the 1977 CAA amendments, many air quality studies were initiated across the United States, and most of those studies had a visibility component as part of their overall objective. A table listing these studies prior to 1990 and their objectives can be found in Trijonis et al. (1990) and an updated list of longer-term studies in Table 8.1.

VISIBILITY MONITORING

A number of authors used the National Weather Service airport observations of visual range data to track changes in haziness over time. Identifying a dark feature against a background sky that is just visible establishes the visual range. Husar et al. (1981) and Trijonis et al. (1990) showed temporal and seasonal trends in haziness from 1948 to 1975. While these analyses were able to show dramatic increases in haziness over time in both the eastern United States and California, the combination of haziness changes and appropriate landscape features at distances sensitive to changes in haziness did not work well for the interior of the western United States.

Little information on visibility changes on the Colorado Plateau existed prior to 1977. A few measurements of extinction were made by Hall et al. (1975) and by O'Dell and Layton (1974). Two of the very first spatially comprehensive studies carried out in the interior western United States were initiated by the EPA Environmental Monitoring Systems Laboratory (EMSL) lab in Las Vegas, Nevada. The first, designated as the Visibility Investigative Experiment in the West (VIEW), started in March 1978 and included teleradiometer measurements of visibility in

TABLE 8.1   Regional, Long-Term Aerosol and Visibility Monitoring Programs

Program	References	Location	Time of Operation	Information Collected	Purpose
National Weather Service (NWS) Airport Visibility	Trijonis and Yuan, 1978; Trijonis, 1979, 1982b; Husar et al., 1981	Airports throughout the country	1918–2004	The farthest landscape feature that can be seen	Assess visibility conditions primarily for aircraft operations. Used by others to assess visibility trends.
National Weather Service (NWS), Federal Aviation Administration (FAA), Department of Defense (DOD)	Bradley and Imbembo, 1985	Airports and other locations	1991–present	Visibility is reported but based on a forward scatter measurement	Aircraft operations and climatological trends
National Air Surveillance Network (NASN)	U.S. EPA, 1976	Urban and rural U.S.	1956–1975	Total Suspended Particles (TSP)	General health concerns of public
Sulfate Regional Experiment (SURE)	Mueller and Hidy, 1983	54 stations, primarily in the Northeast	1977–1978	TSP, fine and coarse fractions, mass, ions, and elements	Focus on sulfate aerosols and their emission sources
Visibility Investigative Experiment in the West (VIEW)	Snelling et al., 1984	14 sites, primarily on Colorado Plateau	1978–1987	Visibility data primarily in the form of contrast measurements of landscape features	Characterize visibility in western United States

(Continued)

TABLE 8.1 Regional, Long-Term Aerosol and Visibility Monitoring Programs (*cont.*)

Program	References	Location	Time of Operation	Information Collected	Purpose
Eastern Regional Air Quality Study (ERAQS)	Mueller and Watson, 1982; Tombach & Allard, 1983	Nine nonurban in northeastern United States	1978–1989	Aerosol and visibility; TSP, fine and coarse fractions, mass, ions, and elements. b_{sat}, b_{ext}, and photography	Characterize visibility and air quality in northeastern United States
Inhalable Particle Study (IPA)	Watson et al., 1981	163 rural and urban sites in United States	1978–1984	TSP, fine and coarse fractions, mass, ions, and elements	Characterize inhalable particulates
Western Particle Characterization Study (WPCS)	Cahill et al., 1984	40 nonurban sites in western United States	1979–1981	TSP, fine and coarse fractions, mass, and elements	Characterize background aerosol levels in western United States
Western Regional Air Quality Study (WRAQS)	Macias et al., 1987	11 nonurban sites in western United States	1981–1982	Aerosol and visibility; TSP, fine and coarse fractions, mass, ions, and elements; b_{sat}, b_{ext}, and photography	Characterize background aerosol and visibility levels in western United States
Ohio River Valley Study (ORV)	Shaw and Paur, 1983	Three rural sites in Ohio River valley	1980–1981	Fine and coarse particulate mass and elements	Characterize particulate levels in the Ohio River valley
Research on Operations Limiting Visual Extinction (RE-SOLVE)	Trijonis and Pitchford, 1986; Trijonis et al., 1988	Seven sites in California Mojave Desert	1983–1985	Aerosol and visibility; TSP, fine and coarse fractions, mass, ions, and elements. b_{sat}, b_{ext}, and photography	Dept. of Defense concerns over flight operations limiting visibility conditions in the R-2508 military air space

Program	Reference	Location/Sites	Time Period	Measurements	Objectives
Subregional Cooperative Electric Utility, Department of Defense, National Park Service, and EPA Study (SCENES)	McDade and Tombach, 1987	Eleven rural and remote sites in southwestern United States	1984–1989	Aerosol and visibility; TSP, fine and coarse fractions, mass, ions, and elements; b_{scat}, b_{ext}, and photography	Understand the source of visibility impairing aerosols on the Colorado Plateau primarily the Grand Canyon
Interagency Monitoring of Protected Visual Environments (IMPROVE)	Joseph et al., 1987; Malm et al., 1994	Initially 20 sites in Class I areas, expanded to 177 sites across United States	1987–present	Aerosol and visibility; fine and coarse fractions, mass, ions, elements, organics, light-absorbing carbon, b_{scat}, b_{ext} and scene monitoring	Establish baseline visibility levels, track progress toward Regional Haze Rule goal of natural conditions, and identify causes and sources of visibility impairing particles
Chemical Speciation Network (CSN)	Rao et al., 2003; Malm et al., 2011	Approximately 200 urban sites across United States	2000–present	Aerosol and visibility; fine and coarse fractions, mass, ions, elements, organics, light-absorbing carbon, b_{scat}, b_{ext} and scene monitoring	Track progress of emission reduction strategies through characterization of trends, validation of air quality modeling and source apportionment activities, support of regulatory efforts such as the Regional Haze Rule, and support of health effects and exposure studies

14 national parks in the Southwest, including parks on the Colorado Plateau. The contrast of a number of scenic vistas at four wavelengths (colors) was measured using a telescope fitted with a detector and four wavelength-dependent filters, three times a day (9:00 am, 12:00 pm, and 3:00 pm). These contrast levels can, with a number of limiting assumptions, be used to estimate atmospheric extinction and visual range (see Chapter 3). Also, a photograph was taken of each landscape feature at the time of the contrast measurement. Some of the pictures showing various types of visibility impairment presented in Chapter 1 were taken during this early study (Snelling et al., 1984).

Fig. 8.3 shows the results of the very first quantitative visibility measurements made in the fall of 1978. The park units with the best visibility were Capitol Reef NP, Bryce Canyon NP, Grand Canyon NP, Chaco Culture National Historical Park, Canyonlands NP, and Navajo National Monument. Visual ranges at these National Park Service (NPS) units were in excess of 200 km (Snelling et al., 1984).

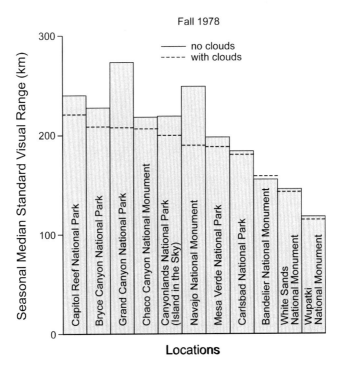

FIGURE 8.3   Geometric mean standard visual range of 11 national parks and monuments during fall 1978 on cloudy and cloudless days, ranked in descending order.

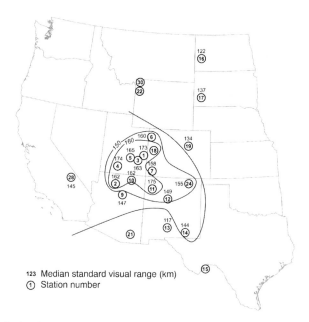

123 Median standard visual range (km)
① Station number

FIGURE 8.4   Median standard visual range (km), summer 1978 through fall 1979.

The median standard visual range for the first year of data is shown in Fig. 8.4. The visual range on the Colorado Plateau was about 160–170 km, which represents the most transparent or clearest air in the United States. This initial finding in 1978 for the most part is still true today, although there are some high mountainous sites in the middle and upper Rocky Mountains and Alaska that have slightly better visibility (Malm et al., 2004).

The teleradiometer network was expanded to include those sites shown in Fig. 8.5 (Malm and Molenar, 1984). Fig. 8.5 shows isopleths of standard visual range (SVR) over the western United States for the summer of 1982. Fig. 8.5 is similar to Fig. 8.4, showing about the same visual range of 160 km to near 180 km on and near the Colorado Plateau. Southern Arizona, New Mexico, and the Front Range area of the Rocky Mountains have SVR = 140 km. The lowest visual range is found in California near the populated urban areas and in areas where there is significant industrial and agricultural activity.

Data on air quality routinely collected in the regions of greatest concern for visibility protection (i.e., where energy-related developments are in otherwise pristine areas) were for the most part not germane to the problem of attributing the haze to specific particulate species. Most existing air monitoring sites were in urban areas subject to local pollution sources. Little information was available on particulate size or composition, factors important both for evaluating natural and man-made contributions to particulate matter and their potential impacts on visibility.

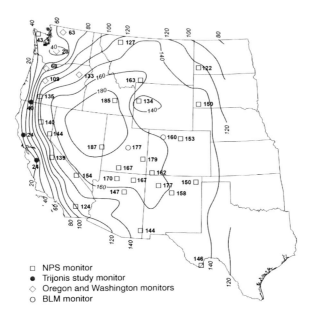

FIGURE 8.5   Standard visual range (SVR) isopleths for the western United States. The symbols correspond to the location of monitoring sites operated by different agencies, while the adjacent numbers are the average SVR values for that monitoring site. □ corresponds to the NPS monitoring networks, ● indicates data from Trijonis (1982a), ◇ represents data gathered by the states of Oregon and Washington, and ○ indicates Bureau of Land Management (BLM) monitoring sites. Trijonis data were obtained from airport human-observer observations, while the BLM and the states of Washington and Oregon used teleradiometers of the same design as those used by the NPS.

Monitoring data for particles were generated on a one-day-in-six protocol that hindered both statistical studies and correlations with meteorology. Finally, relatively few particulate monitoring sites existed in the Colorado Plateau area.

To address this issue, the EPA's VIEW program included a 40-site particulate monitoring network, referred to as the Western Particulate Characterization Study (WPCS), which was deployed starting in August 1979. The sampling stations were located in eight western states: Arizona, New Mexico, Utah, Colorado, Wyoming, Montana, North Dakota, and South Dakota. Stacked filter unit samplers were used in the study to collect both coarse (diameter 15–2.5 μm) and fine (diameter <2.5 μm) particles. The sample was an integrated 72-h sample that was collected twice weekly. Particle-induced X-ray emission (PIXE) analyses were used to estimate elemental concentrations greater in atomic weight than sodium. Samples were also analyzed gravimetrically for mass (Cahill et al., 1984).

Coarse particle concentrations were generally greater than fine particle concentrations. The former are dominated by soil-related elements. Unlike coarse particle parameters, fine mass and fine sulfur concentrations are present in regional patterns on a scale that is easily seen by the monitoring network. It is now known that fine particles, within the WPCS monitoring region, are composed primarily of ammoniated sulfate, organic compounds, and soil-related materials (Malm et al., 2004).

Fig. 8.6 shows the initial elemental sulfur data collected in the network. If sulfur were in the form of ammonium sulfate, the concentrations in Fig. 8.6 would be multiplied by 4.125. Unlike soil-related material, which exhibits concentrations at one or two monitoring sites that are quite different from nearby sites, fine sulfur has clear regional patterns, with elevated or depressed concentrations that extend over large regional sections of the eight-state network (Cahill et al., 1984).

These initial measurements led to an overall visibility monitoring strategy, as outlined in the EPA's Interim Guidance for Visibility Monitoring (U.S. EPA, 1980) and Visibility Monitoring Guidance (U.S. EPA, 1999). It was determined that three types of measurements were necessary to quantify the impairment of aerosols on scenic landscape features: an optical measurement is required to quantify the optical properties of an aerosol; particle measurements are necessary to understand which of the many particulate components are responsible for the visibility impairment; and finally, some type of camera system is needed to quantify the visual appearance of the impaired visibility. Visibility impairment was found to occur as uniform haze and as elevated and ground-based haze layers. Measurements of particle concentrations or their optical characteristics are not able to quantify the chemical and physical characteristics of distributions of haze, which are not uniform in space. However, simple photographic techniques can give some insight as to the layered haze's chemical and physical characteristics.

Also, during 1977–1979, there were intensive, short-term research studies in the Southwest, mainly focused on plume blight and plume chemistry with an eye toward improving plume visibility models. Most notable was the Visibility Impairment due to Sulfur Transport and Transformation in the Atmosphere (VISTTA) study funded by the EPA (Blumenthal et al., 1983; Swicker et al., 1983). The stated objectives of VISTTA were to determine the relationship between emissions and downwind optical properties of coal-fired power plants, determine anthropogenic and natural sources of visibility impairing aerosols, and provide data for the evaluation of plume visibility models. The valuable and detailed information from such studies is limited to relatively short periods of time and limited meteorological conditions. They did not provide information on seasonal factors or baseline values for trend analysis.

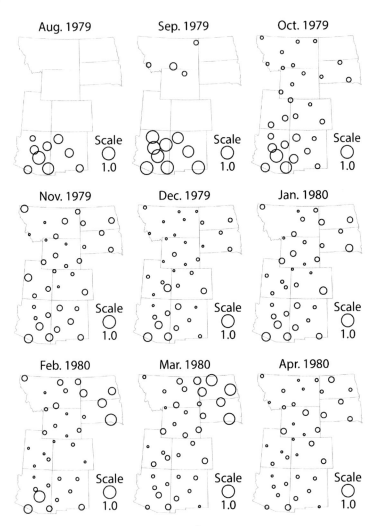

Concentration (micrograms /m³) Proportional to Diameter
Fine Sulfur Maps, 8-79 - 4/80

FIGURE 8.6   Circle size corresponds to the concentrations of elemental sulfur. The first measurements made in the Western Particulate Characterization Study were in August 1979.

Then, in the late 1970s and through the 1980s, over 20 shorter-term visibility and aerosol studies were carried out, many of them in urban areas (Trijonis et al., 1990). Los Angeles, Denver, Detroit, Dallas, and Houston were major urban areas where visibility as an aesthetic issue was addressed.

VISIBILITY MODELING

Subpart B of CAA § 169A, which called for modeling tools capable of predicting the visual appearance of haze resulting from emissions from industrial and/or mobile sources, stimulated the development of two types of models. Because emissions from coal-fired power plants resulted in constrained plumes (known as plume blight), there was an effort to develop models capable of simulating the dispersion and chemical conversion of emissions and the visual appearance of these pollutants as a function of sun–observer geometry. When single or multiple plumes become dispersed and mixed with the background aerosol, the appearance of this mixture appears as a uniform degradation of the scenic appearance of landscape features. From a visibility perspective, modeling of plumes and uniform hazes are uniquely different. In one case it is the appearance of the haze or plume that is of direct concern, while in the other case of uniform haze, the concern is primarily with the degradation of the scene itself, although viewing a scene through a plume will cause the landscape features behind the plume to appear visually degraded.

Fig. 8.7 outlines the submodules in a comprehensive visibility model. Emissions (box 1) are advected, dispersed, and chemically transformed from one gas and particle species (aerosol) to another (box 2). The visible appearance of the aerosol, as well as its effect on landscape features, depends on sun–observer–aerosol geometry and the inherent aerosol optical characteristics as well as the background aerosol concentration (box 3). Finally, the human observer views the plume or layered haze or the effect of the haze on the ability to see a landscape feature (box 4).

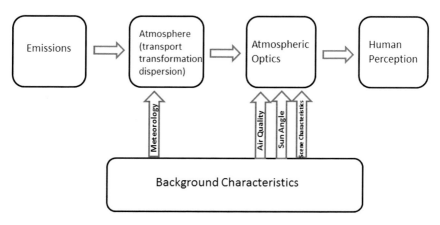

FIGURE 8.7   Components of a visibility model.

The first plume models developed are outlined in Latimer and Samuelsen (1978), Latimer et al. (1978), and Williams et al. (1979). These models were single source oriented and applicable to power plants or other large point sources and had Gaussian-based dispersion modules, simple SO_2- > SO_4 and NOx- > NO_3 chemistry, and first-order radiation transfer algorithms. The human perception indexes associated with the Latimer et al. model were reduction in visual range, contrast of the plume against some background at 0.55 μm (green) or blue (0.4 μm) to red (0.7 μm) ratio (an indication of plume coloration), and color difference between the plume and its background as represented by the color difference parameter (ΔE) (refer to Chapter 4 for more detail on these parameters). The Williams et al. model displayed the modeled appearance of the plume on some background landscape feature on a television monitor.

An example output of the Latimer et al. model for a well-controlled, 2000-MW coal-fired power plant is shown in Fig. 8.8. All parameters are displayed as a function of downwind distance from the source, and the viewing angle is perpendicular to the plume. The first graph shows the reduction in visual range, the second the blue-red ratio, the third plume contrast, and the final graph ΔE. The graphs labeled "East" and "West" correspond to background ambient visibility conditions of 15 km and 130 km, which are representative of eastern and western visibility conditions, respectively. The point of this graph was to show that when emissions are embedded in a hazy background, the plume is less perceptible than in a relatively clean or clear background.

The Latimer et al. model evolved and became known as the PLUVUE (Plume Visibility) model (U.S. EPA, 1992), and a screening tool, similar to PLUVUE but with simpler optics modules to approximate the visibility impact of single sources, known as VISCREEN (Visual Impact Screening), was developed (Latimer and Ireson, 1988).

Eventually, a federal land manager (FLM) working group in conjunction with the EPA came to recommend a Lagrangian variable-trajectory puff superposition model suitable for modeling the transport, transformation, diffusion, and removal of air pollutants from multiple point and area sources at transport distances beyond the range of conventional, straight-line Gaussian plume models (i.e., beyond 10–50 km) (Benkley and Bass, 1979). It allows for flow within and above the boundary layer and incorporates enhanced sulfate and nitrogen chemistry (U.S. EPA, 1993).

The photographic display technique was extended by Malm et al. (1983) and Molenar et al. (1994) to model the visual effect of haze on a variety of scenes throughout the United States. The model, WinHaze, incorporates aerosol and radiative transfer modules with image processing to synthesize images representing air quality and environmental conditions. It is

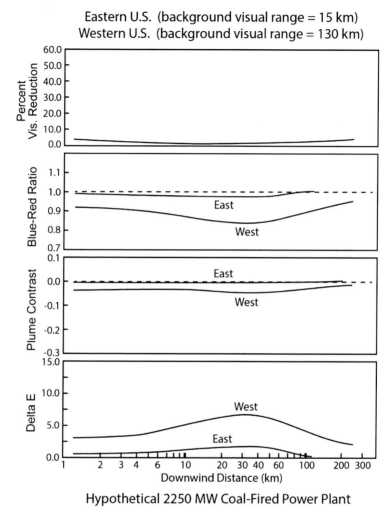

Eastern U.S. (background visual range = 15 km)
Western U.S. (background visual range = 130 km)

Hypothetical 2250 MW Coal-Fired Power Plant

FIGURE 8.8 Comparison of predicted plume visibility indexes for hypothetical eastern and western U.S. cases.

available at http://www.air-resource.com/resources/downloads.htm. This technique and model are discussed further in Chapter 6.

Beyond a few hundred kilometers, plumes and other sources become mixed and typically appear as uniform and, in many cases, regional haze. At the time of the passage of the CAA amendments of 1977, regional-scale model development was in its infancy. Wind field models were not capable of simulating winds in complex terrain such as the mountainous areas found in the western United States, and only very simple chemistry was included in these initial models. It is known that in cloud chemistry there is a mechanism for accelerated gas to particle conversion and particle growth, and none of the early models addressed this issue.

The first EPA report to Congress highlighted the following issues with respect to regional scale models (U.S. EPA, 1979):

> "Regional visibility models are extremely sensitive to the rate of conversion of SO_2 to sulfate and to the wind field, both of which are uncertain and must be assumed or interpolated from limited data.
> Among the additional limitations of regional models are:
>
> 1. Lack of adequate inventory of emission sources and base-line visibility. The visual impact of any source at any location is a function of existing visibility. Any model must be able to handle multiple sources. Urban areas in particular are difficult to characterize.
> 2. Incomplete knowledge of large-scale meteorological processes and uncertainties about boundary conditions. Chemical transformations apparently vary greatly and are largely unknown on regional scales.
> 3. The existing visibility database is insufficient for input to a model and too imprecise for validation of output."

VISIBILITY REGULATIONS

It is worth noting that the 1979 EPA report to Congress (U.S. EPA, 1979) was the first EPA report to obtain permission from the Government Printing Office to be printed with detailed color photography, due to the nature of the resource being protected. Among other things, it interprets the EPA's and states' responsibilities under Section CAA § 169A:

> "EPA must promulgate regulations that (a) provide guidelines to the States on appropriate techniques for implementing the national visibility goal through State Implementation Plans (SIPs) and (b) require affected States to incorporate into their SIPs measures needed to make reasonable progress toward meeting the national visibility goal. The regulations and guidelines must require that certain major stationary sources, likely to impair visibility, install best available retrofit technology (BART). The regulations must also require that the SIPs include a long-term (10–15 year) strategy for making reasonable progress toward the visibility goal. The long term strategy may require control of sources not otherwise addressed by the BART provision. The Act states that costs, energy and non-air environmental impact, and other factors must be considered in determining BART and reasonable progress.
> The language of the national visibility goal and the legislative history of the Act indicate that the national goal of Section 169A mandates, where necessary, control of both existing and new sources of air pollution. It is apparent, however, that adverse visibility impacts from proposed major new or modified sources are to be dealt with through the procedure for prevention of significant deterioration (PSD) mandated in

Section 165(d) of the Clean Air Act. In addition to the activities required of EPA and the States, the Clean Air Act requires that the Federal Land Managers (the Department of Interior and Department of Agriculture, through the National Park Service, the Fish and Wildlife Service, and the Forest Service) play an important role in visibility protection. Land Manager responsibilities include reviewing the adequacy of the state visibility protection strategies and determining whether proposed major air pollution sources have an adverse impact on visibility in Class I areas.

In establishing the national visibility goal, Congress called for explicit recognition of the value of visibility in special class I areas. By requiring consideration of "significant" impairment in BART decisions, "adverse" effects of proposed new sources, and "reasonable progress" in implementing the national goal, Congress has, in effect, mandated that judgments be made on the value of visibility in the context of specific decisions on control and location requirements for sources of visibility impairing air pollution. Preliminary economic studies of the value of visibility and research in recreational psychology and human perception support the notion that visibility is an important value in class I areas and suggest that several approaches are available for estimating the value of an incremental improvement or deterioration in visibility."

The report also brought up the notion of vistas outside the boundary of a Class I area but integral to the enjoyment of the area, or "integral vistas." It stated that

"The preliminary analysis confirms the notion that it is important to consider the impact of air pollution on visibility for vistas that extend beyond class I boundaries. Land Managers in over 90 percent of the class I areas, who provided detailed information on vistas, reported that one or more views from within the area looking outside the area may be, to some extent, important. Moreover, in some areas, these external views appear to be an integral part of the visibility experience in the area. For example, the view from Mesa Verde of Shiprock (New Mexico), a unique natural feature, was reportedly impaired regularly by power plant plumes. To exclude consideration of visibility impairment of this kind of vista appears contrary to the national goal."

In 1980, the EPA issued the first visibility regulations codified under 40 CFR 51.300–51.307. The reader is referred to Appendix 2 for definitions of terms, such as visibility impairment, that are used in CAA § 169A and the 1980 rule.

The 1980 rule characterized visibility impairment as either plumes or layered hazes that could be reasonably attributable to a single source or small group of sources and haze that was uniform and regional in nature and not attributable to one source or small group of sources. The rule addressed the first type of visibility impairment while putting off regional haze for future consideration. It required the 36 states with one or more Class I areas within their borders to develop a plan to address visibility impairment reasonably attributable to sources within their states by September 1981. The regulations required each SIP to have five components: development of a visibility monitoring plan, a review of BARTs for sources identified by states or FLMs to contribute to visibility impairment in one or more Class I areas, a long-term strategy for addressing possible future impairment from any source(s), provisions for FLM review of new or modified source permits, and incorporation of integral vistas.

The integral vistas provision allowed for protection of important views from within a Class I area to landmarks outside of the Class I area. The rule gave the FLM the authority to identify these vistas according to criteria developed by the FLM. The NPS proposed 170 integral vistas in 43 of their mandatory 48 Class I areas. They included such key landscape features as Shiprock, New Mexico, as seen from Mesa Verde NP, the 130-km-distant Navajo Mountain as seen from Bryce Canyon and the Grand Canyon, and the beautiful and scenic La Sal Mountains as viewed from Canyonlands and Arches national parks. In 1985, Secretary of the Interior under President Ronald Reagan, Donald Hodel, made the decision to not finalize the list. He announced in a news release dated October 25, 1985 that "designation of integral vistas would not be good for the parks" (Simon, 1988). Only Roosevelt Campobello International Park has FLM-protected integral vistas, and the states of Alaska and Washington identified integral vistas as part of their SIP development process.

The 1980 rule requires states to make reasonable progress toward natural conditions by reducing emissions from sources that are reasonably attributable to visibility impairment in Class I areas. "Reasonably attributable" refers to a source whose plume can be observed to enter into a Class I area or whose emissions can be attributed to visibility impairment by visual observation or other appropriate techniques, including new monitoring strategies.

IMPLEMENTATION ACTIVITIES

Photographs taken as part of project VIEW of several integral vistas at each of the 14 monitoring sites documented uniform and layered hazes. The film used was Kodachrome 25. The slides were archived, and representative images were digitized and are available at http://vista.cira. colostate.edu/improve/Data/IMPROVE/Data_IMPRPhot.htm (Chinkin et al., 1986). Analysis of these slides showed frequent layered hazes throughout the Colorado Plateau, especially during the winter months. The view of Navajo Mountain as seen from Bryce Canyon had occurrences of layered hazes approximately 70% of the time. The Navajo Generating Station (NGS) is just a few kilometers from Navajo Mountain. Its emissions are released into a mixed layer that is greater than the stack height but below the height of the mountain. It was hypothesized that much of the haze layers observed from Bryce Canyon were associated with NGS emissions. The same type of layered hazes was observed for the view of the Chuska Mountains as seen from Mesa Verde NP. The view is across the basin from where the Four Corners and San Juan coal-fired power plants are located. Even though there were also significant haze layers in or along the Green River, Colorado, and San Juan River basins, individual plumes were not observed to emanate from power plants and extend into any Class I area.

Eight-millimeter, color, time-lapse photography was also used in an effort to identify plumes or haze layers that were transported into Class I areas from known sources. This technique successfully identified emissions from a pulp and paper mill and an asphalt plant outside the Moosehorn National Wildlife Refuge that were transported into the refuge as highly perceptible, coherent plumes. This was the sole Class I area where the identification of visibility impairment that was attributable to a single source was accomplished by using simple monitoring techniques such as photography or human observation. More complex apportionment techniques and analyses of emission sources' contributions to visibility impairment were required, as described later.

The particle measurements made as part of project VIEW showed that there was sufficient sulfate aerosol associated with sulfur dioxide (SO_2) emissions to cause the haze layers that were observed. Analysis also showed that emissions of SO_2 from the coal-fired power plants on the Colorado Plateau were large enough to account for observed sulfate levels. However, because sulfate is a secondary particle formed from the invisible primary molecule, SO_2, the haze was not attributable to a single source through visual observation alone. Moreover, because integral vista protection was not supported by the FLM, in this case Secretary Hodel, the haze layer obscuring landscape features had to be observed within the boundaries of a park to be considered as a source of impairment that required the states to address possible emission reduction of responsible sources as part of the SIP development process.

If a state did not meet its requirement to develop a SIP for protecting visibility, the 1980 rule required the federal government develop a federal implementation plan (FIP) (CAA § 110(c)(1)). In 1982, the Environmental Defense Fund (EDF) sued the EPA for not developing Federal Implementation Plans (FIPs) for those states that had not met the requirement for a SIP to protect visibility (Environmental Defense Fund v. Reilly, No. C82-6850-RPA, N.D. Cal., 1984). From 1984 to 1991, the EPA, acting on behalf of 35 states, promulgated a series of FIPs to implement the 1980 visibility protection regulations. The first actions were to establish the legal basis for FLM review of new sources and major modifications to large industrial sources under the CAA § 162 New Source Review program and to establish a cooperative federal visibility monitoring program between the EPA and the federal land management agencies. The monitoring program was the beginning of the Interagency Monitoring of Protected Visual Environments (IMPROVE) network.

The EPA solicited input from the states and FLMs to certify visibility impairment in any mandatory federal Class I area as "reasonably attributable" to a source or group of sources. The Department of the Interior identified a number of possible sources affecting Class I areas. As a result of that action, the EPA agreed to an FIP schedule that in part stated that by August 31, 1989, they would identify any sources that caused humanly perceptible, reasonably attributable haze to a Class I area. On March 24, 1986, the FLM for the Grand Canyon and Canyonlands NPs certified visibility impairment at those parks that was probably associated with NGS emissions.

The NPS sponsored several modeling efforts to evaluate the possibility that the NGS is partially responsible for the haze. A wind field model was adapted for the area's terrain and winter meteorology to investigate transport and diffusion (Latimer and Samuelsen, 1978; Johnson et al., 1980; Latimer, 1980). In a separate study, ambient nitrogen chemistry was theoretically simulated to estimate the role of particulate nitrates (Latimer et al., 1986). Though these efforts added to the knowledge of the source of the haze, the uncertainties in modeling this situation led to approaching the question with observational studies.

The Subregional Cooperative Electric Utility, Department of Defense, NPS, and EPA Study (SCENES) monitoring program listed in Table 8.1 was part of the response to the need for developing other monitoring techniques to address the reasonable attribution question. The SCENES program consisted of 11 monitoring sites on and around the Colorado Plateau but with a focus mainly on Canyonlands, Mesa Verde, and Grand Canyon NPs and on the large Navajo, Mojave, and Four Corners coal-fired power plants as the primary sources of visibility-reducing particles (EPRI, 1986). The nearest large urban areas are over 300 km away. Only a few smaller urban areas or towns are within the area; these include Moab, Utah, Page, Arizona, and at the most-western end of the study area, Las Vegas, Nevada. There are a few small industrial enterprises in the vicinity such as sawmills, mining, and milling operations.

The stated objectives of SCENES were to, in part, analyze the relationships between haze, air quality, and meteorological conditions, determine the contribution to light extinction made by the types of material suspended in the atmosphere, and identify the sources of those materials. These objectives were to be carried out through a set of routine measurements and special studies. A 1986 special study involving aircraft, along with ground-based measurements, showed that the ground-based haze layers were widespread throughout the canyons and river basins on the Colorado Plateau. These findings were consistent with the extensive photography results of the VIEW monitoring program.

While comprehensive aerosol and visibility measurements were being made in the SCENES program, little progress was being made toward understanding specific source–receptor relationships. The frustration level was high at times among the consultants and participants in the various ongoing monitoring and research programs, leading to some caustic and insightful jingles (recollections of the author):

> We've all heard of the RESOLVE*, IMPROVE, and SCENES measurement programs, but what happened to our resolve to improve the scenes!!
> (Doug Latimer, Environmental Scientist, Boulder, Colorado)

*RESOLVE, Research on Operations Limiting Visual Extinction, was a defense department funded special study on visibility.

and,

> Here, locked in a windowless room,
> Like corpses inside a big tomb,
> We've dreamed of blue skies
> And unfettered eyes
> And who's doing what, and to whom.
> (Warren White, Data Analyst, Davis, California)

In early 1986 while at a SCENES working group meeting at Los Alamos National Laboratory, New Mexico, the author (Malm) was lamenting to Eugene Mroz, a researcher there, over the difficulty of how to establish a definitive and defensible link, if there was one, between coal-fired power plant emissions and visibility impairment within the boundaries of Class I areas. Grand Canyon and Canyonlands NPs are large parks with distant landscape features that exhibit some level of visibility impairment most of the time. It was clear that the impairment was due to emissions from a number of sources and from as far away as Los Angles and other distant urban and industrial areas (Pitchford et al., 1981; Blumenthal et al., 1983). NGS emissions were, on average, upwind of the Grand Canyon and would have to travel about 150 km to enter the canyon proper.

In response to these discussions, Mroz indicated that scientists at Los Alamos had been able to fabricate heavy methane, or deuterated methane (CD_4), which was being used as an atmospheric tracer molecule to test dispersion models. It is detected by mass spectrometry at levels of about one part in 10^{-16} by volume. Mroz indicated that it would be stable at stack temperatures found in coal-fired power plants such as NGS. This tracer material became the core of a reasonable attribution scoping study designed to inform as to the possibility of using tracers to assess the impact of NGS emissions on visibility in Grand Canyon and Canyonlands NPs (Malm et al., 1989, 1990).

It was recognized that the difficulty with using an inert tracer to assess visibility impairment is that the aerosol responsible for the impairment is secondary in nature. SO_2 gaseous emissions, which do not cause visibility impairment, are emitted into the atmosphere where they chemically transform into a sulfate particulate that is ultimately responsible for the scattering of light and reduction in visibility. Therefore the ratio of tracer to secondary particulate will change in time and space and vary as a function of atmospheric conditions such as the presence of oxidizing agents and aqueous aerosols. Hidy et al. (1985; Hidy, 1987) concluded that only massive sulfur isotope releases that can be measured in both their primary and transformed state ($SO_2 \rightarrow SO_4$) would be successful for attributing the primary emissions of sulfur dioxide to the secondary sulfate particle causing visibility impairment. However, the cost of implementing such a program was prohibitive. Therefore, it was decided to investigate how the release of an inert tracer from a large SO_2 source could inform the attribution problem.

The resulting scoping study designed to track the emissions from NGS to possible visibility impairment in Grand Canyon and Canyonlands NPs and Glen Canyon Recreation Area became known as the Winter Haze Intensive Tracer Experiment (WHITEX) (Malm et al., 1989). The study was carried out from January 7 through February 18, 1987. The measurement program consisted of four different types of ground station configurations and one airborne platform. The configurations are classified as major receptor, satellite, gradient, and background sites. Grand Canyon NP, Glenn Canyon Recreation Area (Lake Powell), and Canyonlands NP, classified as major receptors, had full characterizations of fine and coarse particles, along with trace element species; optical measurements, including still and time lapse photography; aerosol physical characteristics, such as size distribution; and, of course, CD_4 concentrations.

Other satellite and gradient sites had a subset of measurements designed to be consistent with the type of analysis used for the data collected at those sites. The purpose of background sites is self-explanatory; however, understanding the background CD_4 concentrations was key to using CD_4 as a tracer of NGS emissions. The analysis protocol called for exercising a number of different receptor and receptor hybrid models and reconciling any differences in the attribution of the NGS contribution to visibility impairment at the receptor sites. For details of the various analysis and modeling activities the interested reader is directed to http://vista.cira.colostate.edu/improve/Studies/WHITEX/Reports/FinalReport/whitexreport.htm, where the full WHITEX report is available, and to a list of publications included in the references at the end of this chapter.

Time-lapse photography in combination with tracer data was essential to understanding transport conditions and chemistry mechanisms that resulted in observed visibility impairment. The photo in Fig. 8.9a is a

FIGURE 8.9   (a) shows a cloud-filled air mass filling the Grand Canyon. The view is from Desert View Watch, located on the east side of Grand Canyon, looking east toward Navajo Mountain and the Navajo Generating Station. (b) is from the same vantage point, looking west down the length of the canyon.

frame capture of time-lapse photography taken on February 12, 1987, of the view from the east end of the Grand Canyon, looking to the east, with Page, Arizona, the NGS, and Navajo Mountain about 130 km distant. It shows clouds flowing down the Grand Canyon and below the rim, as water would fill a bathtub, from the direction of Lake Powell and the NGS, and ultimately filling the canyon with a cloud layer as far as the eye can see. Fig. 8.9b is a view in the opposite direction, looking to the west, down the length of the canyon.

Fig. 8.10 is a split-screen view of Fig. 8.9b; Fig. 8.10b shows how the canyon appeared after the clouds evaporated, leaving a dense haze, and Fig. 8.10a shows the canyon two days later after the haze had cleared out. The haze remaining after the evaporation of clouds (right) was primarily a sulfate haze, a haze that totally obscured canyon scenic elements only a few kilometers distant. The left side of the figure represents nearly particle-free air, or near-zero visibility impairment. During the time period of this extreme haze event, concentrations of CD_4 were near their maximum.

FIGURE 8.10   Split-screen view of the Grand Canyon as seen from Desert View Watch Tower, looking west. (a) was taken under near-Rayleigh, or particle-free, conditions, while (b) shows the canyon filled with a haze composed primarily of sulfate particles.

First-order calculations of the amount of sulfate that could be produced from NGS SO_2 emissions during this time period indicated that those emissions were sufficient to account for all the observed sulfate haze. Receptor modeling further suggested the NGS emissions were responsible for a large fraction of the observed haze during certain wintertime meteorological conditions. The study identified those synoptic meteorological conditions that were conducive to the flow of air masses "down-river," similar to water flow in the Colorado and San Juan river basins. Those conditions tended to occur once or twice a year during winter months and lasted about one to six days. On the other hand, during summer months the airflow is such that the NGS emissions are primarily transported to the northeast, away from the Grand Canyon but toward Canyonlands NP and beyond.

In April 1989, a draft of the WHITEX study was issued by the NPS, indicating that NGS emissions were reasonably attributable under certain wintertime conditions to significant and perceptible visibility impairment in Grand Canyon NP. Based on this report, the EPA initially attributed several episodes of wintertime visibility impairment at Grand Canyon NP to the NGS. The EPA solicited public comment on the merits of its preliminary attribution finding and began an informal rulemaking process to determine the appropriate action to be taken. In December 1989, the NPS released its final WHITEX report, affirming that NGS emissions contributed to visibility impairment in Grand Canyon National Park.

In 1990, Congress again amended the CAA, adding Section 169B, which required, among other things, that the EPA undertake research to identify visibility transport regions and create visibility transport commissions to address interstate transport issues as they relate to visibility impairment in Class I areas. However, until new regulations were promulgated, the 1980 rules were the law of the land. So, the issue of establishing reasonably attributable visibility impairment in the Grand Canyon from NGS emissions proceeded under the 1980 rule.

Because it was recognized that there were large uncertainties in the estimated fraction of sulfate in Grand Canyon during certain winter haze episodes that could be attributed to the NGS, and because the finding of reasonably attributable visibility impairment that was linked to the NGS emissions would result in significant costs associated with BARTs, the findings of the WHITEX study became quite contentious, with industry and the owners of the NGS lined up on one side and the EPA and NPS on the other. Through negotiations between the owners of the NGS and the EPA, it was established that if the owners were not accepting of WHITEX findings, they could carry out their own study. It is interesting to note that prior to WHITEX, the participants in SCENES showed little interest in carrying out an attribution study, but it only took a few months after the final WHITEX report was published for the same participants to formulate an attribution study that became known as the Navajo Generating Station Visibility Study (NGSVS) (Richards et al., 1991a,b).

The design of the NGSVS was similar to WHITEX but with more extensive measurements from January 10 through March 31, 1990. Four perfluorocarbon tracer species were used and were changed daily on a 4-day cycle. Tracer species were measured at 27 sites, along with SO_2, sulfur, and fine mass concentrations. The aerosols were integrated, 4-h samples. A network of 15 surface meteorological stations was operated along with measurements of upper air winds at 12 sites. The analysis plan called for exercising of receptor-oriented models along with deterministic chemical transport modeling.

Ironically, the unique meteorological conditions that transport NGS emissions into and below the rim of Grand Canyon during the presence of cloud layers did not occur during the NGSVS. Despite the fact that meteorological conditions that result in maximal impact of NGS emissions on visibility did not occur an analysis of NGSVS data suggested that sulfate from NGS emissions did exceed regional background by about 30% on approximately 13 days of the 90-day study period. Results of the study also suggested that rapid conversion of SO_2 to SO_4 required the passage of the plume through a fog or cloud and further verified that under dry wintertime conditions, conversion of SO_2 to SO_4 is slow, resulting in small particles that have a relatively low mass scattering efficiency. Despite the extensive measurement and analysis program associated with the NGSVS, there still remained large uncertainties as to the exact contribution of NGS emissions to visibility impairment in the Grand Canyon. It is of interest to point out that another sulfate apportionment study during the same time period also attributed NGS emissions to measured ambient sulfate levels at Canyonlands NP (Eatough et al., 1996).

Because of the difficulty in an exact attribution of NGS emissions to visibility impairment in the Grand Canyon, the National Academy of Sciences (NAS) was asked to review the WHITEX findings and any other information concerning transport of NGS emissions into the Grand Canyon that might affect the EPA's initial ruling of reasonably attributable impairment (National Research Council, 1993). The committee concluded that "at some times during the study period, NGS contributed significantly to haze in the GCNP; however, WHITEX did not quantitatively determine the fraction of sulfate SO_4 particles and resulting haze attributable to NGS emissions."

As a result of the WHITEX study and NAS review, there was a negotiated agreement between interested parties to control NGS SO_2 emissions at the 90% level based on a 30-day averaging period. In October 1991, the EPA issued its final determination that certain visibility impairment episodes at Grand Canyon NP were attributable to NGS emissions (40 CFR 52; 56 Federal Register at 50,172, 1991).

But things were not over yet! In January 1993, the Central Arizona Water Conservation District brought suit against the EPA, challenging the October 3, 1991, Final Rule, which imposed 90% SO_2 emission controls at the NGS. In March 1993, the U.S. 9th Circuit Court of Appeals decided the case (Central Arizona Water Conservation District v. United States Environmental Protection Agency, 990 F.2d 1531, 23 ELR 20678). The court ruled that the water districts did have standing to bring the challenge, but the court further ruled that the EPA's actions in promulgating the rule were reasonable and within its statutory and regulatory authority and were not arbitrary and capricious. In reaching its conclusion, the court addressed the issue of "reasonably attributable." The court wrote

"EPA has acknowledged that 'NGS is not the only source of visibility impairment' at the Grand Canyon, 56 Fed. Reg. at 50,177, and that regional haze also adversely affects visibility there.... Nonetheless, these mere facts hardly mean that EPA is without statutory authority to remedy the impairment attributable to NGS. Even if the Final Rule addresses only a small fraction of the visibility impairment at the Grand Canyon, EPA still has the statutory authority to address that portion of the visibility impairment problem which is, in fact, 'reasonably attributable' to NGS. Congress mandated an extremely low triggering threshold, requiring the installment of stringent emission controls when an individual source 'emits any air pollutant which may reasonably be anticipated to cause or contribute to any impairment of visibility' in a class I Federal area. 42 U.S.C. 7491(b)(2)(a). The National Academy of Sciences correctly noted that Congress has not required ironclad scientific certainty establishing the precise relationship between a source's emission and resulting visibility impairment: The phrase 'may reasonably be anticipated' suggests that Congress did not intend to require EPA to show a precise relationship between a source's emissions and all or a specific fraction of the visibility impairment within a Class I area. Rather, EPA is to assess the risk in light of policy considerations regarding the respective risks of overprotection and underprotection."

Other certifications of visibility impairment that may be reasonably attributable were also made by the EPA in 1985. However, only two reasonable attribution determinations were made, the NGS as discussed above and the Moosehorn National Wildlife Refuge located in Maine. No reasonable attribution decisions were made with respect to the other cases.

The suspected sources of visibility impairment in the Moosehorn National Wildlife Refuge were identified as an asphalt plant and a pulp and paper mill. Starting in October 1987, visibility monitoring was initiated using an 8-mm, time-lapse camera similar to those used in WHITEX. The results dramatically showed that plumes could be seen from the mill almost every day and that the plume appeared to enter the Moosehorn area. The EPA attributed this visibility impairment to the Georgia-Pacific Corporation's Woodland Mill on September 5, 1989 (see Federal Register, Volume 54, No. 170).

The NAS report (National Research Council, 1993) states

"Progress towards the national goal of remedying and preventing anthropogenic visibility impairment in Class I areas will require regional programs that operate over large geographic areas. Because most visibility impairment in Class I areas results from the transport by winds of emissions and secondary airborne particles over great distances, focusing only on sources immediately adjacent to Class I areas—as under the current program—is unlikely to improve visibility effectively. A program that focuses solely on determining the contribution of individual emission sources to visibility impairment is doomed to failure. Instead, strategies should be adopted that consider many sources simultaneously on a regional basis."

In the 1980s, the EPA was sued for failure to issue the next phase of the visibility protection regulations. The EPA defended its position in two ways (Maine v. Thomas, 874 F.2d 883, 1st Cir. 1989). The primary rationale was process based, in that the EPA noted that the 1980 rules did not set any firm timetable for issuing future regulations to address the main source of human-caused visibility impairment at mandatory federal Class I areas: regional haze. The EPA argued that the lack of a firm deadline in 1980 was no longer challengeable. In addition, the EPA claimed there was insufficient science to develop the implementation tools needed to require states to assess long-range contributions of their sources to levels of impairment. The EPA's position was sustained by the D.C. circuit court (GCVTC, 1996).

As the 1990 CAA amendments were being formulated, there was a political push to include requirements for the EPA to issue regulations to address regional haze. Recognizing that most visibility impairment is regional, the 1990 CAA amendments acknowledged the issue by adding additional visibility protection provisions in Section 169B that explicitly required the identification of source regions and transport pathways, both "clean" and "hazy," to Class I areas with visibility protection. Within the context of source regions and transport pathways, the amendments called for the establishment of interstate commissions to report to the EPA on the promulgation of regulations to address the regional haze problem.

Section 169(B) states that

> (1) "The Administrator, in conjunction with the National Park Service and other appropriate Federal agencies, shall conduct research to identify and evaluate sources and source regions of both visibility impairment and regions that provide predominantly clean air in class I areas. A total of $8,000,000 per year for 5 years is authorized to be appropriated for the Environmental Protection Agency and the other Federal agencies to conduct this research. The research shall include—
>
> (A) expansion of current visibility related monitoring in class I areas;
> (B) assessment of current sources of visibility impairing pollution and clean air corridors;
> (C) adaptation of regional air quality models for the assessment of visibility;
> (D) studies of atmospheric chemistry and physics of visibility.
>
> (2) Based on the findings available from the research required in subsection (a)(1) of this section as well as other available scientific and technical data, studies, and other available information pertaining to visibility source-receptor relationships, the Administrator shall conduct an assessment and evaluation that identifies, to the extent possible, sources and source regions of visibility impairment including natural sources as well as source regions of clear air for class I areas. The Administrator shall produce interim findings from this study within 3 years after November 15, 1990."

Furthermore, the 1990 amendments explicitly called for a Grand Canyon visibility transport commission that was to include a region that could affect visibility in Grand Canyon NP. This commission was expanded to include additional Class I areas, including wilderness areas found on the Colorado Plateau, and included the following states: Arizona, New Mexico, Colorado, Utah, California, Idaho, Nevada, Oregon, and Wyoming. The Grand Canyon Visibility Transport Commission (GCVTC) was directed to report to the EPA on long-term strategies for reducing regional haze from anthropogenic sources (GCVTC, 1996). The report was delivered in June 1996 and included the following recommendations:

- "**Air Pollution Prevention.** Air pollution prevention and reduction of per capita pollution is a high priority for the Commission. The Commission recommends policies based on energy conservation, increased energy efficiency and promotion of the use of renewable resources for energy production.

- **Clean Air Corridors.** Clean air corridors are key sources of clear air at Class I areas, and the Commission recommends careful tracking of emissions growth that may affect air quality in these corridors.
- **Stationary Sources.** For stationary sources, the Commission recommends closely monitoring the impacts of current requirements under the Clean Air Act and ongoing source attribution studies. Regional targets for SO_2 emissions from stationary sources will be set, starting in 2000. If these targets are exceeded, this would trigger a regulatory program, probably including a regional cap and market-based trading. During the next year, participants in the Commission's process will develop a detailed plan for an emissions cap and market trading program.
- **Areas in and Near Parks.** The Commission's research and modeling show that a host of identified sources adjacent to parks and wilderness areas, including large urban areas, have significant visibility impacts. However, the Commission lacks sufficient data regarding the visibility impacts of emissions from some areas in and near parks and wilderness areas. In general, the models used by the Commission are not readily applicable to such areas. Pending further studies of these areas, the Commission recommends that local, state, tribal, federal, and private parties cooperatively develop strategies, expand data collection, and improve modeling for reducing or preventing visibility impairment in areas within and adjacent to parks and wilderness areas.
- **Mobile Sources.** The Commission recognizes that mobile source emissions are projected to decrease through about 2005 due to improved control technologies. The Commission recommends capping emissions at the lowest level achieved and establishing a regional emissions budget, and also endorses national strategies aimed at further reducing tailpipe emissions, including the so-called 49-state low emission vehicle, or 49-state LEV.
- **Road Dust.** The Commission's technical assessment indicates that road dust is a large contributor to visibility impairment on the Colorado Plateau. As such, it requires urgent attention. However, due to considerable skepticism regarding the modeled contribution of road dust to visibility impairment, the Commission recommends further study to resolve the uncertainties regarding both near-field and distant effects of road dust, prior to taking remedial action. Because this emissions source is potentially such a significant contributor, the Commission feels that it deserves high priority attention and, if warranted, additional emissions management actions.

- **Emissions from Mexico.** Mexican sources are also shown to be significant contributors, particularly of SO_2 emissions. However, data gaps and jurisdictional issues make this a difficult issue for the Commission to address directly. The Commission recommendations call for continued binational collaboration to work on this problem, as well as additional efforts to complete emissions inventories and increase monitoring capacities. These matters should receive high priority for regional and national action.
- **Fire.** The Commission recognizes that fire plays a significant role in visibility on the Plateau. In fact, land managers propose aggressive prescribed fire programs aimed at correcting the buildup of biomass due to decades of fire suppression. Wildfire levels are projected to increase significantly during the studied period. The Commission recommends the implementation of programs to minimize emissions and visibility impacts from prescribed fire, as well as to educate the public.
- **Future Regional Coordinating Entity.** Finally, the Commission believes there is a need for an entity like the Commission to oversee, promote, and support many of the recommendations in this report. To support that entity, the Commission has developed a set of recommendations addressing the future administrative, technical and funding needs of the Commission or a new regional entity and has asked the Operations Committee to complete detailed plans by September, 1996. The Commission strongly urges the EPA and Congress to provide funding for these vital functions and give them a priority reflective of the national importance of the Class I areas on the Colorado Plateau."

The GCVTC report provided the technical and policy underpinnings for the national Regional Haze Rule and put the EPA under an enforceable deadline to promulgate a rule (CAA § 169B). In 1997, the EPA issued a notice of proposed rulemaking for regional haze. In 1998, Congress enacted the Transportation Equity Act for the 21st Century (TEA-21, Public Law 105-178), which among other things delayed the requirement for states to submit regional haze SIPs until after the EPA determined $PM_{2.5}$ nonattainment areas and had time to develop attainment plans. This new schedule was reflected in the EPA's final Regional Haze Rule.

In 1999, the EPA promulgated a regional haze rule (40 CFR 51, 64 Federal Register 126, Regional Haze Regulations, Final Rule, 1999). The rule required that all states submit an implementation plan that provides for reasonable progress by 2064 toward achieving "natural background conditions" in national parks and wilderness areas on the average of the worst 20% visibility days while maintaining the air clarity of the best 20% days. The first plans were due in 2008 and were to cover the long-term strategy period of 2008 to 2018. New plans are then required by the end of 2018 and every

10 years thereafter. The GCVTC states were given the option of implementing the regional haze program planning requirements or the GCVTC recommendations for the first round of planning that extends to 2018. After 2018 all states are to manage regional haze under their own regional haze SIPs.

The scope of developing SIPs to address visibility protection on a regional scale was much larger than any previous effort to address pollution impacts in previous SIP requirements. Even though some states had cooperated to address ozone on a regional basis, they had not developed the techniques to address the complex nature of visibility impairment. The need to understand not only the magnitude of fine particulate matter but its composition and distribution in remote areas of the country was key to developing technically sound regulatory programs. New tools and inventories of emissions sources were required. In line with that effort, in 1990 the EPA, through monetary grants, supported the development of the five regional planning organizations (CAA § 165(B) 2) shown in Fig. 8.11.

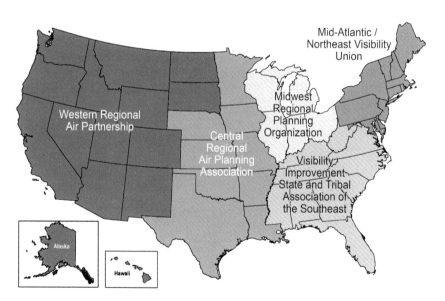

FIGURE 8.11 The five regional planning organizations formed to address interstate transport of haze across state boundaries (http://www.epa.gov/visibility/regional.html).

These organizations evaluated technical information to better understand how their states and tribes impact national parks and wilderness areas (Class I areas) across the country, and they then pursued the development of regional strategies to reduce emissions of particulate matter and other pollutants leading to regional haze (http://www.epa.gov/air/visibility/regional.html).

EPA funding lasted from 2001 to 2009 and supported new emissions inventory efforts, including rural ammonia in the Midwest, and coordinated air quality model development across the nation.

Through the regional planning organizations, all states were provided with technical assessments for their contributions to haze in Class I areas affected by emissions from their state. During this same time, the EPA expanded funding for monitoring of fine particulate matter in most of the Class I areas to further support the efforts to develop regional haze SIPs.

As noted before, the CAA establishes a national goal of returning the visual air quality in mandatory federal Class I areas to "natural conditions." This goal was the premise for the glide-path approach promulgated in EPA regional haze regulations. To start implementation of the Regional Haze Rule, the EPA needed to establish the natural conditions goal for each Class I area. The regulations then allow states, in cooperation with the FLMs, to adjust the natural conditions estimates as scientific research advances. Of course, natural background is unique to every Class I area and will vary from day to day or even hour to hour. The interior West is different from the interior East, the Southeast is different from the Northeast, the Northwest is quite different from the desert Southwest, and so forth. And the average background conditions are made up of a distribution of background values, so the 20% haziest background conditions are not the same as the average or the 20% clearest background conditions. Estimating the 20% haziest background conditions consistent with tracking progress of the 20% worst or haziest days presents a unique calculation problem.

The EPA has actually published guidance on estimating natural visibility conditions, primarily based on the works of Trijonis et al. (1990) and Ames and Malm (2001). Trijonis et al. developed average background conditions on the basis of a 1987–1988 literature review that is included here for reference in Table 8.2. Implicit in the table are some atmospheric aerosol and optics concepts, such as names of certain aerosol species, extinction, and extinction efficiencies that are discussed in some detail in Chapters 2 and 3.

The extinction efficiencies are based on the literature review by Trijonis et al. (1990). All the extinction efficiencies represent particle scattering, except for elemental carbon where the 10.5 m^2/g value is assumed to consist of 9 m^2/g absorption and 1.5 m^2/g scattering. Note that the 0.6 m^2/g value for coarse particles is a "pseudo-coarse scattering efficiency" representing the total scattering by all ambient coarse particles divided by the coarse particle mass between 2.5 and 10 μm.

The background levels were based on emission inventories of natural versus anthropogenic sources, ambient background measurements,

TABLE 8.2   Average Natural Background Levels of Aerosols and Light Extinction

	Average Concentration			Extinction Efficiencies	Extinction Contributions	
	East (μg/m^3)	West (μg/m^3)	Error Factor	(m^2/g)	East (Mm^{-1})	West (Mm^{-1})
Fine particles (<2.5 μm)						
Sulfates (as NH$_4$ HSO$_4$)	0.2	0.1	2	2.5	0.5	0.2
Organics	1.5	0.5	2	3.75	5.6	1.9
Light absorbing carbon	0.02	0.02	2–3	10.5	0.2	0.2
Ammonium nitrate	0.1	0.1	2	2.5	0.2	0.2
Soil dust	0.5	0.5	1.5–2	1.25	0.6	0.6
Water	1.0	0.25	2	5	5.0	1.2
Coarse particles (2.5–10 μm)	3.0	3.0	1.5–2	0.6	1.8	1.8
Rayleigh scatter					12	11
				Total	26 ± 7	17 ± 2.5

especially in the Southern Hemisphere, and receptor-type models with natural versus anthropogenic tracers.

Notice the difference in background concentration for fine particle species (<1.5 μm) between the eastern and western United States. Sulfates and organics are judged to be higher in the East than the West, while elemental carbon, nitrate, soil dust, and coarse particles are all judged to be about the same. The water associated with hygroscopic aerosols is higher in the East primarily because the relative humidity is higher. The error factors for all species are about 2. Mass extinction efficiencies are reviewed in some detail in Chapter 2, and the current best estimates are not all that different from those shown in Table 8.2. Using the mass extinction efficiencies listed in the table, the extinction associated with each species can be calculated and is also presented in Table 8.2.

By apportioning water scattering between sulfates and nitrates, it is evident that these two species contribute about 60% of the extinction budget and are about equal in the East, while in the West coarse particles and organics are all about equal in their contribution to background extinction levels, as are sulfates, nitrates, and soil dust, but at a somewhat lower level.

The challenge is to extend this analysis to try to account for differences in background conditions that make up the mean background level and to account for the extreme differences in relative humidity conditions that occur in different seasons and across the country. No attempt was made to extend background aerosol levels to regions smaller than the East/West subdivisions carried out by Trijonis (1982b). However, the IMPROVE extinction algorithm (Chapter 2) was used to calculate background extinction levels from the estimated background aerosol concentrations. The IMPROVE extinction equation accounts for the hygroscopicity of inorganic aerosols using an $f(RH)$ factor that adjusts the sulfate and nitrate mass extinction efficiencies as a function of relative humidity. A monthly average $f(RH)$ factor was calculated for each monitoring site. These $f(RH)$ factors were used to understand the distributions of background extinction as a function of RH as it changes seasonally and across the country. These distributions were then used to estimate the visibility on the best and worst days; it is these values that are used on a site-by-site basis for establishing the natural background visibility levels that are to be achieved by 2064. In the EPA guidance for estimating natural visibility conditions, there is a discussion concerning using better estimates of background conditions as visibility and aerosol science progress (U.S. EPA, 2003). For instance, if better estimates of background or natural emissions were available, chemical transport models could be used to estimate background aerosol concentrations as opposed to relying on only a few background measurements as was done previously.

The Regional Haze Rule extended the haze program to all states, set reasonable progress goals, revised SIPs as needed depending on whether reasonable progress goals are achieved, expanded the monitoring networks to include representation of all Class I areas, and incorporated emission reduction associated with revisions of NAAQS standards into reasonable progress goals. The rule also has a new source review provision to prevent potential future visibility impairment.

States are required to determine whether emission sources within the state are reasonably expected to contribute to visibility impairment not only in Class I areas within the state but also to Class I areas downwind of the state, even in states that do not contain any Class I areas. The rule explicitly states that "emissions from each of the 48 contiguous states may be reasonably anticipated to cause or contribute to visibility impairment in a Class I area" (40 CFR 51, 64 Federal Register 126, Regional Haze Regulations; Final Rule, 1999). It also states that "a BART-eligible source is reasonably anticipated to cause or contribute to regional haze if it can be shown that the source emits pollutants within a geographic area from which pollutants can be emitted and transported downwind to a class I area." This rule essentially puts every source built between

1962 and 1977 under BART emission limit requirements. Another twist to the rule is that when a state does a BART analysis to assess its impact on visibility impairment, it is to assume that all other sources have implemented BART and reasonable progress measures to eliminate degradation to visibility. This assures that the analysis will be done under the cleanest conditions and hence the most sensitive conditions to an incremental change in emissions. States can choose alternatives to BART, such as implementing alternative energy production strategies, as long as the alternatives equal or exceed the visibility improvement associated with BART implementation.

To assist states in determining which BART is needed to address impacts to regional haze, an examination of BART determinations to address reasonably attributable impairment (RA BART) under the 1980 visibility protection rules was conducted. Lebans (2001) reviewed six BART and RA-BART-like case studies: Healy Clean Coal Project, Mohave Generating Station, Navajo Generating Station (NGS), Georgia-Pacific Pulp and Paper Mill, Centralia Power, and Craig and Hayden power plants. They state, "Following certification by a FLM, and attribution to specific source(s) by the state, federal regulation prescribes a process of establishing appropriate source controls on a case-by-case basis." All but one of the cases examined are coal-fired power plants. The exception is the Georgia-Pacific Pulp and Paper Mill case. Though not subject to BART, this source is subject to prevention of significant deterioration (PSD) and new source review (NSR) procedures and general state authority to make reasonable progress toward the visibility goal.

Attribution studies related to the above-described RA-BART analysis are listed in Table 8.3. The interested reader is directed to http://vista.cira.colostate.edu/improve/Studies/Specialstudies.htm for full descriptions and findings of the studies.

Among the other cases, the NGS is the only one in which a formal BART analysis was completed under the 1980 visibility protection regulations. However, the controls identified as BART by the EPA in February 1991 (70% SO_2 reduction) were replaced by "better than BART" controls (90% SO_2 reduction) that were determined to be more cost effective. Since the cases involving Centralia, Healy, and Georgia-Pacific were precipitated by intervention in the state PSD permit process or through reasonably available control technologies (RACT) requirements, these other regulatory processes served as the major basis for establishing controls. However, in the Centralia case, a BART analysis was included in the technical support document, and the Southwest Air Pollution Control Agency (SWAPCA), the local air regulatory authority, asserted that the final RACT order emission limits met or exceeded the BART requirements.

TABLE 8.3   Visibility Attribution Studies Focusing on the Contribution of Coal-Fired Power Plant Emissions to Visibility Reduction in a Number of Class I Areas

Abbreviation	Name	Year/Receptor	Source of Concern	Comments	Reference
WHITEx	Winter Haze Intensive Tracer Experiment	1987-Grand Canyon and Canyonlands NPs	Navajo Generating Station	Release of CD_4 tracer from NGS	Malm et al., 1989
PReVEnt	Pacific Northwest Regional Visibility Experiment	1990-Class I areas in northwestern United States	Centralia Power Plant	Natural tracers	Malm and Gebhart, 1996; Henry, 1997
Moosehorn	Moosehorn National Wildlife Refuge	1987-Moosehorn Wildlife Refuge	Pulp and paper mill	Time-lapse photography	None
NGS	Navajo Generating Station Visibility Study	1990-Grand Canyon NP	Navajo Generating Station	Release of perfluorocarbon tracers	Lebans, 2001
MoHaVe	Measurement of Haze and Visibility Effects	1992-Grand Canyon	Mojave Power Plant	Natural tracers	Pitchford et al., 1999
Mt. Zirkel	Mt. Zirkel Reasonable Attribution Study	1995-Mt. Zirkel Wilderness, CO	Craig and Hayden power plants	Natural tracers	Watson et al., 1996
BRAVO	Big Bend Regional Aerosol and Visibility Observational Study	1999-Big Bend NP	Mexico's Carbon I and II power plants and other sources	Release of perfluorocarbon tracers near Carbon power plants	Pitchford et al., 2004

Studies were carried out from 1987 to 1999.

In 2002, the United States Court of Appeals for the District of Columbia Circuit found portions of the EPA's Regional Haze Rule contrary to the CAA in important respects, including that it allowed the states to require BART controls at sources, without any empirical evidence of the particular source's contribution to visibility impairment in a Class I area.

In 2005, the EPA revised the Regional Haze Rule and issued guidelines for making BART determinations. The revised rule calls for imposition of BART if a source "may reasonably be anticipated to cause or contribute to any impairment of visibility in any mandatory class I Federal area." In determining what constitutes BART for a particular source, the states (and the EPA) must consider five factors, including the cost of compliance and the degree of visibility improvement that may reasonably be anticipated to result from the use of the technology. The guidelines include presumptive limits for nitrogen oxide (NOx) emissions from large coal-fired power plants and state that these presumptive limits "are extremely likely to be appropriate" for such plants. The EPA said, "Based on our analysis of emissions from power plants, we believe that applying these highly cost-effective controls at the large power plants covered by the guidelines would result in significant improvements in visibility and help to ensure reasonable progress toward the national visibility goal." The EPA's presumptive limits are based on the use of combustion controls, such as low-NOx burners, at such plants (40 CFR 51, Appendix Y, Guidelines for BART Determinations under the Regional Haze Rule).

Since the promulgation of the 1999 Regional Haze Rule and the 2005 revisions to the BART regulations, many sources have undergone a BART review for their contributions to regional haze, and reductions in emissions have been required. This round of BART reviews has been focused on western states and on control of NOx, as the regional programs discussed later addressed many of the similar sources in the eastern United States.

PARALLEL REGULATORY PROGRAMS AFFECT CLASS I VISIBILITY

While the Regional Haze Rule sets forth a goal of reaching natural visibility conditions by 2064, other regulatory provisions designed to reduce the impacts of air pollution on health also affect visibility. On March 10, 2005, the EPA issued the Clean Air Interstate Rule (CAIR) (40 CFR 96; 40 CFR 97). This rule provides states with a solution to the problem of power plant pollution that is transported from one state to another. The CAIR covers the 27 eastern states, shown in Fig. 8.12, and the District of Columbia. The rule uses a "cap and trade" system to reduce the target pollutants—sulfur dioxide (SO_2) and nitrogen oxides (NOx)—by 70%.

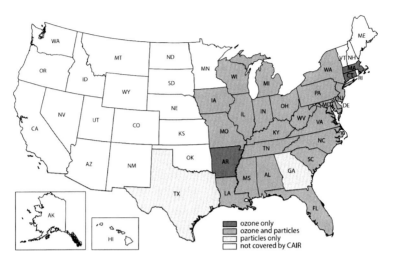

FIGURE 8.12   States covered by the Clean Air Interstate Rule (CAIR).

States must achieve the required emission reductions using one of two compliance options: (1) meet the state's emission budget by requiring power plants to participate in an EPA-administered interstate cap and trade system that caps emissions in two stages, or (2) meet an individual state emissions budget through measures of the state's choosing (http://www.epa.gov/airmarkets/programs/cair/).

The 2005 CAIR rule was subject to a prolonged legal challenge. In response to that challenge and to changing needs to address cross-state air pollution in the eastern United States, the EPA supplanted the CAIR with the Cross-State Air Pollution Rule (CSAPR). The CSAPR requires states to "significantly improve air quality by reducing power plant emissions that contribute to ozone and/or fine particle pollution in other states." In a separate but related regulatory action, the EPA finalized a supplemental rulemaking on December 15, 2011, to require five states—Iowa, Michigan, Missouri, Oklahoma, and Wisconsin—to make summertime NOx reductions under the CSAPR ozone season control program. The CSAPR requires a total of 28 states to reduce annual SO_2 emissions, annual NOx emissions, and/or ozone season NOx emissions to assist in attaining the 1997 ozone and fine particle requirements and the 2006 fine particle NAAQS. On February 7, 2012, and June 5, 2012, the EPA issued two sets of minor adjustments to the CSAPR. The revised rule requires significant reductions in SO_2 and NOx emissions that cross state lines. These pollutants react in the atmosphere to form fine particles and ground-level ozone and are transported long distances, making it difficult for other states to achieve NAAQS. Emission reductions were to take effect quickly, starting January 1, 2012, for SO_2 and annual NOx reductions, and May 1, 2012, for ozone season NOx reductions. By 2014, combined with other final state

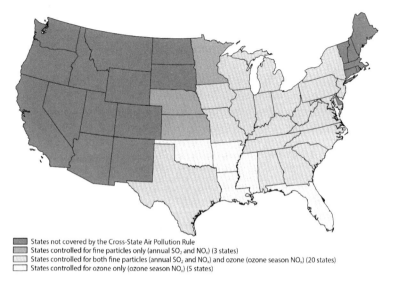

States not covered by the Cross-State Air Pollution Rule
States controlled for fine particles only (annual SO₂ and NOₓ) (3 states)
States controlled for both fine particles (annual SO₂ and NOₓ) and ozone (ozone season NOₓ) (20 states)
States controlled for ozone only (ozone season NOₓ) (5 states)

FIGURE 8.13   States affected by the cross-state transport rule (http://www.epa.gov/crossstaterule/). States in the West are not affected by the rule.

and EPA actions, the CSAPR would have reduced power plant SO_2 emissions by 73% and NOx emissions by 54% from 2005 levels in the CSAPR region. This rule was challenged and vacated in late 2012. At the time of this writing, the EPA is in the process of appealing to the U.S. Supreme Court (http://www.epa.gov/crossstaterule/). The regions subject to the CSAPR are shown in Fig. 8.13.

PROGRESS IS UNDERWAY

As with the CAIR rule, other provisions of the 1990 CAA amendments that reduce the impacts of air pollution on health also affect visibility. Title I addressed the designation of NAAQS nonattainment areas as well as interstate and intercontinental transport, and Title II addressed motor vehicle emissions and fuel standards. Title IV (Acid Deposition Control) established regulatory mandates to reduce electric utility emissions of SO_2 and NOx, both of which form particulates in the atmosphere that contribute to visibility degradation and to acid deposition (U.S.C. Title 42, Chapter 85, Subchapter IV § 7651).

The combined effect of all regulatory control strategies for SO_2, the precursor gas to visibility-reducing sulfate particles, and nitrogen oxides, the precursors to particulate nitrate, is shown in Fig. 8.14 (Hand et al., 2014). SO_2 emissions have decreased from 23 Mtons yr^{-1} in 1990 to 6 Mtons yr^{-1} in 2012, and NOx emissions have decreased from 26 to 11 Mtons yr^{-1}

during that same time period. Also shown in Fig. 8.14 is the total U.S. wildfire burn area (right axis), compiled from annual total wildland fire statistics by the National Interagency Fire Center (https://www.nifc. gov/fireInfo/fireInfo_stats_totalFires.html). Area burned is an indicator of wildfire activity and therefore smoke emissions (Hand et al., 2014). Fig. 8.14 is suggestive of increased fire emissions over the 1990–2012 time period and could offset any visibility improvement for SO_2 and NOx emission reductions.

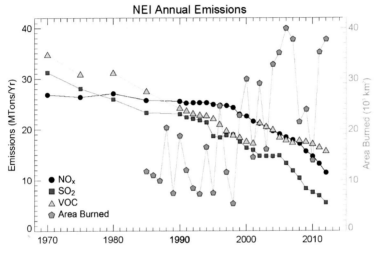

FIGURE 8.14   National Emission Inventory (NEI) total U.S. emissions (Mtons yr^{-1}) (left axis). Area burned (10^3 km^2) on right axis (Hand et al., 2014).

The combined effect of emission reductions of SO_2 and NOx and increases in carbon emissions from increased fire activity is shown in Fig. 8.15, a graph of regionally averaged 10- and 20-year trends in the 20% highest derived atmospheric extinction values, a measure of atmospheric haziness. Derived extinction from aerosol concentrations is covered in some detail in Chapter 2. The small circles in the maps of the United States presented in Fig. 8.15 represent the locations of some 140 monitoring sites in remote and semiremote areas of the United States, many of which are national parks and wilderness areas. The IMPROVE monitoring network underwent an expansion in the 2001–2002 time frame, from about 40 sites to approximately 140 sites, so a change in extinction over a 20-year time frame can be shown for the original 40 sites, while a 10-year trend can be shown for all 140 sites. The number representing the yearly average extinction represents the number of sites in the yearly average.

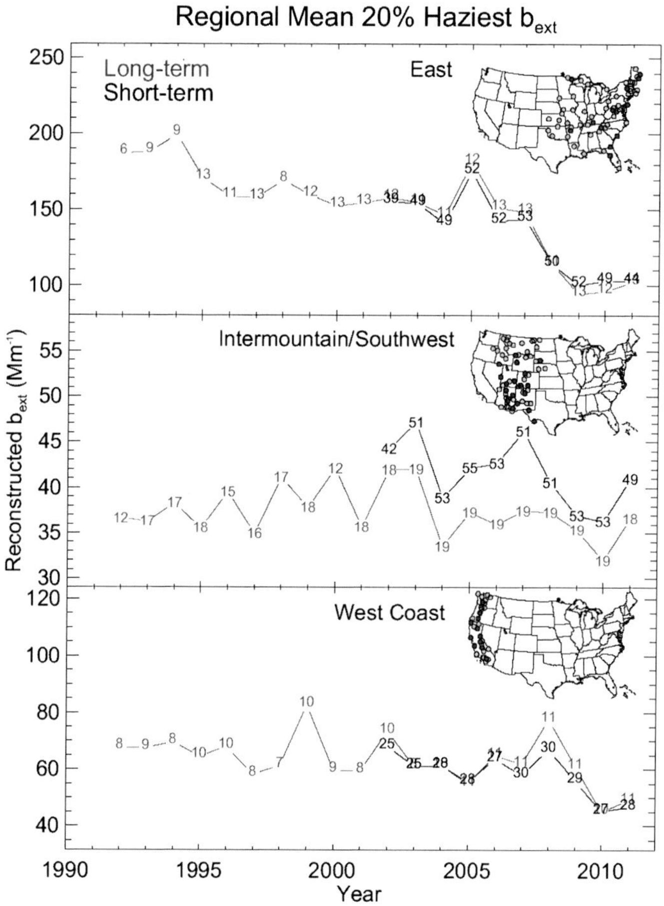

FIGURE 8.15 Top panel: Eastern mean reconstructed 20% haziest ambient light extinction coefficient (b_{ext}, Mm^{-1}, 550 nm), including contributions from Rayleigh scattering from long-term (LT) and short-term (ST) IMPROVE sites. The number of complete sites with valid data available for a given year is used as the plot symbol for each time series. Middle panel: Intermountain/Southwest regional mean 20% haziest b_{ext}. Bottom panel: West Coast regional mean 20% haziest b_{ext}. The inset maps show regionally aggregated ST sites in gray and LT sites in red (Hand et al., 2014).

The top panel in Fig. 8.15 represents yearly 20% highest extinction averaged across the Midwest and East monitoring sites for both the 10- and 20-year datasets, while the middle and bottom panels present the yearly average data for the Intermountain/Southwest and West Coast sites. The Midwest/East 10- and 20-year trends track each other over the time period in which they overlap. There is a continual decrease in haziness or increase in visibility over the 20-year time frame consistent with SO_2 and NOx reductions shown in Fig. 8.15. The greatest reduction in haziness, or improvement in visibility, occurred from 2005 to 2009 when the average 20% highest extinction days dropped from about 175 Mm^{-1} to 100 Mm^{-1} or about 15% $year^{-1}$. The overall average decrease in haziness over the 20-year time period is on the order of about 3% $year^{-1}$.

In the Intermountain/Southwest, the 10- and 20-year datasets differ because the 20-year dataset includes some monitoring locations that represent locations that are inherently hazier than the original 40 monitoring sites, which were primarily in more remote western national parks. Because the emission reductions of SO_2 and NOx have been substantially less in the West, primarily because the largest SO_2 emitters and the highest population density are in the East, the decrease in haziness has been substantially less. In 1990, the average 20% highest extinction values were around 50 Mm^{-1}, and for 2010–2011 they were approximately 40 Mm^{-1}, a decrease of only 10 Mm^{-1} over a 20-year time frame. Also, in the West, any decreases in SO_2 and NOx emissions have been to some degree offset by an increase in wildfire activity and its associated emissions.

As in the East, the West Coast 10- and 20-year time frame trends track each other quite nicely. Little change in extinction is observed from the early 1990s to about 2008, while from about 2008 to 2011 extinction has decreased from about 60 Mm^{-1} to 45 Mm^{-1}. Over the last three years, this corresponds to about a 10% per year decrease in extinction.

Fig. 8.16 shows the same set of plots for the lowest 20% b_{ext} days. The downward trend on the lowest 20% extinction days is even more pronounced than on the 20% haziest days. Both the Intermountain/Southwest and West Coast show a continual decrease in extinction from 1992 to 2011. All regions of the country show about a 1.3–1.5% per year decrease in extinction, which is a 20–30% decrease in extinction over a 20-year period.

The decrease in haziness can be pictorially represented using the computer software embedded in WinHaze. The use and description of WinHaze are covered in Chapter 6; examples are presented here to illustrate the improvements in visibility during the haziest conditions over the last 20 years in two Class I scenic areas. Fig. 8.17 shows an eastern landscape feature in Great Smoky Mountains NP, Tennessee, as it would have appeared in 1990 under the average 20% haziest conditions (302 Mm^{-1}) and how it appeared on the average 20% haziest conditions in 2011 (104 Mm^{-1}). The improvement in visibility is primarily a result of reductions in sulfate particles that are linked to a reduction in SO_2 gas from coal-fired power plants. The improvements are dramatic. In 1990, the more distant features in this landscape are not visible, while the clarity in the overall scene is significantly improved by 2011.

Fig. 8.18 shows a landscape feature in San Gorgonio Wilderness Area, California, a scene in the western United States. The improvement in the average haziest 20% days from 1990 to 2011 corresponds to a change in extinction from 148 Mm^{-1} to 50 Mm^{-1}. Whereas the improvement in visibility

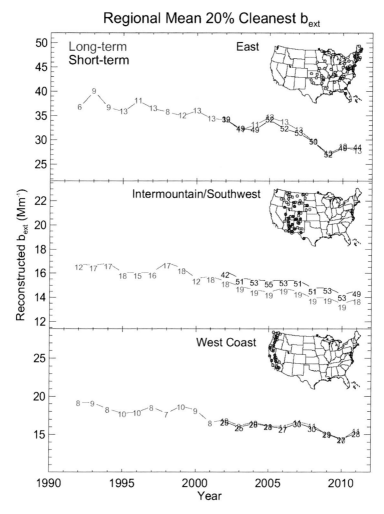

FIGURE 8.16   Mean 20% cleanest ambient light extinction coefficient (b_{ext}, Mm^{-1}, 550 nm), including contributions from Rayleigh scattering from long-term (LT) and short-term (ST) IMPROVE sites. The number of complete sites with valid data available for a given year is used as the plot symbol for each time series. Top panel: Eastern regional mean 20% cleanest b_{ext}. Middle panel: Intermountain/Southwest regional mean 20% cleanest b_{ext}. Bottom panel: West Coast regional mean 20% cleanest b_{ext}. The inset maps show regionally aggregated LT sites in red and ST sites in gray.

at Great Smoky Mountains NP was primarily due to reductions in sulfate particles, the increase in visibility at San Gorgonio Wilderness Area is primarily linked to reductions in particle nitrate concentration reduction, which in turn reflects deceases in NOx emissions primarily from mobile sources (Hand et al., 2014).

FIGURE 8.17   Simulations of the view at Great Smoky Mountains National Park, TN, corresponding to the mean 20% haziest b_{ext} in 1990 (left, b_{ext} = 302 Mm^{-1}) and 2011 (right, b_{ext} = 114 Mm^{-1}). Contributions from Rayleigh scattering are included.

FIGURE 8.18   Simulations of the view at San Gorgonio Wilderness Area, CA, corresponding to the mean 20% haziest b_{ext} in 1990 (left, b_{ext} = 148 Mm^{-1}) and 2011 (right, b_{ext} = 50 Mm^{-1}). Contributions from Rayleigh scattering are included.

Clearly, emission reductions associated with various regulatory programs are having a dramatic effect on improving viewing conditions and the ability to enjoy the scenic national parks and wilderness areas in the United States.

References

Allee, P., Pueschel, R.F., Van Valin, C.C., Roberts, W.F., 1978. Air quality studies in Carbon and Emory counties, Utah (Draft), NOAA Environmental Research Laboratories, Atmospheric Physics and Chemistry Laboratory, Boulder.

Ames, R.B., Malm, W.C., 2001. Recommendations for natural condition deciview variability: an examination of IMPROVE data frequency distributions, presented at the Air & Waste Management Association and American Geophysical Union specialty conference on Regional Haze and Global Radiation Balance – Aerosol Measurements and Models: Closure Reconciliation and Evaluation, Bend, Oregon, October 2–5, 2001.

Anderson, O.L., 1971. Collaborative Research on Assessment of Man's Activities on the Lake Powell Region, project funded by the National Science Foundation – Research Applied to Nation Needs (RANN), (First Progress Report), available at http://arizona.openrepository.com/arizona/bitstream/10150/303129/1/ltrr-0041.pdf.

Bachman, J., 2007. Will the circle be unbroken: a history of the U.S. National Ambient Air Quality Standards. J. Air Waste Manage. Assoc. 57, 652–697.

Bechtel Corporation, 1970. The Navajo Project. Chapter I. Introduction and History of the Navajo Project, in Environmental Planning for the Navajo Generating Station, Status Report, pp. 1–3, OCLC 4711442.

Benkley, C.W., Bass, A., 1979. Development of Mesoscale Air Quality Simulation Model: Vol. 3. User's Guide to MESOPUFF, EPA-600/7-80-058, U.S. Environmental Protection Agency, Atmospheric Research and Exposure Assessment Laboratory, Research Triangle Park, NC.

Blumenthal, D.L., Richards, L.W., Macias, E.S., Bergstrom, R.W., Bhardwaja, P.S., 1983. The Chemistry, Physics, and Optical Properties of Plumes and Background Air in the Southwest United States, EPA-600/S3-83-085, U.S. EPA Environmental Sciences Research Laboratory, Research Triangle Park, NC.

Bradley, J.T., Imbembo, S.M., 1985. Automated visibility measurements for airports, Paper AIAA-85-0191 presented at the American Institute of Aeronautics and Astronautics (AIAA) 23rd Aerospace Sciences Meeting, Reno, Nevada, Jan. 14–17, 1985.

Cahill, T.A., Flocchini, R.G., Eldred, R.A., Feeney, P.J., 1984. Western Particulate Characterization Study, EPA-600/ 4-84-059, U.S. EPA Environmental Monitoring Systems Laboratory, Las Vegas, NV.

Chinkin, L.R., Latimer, D.A., Hogo, H., 1986. Layered haze observed at Bryce Canyon National Park: a statistical evaluation of the phenomenon. In: Bhardwaja, P.S. (Ed.), Proceedings of the Air Pollution Control Association International Specialty Conference, Visibility Protection: Research and Policy Aspects. Air and Waste Management Association, Pittsburgh, 1986, pp. 709–719.

Eatough, D.J., Eatough, M., Eatough, N.L., 1996. Apportionment of sulfur oxides at Canyonlands during the winter of 1990. III. Source apportionment of SOx and sulfate and the conversion of SO2 to sulfate in the Green River Basin. Atmos. Environ. 30, 295–308.

EPRI (Electric Power Research Institute), 1986. The Subregional Cooperative Electric Utility, Department of Defense, National Park Service, and EPA Study (SCENES) on Visibility: An Overview, EPRI EA-4664-SR, Palo Alto, CA.

Geue, J., 1971. Coal-Slurry Pipeline to Mohave Power Plant in Nevada, Research Establishment, Australian Atomic Energy Commission.

GCVTC (Grand Canyon Visibility Transport Commission), 1996. Recommendations for Improving Western Vistas, http://www.wrapair.org/WRAP/reports/GCVTCFinal.PDF.

Haagen-Smit, A.J., 1970. A lesson from the smog capital of the world. Proc. Natl. Acad. Sci. USA 67 (2), 887–897.

Hall, J.S., Jerzykiewicz, M., Riley, L., 1975. Extinction measurements for sensitive assessment of air quality. J. Air Poll. Control Assoc. 25 (10), 1045–1048.

Hand, J.L., Schichtel, B.A., Malm, W.C., Copeland, S.A., Molenar, J.V., Frank, N., Pitchford, M.L., 2014. Widespread reductions in haze across the United States from the early 1990s through 2011. Atmos. Environ. 94, 671–679.

Henry, R.C., 1997. Receptor modeling applied to patterns in space (RMAPS) III: apportionment of airborne particulate sulfur in western Washington state. J. Air Waste Manage. Assoc. 47 (2), 226–230.

Hidy, G.M., 1987. Conceptual design of a massive aerometric tracer experiment (MATEX). J. Air Poll. Control Assoc. 37 (10), 1137–1157.

Hidy, G.M., Hansen, D.A., Bass, A., 1985. Feasibility and Design of the Massive Aerometric Tracer Experiment (MATEX), Vols. I and II, Report EA-4305, Electric Power Research Institute, Palo Alto, CA.

Hundley, N., 2009. Water and the West: The Colorado River Compact and the Politics of Water in the American West, second ed. University of California Press, Berkeley, CA.

Husar, R.B., Holloway, J.M., Patterson, D.E., Wilson, W.E., 1981. Spatial and temporal patterns of eastern United States haziness: a summary. Atmos. Environ. 15, 1919–1928.

Johnson, C.D., Latimer, D.A., Bergstrom, R.W., Hugo, H., 1980. Users Manual for the Plume Visibility Model (PLUVUE), EPA-450/4-80-032, U.S. EPA Office of Air Quality Planning and Standards, Research Triangle Park, NC.

Joseph, D.B., Metza, J., Malm, W.C., Pitchford, M.L., 1987. Plans for IMPROVE: a federal program to monitor visibility in Class I areas. In: Bhardwaja, P.S. (Ed.), Visibility Protection: Research and Policy Aspects. Air Pollution Control Assoc., Pittsburgh, PA.

Latimer, D.A., 1980. Power plant impacts on visibility in the West: Siting and emissions control implications. J. Air Poll. Control Assoc. 30, 142–146.

Latimer, D.A., Ireson, R.G., 1988. Workbook for Plume Visual Impact Screening and Analysis, EPA-450/4-88-015, U.S. Environmental Protection Agency, Office of Air Quality Planning and Standards, Technical Support Division, Research Triangle Park, NC.

Latimer, D.A., Samuelsen, G.S., 1978. Visual impact of plumes from power plants. Atmos. Environ. 12, 1455–1465.

Latimer, D.A., Bergstrom, R.W., Hayes, S.R., Lui, M.K., Seinfeld, J.H., Whitten, G.Z., Wojcik, M.A., Hillyer, M.J., 1978. The Development of Mathematical Models for the Prediction of Anthropogenic Visibility Impairment, EPA-450/3-78-110a,b,c, U.S. Environmental Protection Agency, Office of Air Quality Planning and Standards, Technical Support Division, Research Triangle Park, NC.

Latimer, D.A., Gery, M.W., Hogo, H., 1986. A Theoretical Evaluation of the Role of Night Time Nitrate Formation in the Formation of Layered Haze, Report SYSAPP-86/167, Systems Applications Inc., San Rafael, CA.

Lebans, B., 2001. Westar Council; WESTAR Council RA BART Working Group. RA BART and RA BART-like Case Studies, http://www.wrapair.org/forums/amc/projects/ra_bart_case/, June 2001.

Macias, E.S., Vossler, T.L., White, W.H., 1987. Carbon and sulfate fine particles in the western United States. In: Bhardwaja, P.S. (Ed.), Visibility Protection: Research and Policy Aspects. Air Pollution Control Association, Pittsburgh, PA.

Malm, W.C., 1974. Air movement in the Grand Canyon. Plateau 46 (4), 132–135.

Malm, W.C., Gebhart, K.A., 1996. Source apportionment of organic and light absorbing carbon using receptor modeling techniques. Atmos. Environ. 30, 843–855.

Malm, W.C., Molenar, J.V., 1984. Visibility measurements in national parks in the western United States. J. Air Poll. Control Assoc. 34 (9), 899–904.

Malm, W.C., Molenar, J.V., Chan, L.Y., 1983. Photographic simulation techniques for visualizing the effect of uniform haze on a scenic resource. J. Air Pollut. Control Assoc. 33, 126–129.

Malm, W.C., Pitchford, M.L., Iyer, H.K., 1989. Design and implementation of the Winter Haze Intensive Tracer Experiment. In: Proceedings of the 81st Annual Meeting and Exhibition of the APCA, Air Pollution Control Association, Pittsburgh, PA.

Malm, W.C., Iyer, H.K., Gebhart, K.A., 1990. Application of tracer mass balance regression to WHITEX data. In: Mathai, C.V. (Ed.), Visibility and Fine Particles. Air & Waste Management Association, Pittsburgh, PA.

Malm, W.C., Sisler, J.F., Huffman, D., Eldred, R.A., Cahill, T.A., 1994. Spatial and seasonal trends in particle concentration and optical extinction in the United States. J. Geophys. Res. 99 (D1), 1347–1370.

Malm, W.C., Schichtel, B.A., Pitchford, M.L., Ashbaugh, L.L., Eldred, R.A., 2004. Spatial and monthly trends in speciated fine particle concentration in the United States. J. Geophys. Res. 109, D03306.

Malm, W.C., Schichtel, B.A., Pitchford, M.L., 2011. Uncertainties in PM2.5 gravimetric and speciation measurements and what we can learn from them. J. Air Waste Manage. Assoc. 61, 1131–1149.

McDade, C.E., Tombach, I.H., 1987. Goals and initial findings from SCENES. In: Bhardwaja, P.S. (Ed.), Visibility Protection: Research and Policy Aspects. Air Pollution Control Assoc., Pittsburgh, PA.

Molenar, J.V., Malm, W.C., Johnson, C.E., 1994. Visual air quality simulation techniques. Atmos. Environ. 28, 1055–1063.

Mueller, P.K., Watson, J.G., 1982. Eastern Regional Air Quality Measurements, Vol. I., Report #EA-1914, Electric Power Research Institute, Palo Alto, CA.

Mueller, P.K., Hidy, G.M., 1983. The Sulfate Regional Experiment: Report of Findings, Report #EA-1901, Vols. 1, 2, & 3, Electric Power Research Institute, Palo Alto, CA.

National Research Council, 1993. Protecting Visibility in National Parks and Wilderness Areas, Vol. I. National Academy Press, Washington, DC.

O'Dell, K.D., Layton, R.G., 1974. Visibility studies in the Grand Canyon. Plateau 46 (4), 133–134.

Patton, V.L., 1998. The new air quality standards, regional haze, and interstate air pollution transport. Environ. Law Rep. 28, 10155.

Patton, V., 2004. The visibility protection program. In: Martineau, Jr., R.J., Novello, D.P. (Eds.), The Clean Air Act Handbook, second ed. American Bar Association, Chicago, IL, (Chapter 7).

Pitchford, A., Pitchford, M.L., Malm, W.C., Flocchini, R., Cahill, T., Walther, E., 1981. Regional analysis of factors affecting visual air quality. Atmos. Environ. 15 (10), 2043–2054.

Pitchford, M.L., Green, M., Tombach, I., Malm W.C., Farber, R., Mirabella, V., 1999. Project Mohave Final Report, Measurement of Haze and Visual Effects, http://vista.cira.colostate.edu/improve/Studies/MOHAVE/Reports/FinalReport/mohavereport.htm.

Pitchford, M.L., Tombach, I., Barna, M., Gebhart, K.A., Green M.C., Knipping, E., Kumar, N., Malm, W.C., Pun, B., Schichtel, B.A., Seigneur, C., 2004. Big Bend Regional Aerosol and Visibility Observational Study, http://vista.cira.colostate.edu/improve/Studies/BRAVO/reports/FinalReport/BRAVO/BRAVOFinalReport.pdf.

Rao, V., Frank, N., Rush, A., Dimmick, F., 2003. Chemical Speciation of PM2.5 in Urban and Rural Areas, National Air Quality and Emissions Trends Report, available at http://www.epa.gov/airtrends/aqtrnd03/pdfs/2_chemspecofpm25.pdf.

Reitze, A.W., 2001. Air Pollution Control Law: Compliance and Enforcement. Environmental Law Institute, Washington, DC.

Richards, L.W., Blanchard, C., Blumenthal, D., 1991a. Navajo Generating Station Visibility Study, Final Report, Environmental Services Dept., Salt River Project, Phoenix, AZ.

Richards, L.W., Blanchard, C., Blumenthal, D., 1991b. Navajo Generating Station Visibility Study, Executive Summary, Sonoma Technology Inc. Report STI-90200-1124-FRD2, prepared for Salt River Project, Phoenix, AZ.

Shaw, Jr., R.W., Paur, R.J., 1983. Composition of aerosol particles collected at rural sites in the Ohio River Valley. Atmos. Environ. 17, 2031–2044.

Simon, D.J. (Ed.), 1988. Our Common Lands, Defending the National Parks. National Parks and Conservation Association, Island Press, Washington, DC.

Snelling, R.N., Pitchford, M.L., Pitchford, A., 1984. Visibility Investigative Experiment in the West (VIEW), EPA-600/s4-84-060, U.S. Environmental Protection Agency, Environmental Systems Monitoring Laboratory, Las Vegas, CA.

Stern, A.C., 1982. History of air pollution legislation in the United States. J. Air Poll. Control Assoc. 32, 44–61.

Stevens, J.E., 1990. Hoover Dam: An American Adventure. University of Oklahoma Press, Norman, OK.

Summitt, A.R., 2013. Contested Waters: An Environmental History of the Colorado River. University Press of Colorado, Boulder, CO.

Swicker, J.O., Macias, E.S., Anderson, J.A., Blumenthal, D.L., Ouimette, J.R., 1983. Chemistry and Visual Impact of the Plumes from the Four Corners Power Plant and San Manuel Copper Smelter, U.S. Environmental Protection Agency, EPA-600/S3-83-093, Environmental Sciences Research Laboratory, Research Triangle Park, NC.

Tombach, I., Allard, D., 1983. Comparison of Visibility Measurement Techniques: Eastern United States, Report #EA-3292, Electric Power Research Institute, Palo Alto, CA.

Trijonis, J.C., 1979. Visibility in the Southwest: an exploration of the historical data base. Atmos. Environ. 13, 833.

Trijonis, J.C., 1982a. Visibility in California. J. Air Poll. Control Assoc. 32, 165–169.

Trijonis, J.C., 1982b. Existing and natural background levels of visibility and fine particles in the rural East. Atmos. Environ. 16, 2431–2445.

Trijonis, J.E., Pitchford, M., 1986. Preliminary extinction budget results from the RESOLVE program. In: Bhardwaja, P.S. (Ed.), Visibility Protection: Research and Policy Aspects. Air & Waste Management Association, Pittsburgh, PA.

Trijonis, J.C., Yuan, K., 1978. Visibility in the Northeast: Long-Term Visibility Trends and Visibility/Pollutant Relationships, EPA-600/3-78-075, U.S. Environmental Protection Agency, Environmental Sciences Research Laboratory, Research Triangle Park, NC.

Trijonis, J.E., McGown, M., Pitchford, M., Blumenthal, D., Roberts, P., White, W., et al., 1988. RESOLVE Project Final Report: Visibility Conditions and Causes of Visibility Degradation in the Mojave Desert of California, NWC TP #6869, Naval Weapons Center, China Lake, CA.

Trijonis, J.C., Pitchford, M.L., Malm, W.C., White, W., Husar, R., 1990. Causes and Effects of Visibility Reduction: Existing Conditions and Historical Trends, National Acid Precipitation Assessment Program, State of Science and Technology Report 24, Government Printing Office, Washington, DC.

U.S. EPA, 1976. National Air Surveillance Network Annual Report 1975, U.S. Environmental Protection Agency, Region VI Surveillance and Analysis Division, Houston, TX.

U.S. EPA, 1979. Protecting Visibility: An EPA Report to Congress, EPA-450-5-79-008, http://nepis.epa.gov/Exe/ZyPURL.cgi?Dockey=P1003OJ9.txt.

U.S. EPA, 1980. Interim Guidance for Visibility Monitoring, EPA-450/2-80-082, U.S. Environmental Protection Agency, Office of Air Quality Planning and Standards, Research Triangle Park, NC.

U.S. EPA, 1992. User's Manual for the Plume Visibility Model, PLUVUE II, EPA-454/B-92-008, Environmental Protection Agency, Office of Air Quality Planning and Standards, Research Triangle Park, NC.

U.S. EPA, 1993. Interagency Workgroup on Air Quality Modeling (IWAQM) Phase 1 Report: Interim Recommendation for Modeling Long Range Transport and Impacts on Regional Visibility, EPA-454/R-93-015, Environmental Protection Agency, Office of Air Quality Planning and Standards, Research Triangle Park, NC.

U.S. EPA, 1999. Visibility Monitoring Guidance, EPA-454/R-99-003, Environmental Protection Agency, Office of Air Quality Planning and Standards, Research Triangle Park, NC.

U.S. EPA, 2003. Guidance for Estimating Natural Visibility Conditions under the Regional Haze Rule, EPA-454/B-03-005, Environmental Protection Agency, Office of Air Quality Planning and Standards, Research Triangle Park, NC.

U.S. EPA, 2009. Environmental Protection Agency, Facility Registry System, http://www.epa.gov/enviro/, accessed Jan. 2009.

Vanden Berg, M.D., 2010. Annual Review and Forecast of Utah Coal Production and Distribution 2008, Final 2008 Numbers and Preliminary 2009 Data, Circular 110, ISBN 978-1-55791-827-7, Utah Geological Survey, Utah Department of Natural Resources, Salt Lake City, UT.

Walther, E.G., Malm, W.C., Cudney, R.A., 1977. The Excellent but Deteriorating Air Quality in the Lake Powell Region, Lake Powell Research Project Bulletin No. 52, National Science Foundation, Research Applied to National Needs Program, University of California, Los Angeles, CA.

Watson, J.G., Chow, J., Shah, J.J., 1981. Analysis of Inhalable and Fine Particulate Matter Measurements, EPA-450/ 4-81-035, U.S. Environmental Protection Agency, Office of Air Quality Planning and Standards, Research Triangle Park, NC.

Watson, J.G., Blumenthal, D.B., Chow, J., Cahill, C., Richards, W., Dietrich, D., et al., 1996. Mt. Zirkel Wilderness Area Reasonable Attribution Study of Visibility Impairment, Final Report, Colorado Department of Public Health and Environment, Denver, CO.

Williams, M.D., Treiman, E., Wecksung, M., 1979. Utilization of a Simulated Photograph Technique as a Tool for the Study of Visibility Impairment, LA-UR-79-1741, Los Alamos Scientific Laboratory, Los Alamos, NM.

Zuniga, J.E., 2000. Project History: The Central Arizona Project, Bureau of Reclamation History Program. U.S. Bureau of Reclamation, http://www.usbr.gov/projects/Project.jsp?proj_Name=Central Arizona Project&pageType=ProjectHistoryPage.

APPENDIX

1

Units and Definitions

There are a myriad of units that can be associated with each of the variables in Table A1. Units of the fundamental variables such as length or time can be expressed for each order of magnitude. Each combination of units creates another unit of the derived variables shown in the table. Therefore, only conversions between the most popular variables will be presented later (Table A2).

TABLE A1   This Table is an Extension of Table 2.1. It Mathematically Defines the Various Radiometric and Photometric Variables and gives a Brief Description of Each.

Radiometric Variable	Symbol	Definition	Units	Description
Radiant energy	U	$U = Pdt$ $U = Nm$	Joule (watt sec)	Total quantity of energy emitted at the specified wavelength
Radiant flux	P	$P = dU/dt$	Watt	Total energy emitted per unit time
Radiant intensity	J	$J = dP/d\omega = dU/(dtd\omega)$	Watt/ steradian	Total energy emitted per unit time per unit solid angle
Radiance	N	$N = dJ/dA cos\,(\theta)$	Watt/m²/ steradian	Energy emitted or reflected per unit time from an extended source in some direction
Irradiance	H	$H = dP/dA$	Watt/m²	Energy emitted from a source per unit time per unit area

(Continued)

Visibility: The Seeing of Near and Distant Landscape Features. http://dx.doi.org/10.1016/B978-0-12-804450-6.00009-7

TABLE A1 This Table is an Extension of Table 2.1. It Mathematically Defines the Various Radiometric and Photometric Variables and gives a Brief Description of Each. (*cont.*)

Photometric Variable	Symbol	Definition	Units	Description
Luminous energy	Q	$Q = Fdt$	Talbot	Total quantity of energy emitted weighted to eye sensitivity to wavelength
Luminous flux	F	$F = 683 \int_{380}^{730} PVd\lambda$	Lumen	Weighted energy emitted per unit time
Luminous intensity	I	$I = dF/d\omega$	Lumen/ steradian	Weighted energy emitted per unit time per unit solid angle
Luminance	B	$B = dI/dA\cos(\theta)$	Lumen/ m^2/ steradian	Weighted energy emitted or reflected per unit time from an extended source in some direction
Illuminance	E	$H = dF/dA$	Lumen/m^2	Weighted energy emitted from a source per unit time per unit area

ω = solid angle; θ = angle between a given direction and the normal to the emitting surface; A = surface area of emitting or receiving surface; V = CIE spectral luminosity function; λ = wavelength.

TABLE A2 List of the Most Popular Photometric Variables

Luminous energy	Lumen. sec	Lumber	Lumberg			
Talbot (Q)	1.0	1.0	1×10^7			
Luminous intensity	Candle	Candle (UK)	Candle (German)	Carcel	Hefener Candle	Pentane Candle
Lumen/ steradian (I)	1.0	0.96	0.95	0.104	1.11	0.1
Illuminance	Lux	Meter candle	Foot candle	phot	nox	
Lumen/m^2	1.0	1.0	0.093	0.0001	1000	
Luminance	nit	stilb	Lambert	apostilb	blondal	bril
Lumen/m^2/ steradian (B)	1.0	0.0001	0.000314	3.1416	3.1416	31415926.54

Clean Air Act 169A & B

42 U.S.C.
United States Code, 2010 Edition
Title 42 – THE PUBLIC HEALTH AND WELFARE
CHAPTER 85 – AIR POLLUTION PREVENTION AND CONTROL
SUBCHAPTER I – PROGRAMS AND ACTIVITIES
Part C – Prevention of Significant Deterioration of Air Quality
subpart ii – visibility protection
Sec. 7491 – Visibility protection for Federal class I areas
Sec. 7492 – Visibility
Clean Air Act 169A

§7491. Visibility protection for Federal class I areas
(a) Impairment of visibility; list of areas; study and report

(1) Congress hereby declares as a national goal the prevention of any future, and the remedying of any existing, impairment of visibility in mandatory class I Federal areas which impairment results from manmade air pollution.

(2) Not later than 6 months after August 7, 1977, the Secretary of the Interior in consultation with other Federal land managers shall review all mandatory class I Federal areas and identify those where visibility is an important value of the area. From time to time the Secretary of the Interior may revise such identifications. Not later than 1 year after August 7, 1977, the Administrator shall, after consultation with the Secretary of the Interior, promulgate a list of mandatory class I Federal areas in which he determines visibility as an important value.

(3) Not later than 18 months after August 7, 1977, the Administrator shall complete a study and report to Congress on available methods for implementing the national goal set forth in paragraph (1). Such report shall include recommendations for—

(A) methods for identifying, characterizing, determining, quantifying, and measuring visibility impairment in Federal areas referred to in paragraph (1), and

Visibility: The Seeing of Near and Distant Landscape Features. http://dx.doi.org/10.1016/B978-0-12-804450-6.00010-3

(B) modeling techniques (or other methods) for determining the extent to which manmade air pollution may reasonably be anticipated to cause or contribute to such impairment, and

(C) methods for preventing and remedying such manmade air pollution and resulting visibility impairment.

Such report shall also identify the classes or categories of sources and the types of air pollutants which, alone or in conjunction with other sources or pollutants, may reasonably be anticipated to cause or contribute significantly to impairment of visibility.

(4) Not later than 24 months after August 7, 1977, and after notice and public hearing, the Administrator shall promulgate regulations to assure (A) reasonable progress toward meeting the national goal specified in paragraph (1), and (B) compliance with the requirements of this section.

(b) Regulations

Regulations under subsection (a)(4) of this section shall—

(1) provide guidelines to the States, taking into account the recommendations under subsection (a)(3) of this section on appropriate techniques and methods for implementing this section (as provided in subparagraphs (A) through (C) of such subsection (a)(3)), and

(2) require each applicable implementation plan for a State in which any area listed by the Administrator under subsection (a)(2) of this section is located (or for a State the emissions from which may reasonably be anticipated to cause or contribute to any impairment of visibility in any such area) to contain such emission limits, schedules of compliance, and other measures as may be necessary to make reasonable progress toward meeting the national goal specified in subsection (a) of this section, including—

(A) except as otherwise provided pursuant to subsection (c) of this section, a requirement that each major stationary source which is in existence on August 7, 1977, but which has not been in operation for more than 15 years as of such date, and which, as determined by the State (or the Administrator in the case of a plan promulgated under section 7410(c) of this title) emits any air pollutant which may reasonably be anticipated to cause or contribute to any impairment of visibility in any such area, shall procure, install, and operate, as expeditiously as practicable (and maintain thereafter) the best available retrofit technology, as determined by the State (or the Administrator in the case of a plan promulgated under section 7410(c) of this title) for controlling emissions from such source for the purpose of eliminating or reducing any such impairment, and

(B) a long-term (10 to 15 years) strategy for making reasonable progress toward meeting the national goal specified in subsection (a) of this section.In the case of a fossil-fuel fired generating powerplant having a total generating capacity in excess of 750 megawatts, the emission limitations required under this paragraph shall be determined pursuant to guidelines, promulgated by the Administrator under paragraph (1).

(c) Exemptions

(1) The Administrator may, by rule, after notice and opportunity for public hearing, exempt any major stationary source from the requirement of subsection (b)(2)(A) of this section, upon his determination that such source does not or will not, by itself or in combination with other sources, emit any air pollutant which may reasonably be anticipated to cause or contribute to a significant impairment of visibility in any mandatory class I Federal area.

(2) Paragraph (1) of this subsection shall not be applicable to any fossil-fuel fired powerplant with total design capacity of 750 megawatts or more, unless the owner or operator of any such plant demonstrates to the satisfaction of the Administrator that such powerplant is located at such distance from all areas listed by the Administrator under subsection (a)(2) of this section that such powerplant does not or will not, by itself or in combination with other sources, emit any air pollutant which may reasonably be anticipated to cause or contribute to significant impairment of visibility in any such area.

(3) An exemption under this subsection shall be effective only upon concurrence by the appropriate Federal land manager or managers with the Administrator's determination under this subsection.

(d) Consultations with appropriate Federal land managers

Before holding the public hearing on the proposed revision of an applicable implementation plan to meet the requirements of this section, the State (or the Administrator, in the case of a plan promulgated under section 7410(c) of this title) shall consult in person with the appropriate Federal land manager or managers and shall include a summary of the conclusions and recommendations of the Federal land managers in the notice to the public.

(e) Buffer zones

In promulgating regulations under this section, the Administrator shall not require the use of any automatic or uniform buffer zone or zones.

(f) Nondiscretionary duty

For purposes of section 7604(a)(2) of this title, the meeting of the national goal specified in subsection (a)(1) of this section by any specific date or dates shall not be considered a "nondiscretionary duty" of the Administrator.

(g) Definitions

For the purpose of this section—

(1) in determining reasonable progress there shall be taken into consideration the costs of compliance, the time necessary for compliance, and the energy and nonair quality environmental impacts of compliance, and the remaining useful life of any existing source subject to such requirements;

(2) in determining best available retrofit technology the State (or the Administrator in determining emission limitations which reflect such technology) shall take into consideration the costs of compliance, the energy and nonair quality environmental impacts of compliance, any existing pollution control technology in use at the source, the remaining useful life of the source, and the degree of improvement in visibility which may reasonably be anticipated to result from the use of such technology;

(3) the term "manmade air pollution" means air pollution which results directly or indirectly from human activities;

(4) the term "as expeditiously as practicable" means as expeditiously as practicable but in no event later than 5 years after the date of approval of a plan revision under this section (or the date of promulgation of such a plan revision in the case of action by the Administrator under section 7410(c) of this title for purposes of this section);

(5) the term "mandatory class I Federal areas" means Federal areas which may not be designated as other than class I under this part;

(6) the terms "visibility impairment" and "impairment of visibility" shall include reduction in visual range and atmospheric discoloration; and

(7) the term "major stationary source" means the following types of stationary sources with the potential to emit 250 tons or more of any pollutant: fossil-fuel fired steam electric plants of more than 250 million British thermal units per hour heat input, coal cleaning plants (thermal dryers), kraft pulp mills, Portland Cement plants, primary zinc smelters, iron and steel mill plants, primary aluminum ore reduction plants, primary copper smelters, municipal incinerators capable of charging more than 250 tons of refuse per day, hydrofluoric, sulfuric, and nitric

acid plants, petroleum refineries, lime plants, phosphate rock processing plants, coke oven batteries, sulfur recovery plants, carbon black plants (furnace process), primary lead smelters, fuel conversion plants, sintering plants, secondary metal production facilities, chemical process plants, fossil-fuel boilers of more than 250 million British thermal units per hour heat input, petroleum storage and transfer facilities with a capacity exceeding 300,000 barrels, taconite ore processing facilities, glass fiber processing plants, charcoal production facilities.

(July 14, 1955, ch. 360, title I, §169A, as added Pub. L. 95–95, title I, §128, Aug. 7, 1977, 91 Stat. 742.)

Clean Air Act 169B

§7492. Visibility

(a) Studies

(1) The Administrator, in conjunction with the National Park Service and other appropriate Federal agencies, shall conduct research to identify and evaluate sources and source regions of both visibility impairment and regions that provide predominantly clean air in class I areas. A total of $8,000,000 per year for 5 years is authorized to be appropriated for the Environmental Protection Agency and the other Federal agencies to conduct this research. The research shall include—

   (A) expansion of current visibility related monitoring in class I areas;

   (B) assessment of current sources of visibility impairing pollution and clean air corridors;

   (C) adaptation of regional air quality models for the assessment of visibility;

   (D) studies of atmospheric chemistry and physics of visibility.

(2) Based on the findings available from the research required in subsection (a)(1) of this section as well as other available scientific and technical data, studies, and other available information pertaining to visibility source–receptor relationships, the Administrator shall conduct an assessment and evaluation that identifies, to the extent possible, sources and source regions of visibility impairment including natural sources as well as source regions of clear air for class I areas. The Administrator shall produce interim findings from this study within 3 years after November 15, 1990.

(b) Impacts of other provisions

Within 24 months after November 15, 1990, the Administrator shall conduct an assessment of the progress and improvements in visibility in class I areas that are likely to result from the implementation of the provisions of the Clean Air Act Amendments of 1990 other than the provisions of this section. Every 5 years thereafter the Administrator shall conduct an assessment of actual progress and improvement in visibility in class I areas. The Administrator shall prepare a written report on each assessment and transmit copies of these reports to the appropriate committees of Congress.

(c) Establishment of visibility transport regions and commissions

    (1) **Authority to establish visibility transport regions**

Whenever, upon the Administrator's motion or by petition from the Governors of at least two affected States, the Administrator has reason to believe that the current or projected interstate transport of air pollutants from one or more States contributes significantly to visibility impairment in class I areas located in the affected States, the Administrator may establish a transport region for such pollutants that includes such States. The Administrator, upon the Administrator's own motion or upon petition from the Governor of any affected State, or upon the recommendations of a transport commission established under subsection (b) of this section may—

    (A) add any State or portion of a State to a visibility transport region when the Administrator determines that the interstate transport of air pollutants from such State significantly contributes to visibility impairment in a class I area located within the transport region, or

    (B) remove any State or portion of a State from the region whenever the Administrator has reason to believe that the control of emissions in that State or portion of the State pursuant to this section will not significantly contribute to the protection or enhancement of visibility in any class I area in the region.

    (2) **Visibility transport commissions**

Whenever the Administrator establishes a transport region under subsection (c)(1) of this section, the Administrator shall establish a transport commission comprised of (as a minimum) each of the following members:

    (A) the Governor of each State in the Visibility Transport Region, or the Governor's designee;

    (B) The Administrator or the Administrator's designee; and

    (C) A representative of each Federal agency charged with the direct management of each class I area or areas within the Visibility Transport Region.

    (3) **Ex officio members**

All representatives of the Federal Government shall be ex officio members.

(4) **Federal Advisory Committee Act**
The visibility transport commissions shall be exempt from the requirements of the Federal Advisory Committee Act [5 U.S.C. App.].

(d) **Duties of visibility transport commissions**
A Visibility Transport Commission—
(1) shall assess the scientific and technical data, studies, and other currently available information, including studies conducted pursuant to subsection (a)(1) of this section, pertaining to adverse impacts on visibility from potential or projected growth in emissions from sources located in the Visibility Transport Region; and
(2) shall, within 4 years of establishment, issue a report to the Administrator recommending what measures, if any, should be taken under this chapter to remedy such adverse impacts. The report required by this subsection shall address at least the following measures:
    (A) the establishment of clean air corridors, in which additional restrictions on increases in emissions may be appropriate to protect visibility in affected class I areas;
    (B) the imposition of the requirements of part D of this subchapter affecting the construction of new major stationary sources or major modifications to existing sources in such clean air corridors specifically including the alternative sitting analysis provisions of section 7503(a)(5) of this title; and
    (C) the promulgation of regulations under section 7491 of this title to address long-range strategies for addressing regional haze which impairs visibility in affected class I areas.

(e) **Duties of Administrator**
(1) The Administrator shall, taking into account the studies pursuant to subsection (a)(1) of this section and the reports pursuant to subsection (d)(2) of this section and any other relevant information, within 18 months of receipt of the report referred to in subsection (d)(2) of this section, carry out the Administrator's regulatory responsibilities under section 7491 of this title, including criteria for measuring "reasonable progress" toward the national goal.
(2) Any regulations promulgated under section 7491 of this title pursuant to this subsection shall require affected States to revise within 12 months their implementation plans under section 7410 of this title to contain such emission limits, schedules of compliance, and other measures as may be necessary to carry out regulations promulgated pursuant to this subsection.

(f) Grand Canyon visibility transport commission
The Administrator pursuant to subsection (c)(1) of this section shall, within 12 months, establish a visibility transport commission for the region affecting the visibility of the Grand Canyon National Park.

(July 14, 1955, ch. 360, title I, §169B, as added Pub. L. 101–549, title VIII, §816, Nov. 15, 1990, 104 Stat. 2695.)

Definitions of terms used in CAA § 169A and the 1980 visibility regulations (40 CFR 51.300–51.307)

Adverse impact on visibility means visibility impairment that interferes with the management, protection, preservation, or enjoyment of the visitor's visual experience of the federal Class I area. This determination must be made on a case-by-case basis, taking into account the geographic extent, intensity, duration, frequency, and time of visibility impairments, and how these factors correlate with (1) the times of visitor use of the federal Class I area, and (2) the frequency and timing of natural conditions that reduce visibility. This term does not include effects on integral vistas.

Best available retrofit technology (BART) means an emission limitation based on the degree of reduction achievable through the application of the best system of continuous emission reduction for each pollutant that is emitted by an existing stationary facility. The emission limitation must be established on a case-by-case basis, taking into consideration the technology available, the costs of compliance, the energy and non-air-quality environmental impacts of compliance, any pollution control equipment in use or in existence at the source, the remaining useful life of the source, and the degree of improvement in visibility that may reasonably be anticipated to result from the use of such technology.

Existing stationary facility means any stationary sources of air pollutants, including any reconstructed source, that were not in operation prior to August 7, 1962, were in existence on August 7, 1977, and had the potential to emit 250 tons per year or more of any air pollutant. In determining potential to emit, fugitive emissions, to the extent quantifiable, must be counted.

Federal Class I area means any federal land that is classified or reclassified Class I.

Federal land manager (FLM) means the secretary of the department with authority over the federal Class I area (or the secretary's designee) or, with respect to Roosevelt-Campobello International Park, the chairman of the Roosevelt-Campobello International Park Commission.

Implementation plan means any state implementation plan, federal implementation plan, or tribal implementation plan.

Integral vista means a view perceived from within the mandatory Class I federal area of a specific landmark or panorama located outside the boundary of the mandatory Class I federal area.

Mandatory Class I federal area means any area identified in 40 CFR, part 81, subpart D.

Natural conditions includes naturally occurring phenomena that reduce visibility as measured in terms of light extinction, visual range, contrast, or coloration.

Reasonably attributable means attributable by visual observation or any other technique a state deems appropriate.

Reasonably attributable visibility impairment means visibility impairment that is caused by the emission of air pollutants from one or a small number of sources.

Regional haze means visibility impairment that is caused by the emission of air pollutants from numerous sources located over a wide geographic area. Such sources include, but are not limited to, major and minor stationary sources, mobile sources, and area sources.

Secondary emissions means emissions that occur as a result of the construction or operation of an existing stationary facility but do not come from the existing stationary facility. Secondary emissions may include, but are not limited to, emissions from ships or trains coming to or from the existing stationary facility.

Significant impairment means visibility impairment that interferes with the management, protection, preservation, or enjoyment of a visitor's visual experience of the mandatory Class I area. This determination must be made on a case-by-case basis, taking into account the geographic extent, intensity, duration, frequency, and time of the visibility impairment, and how these factors correlate with (1) the times of visitor use of the mandatory Class I area, and (2) the frequency and timing of natural conditions that reduce visibility.

Stationary source means any building, structure, facility, or installation that emits or may emit any air pollutant.

Visibility impairment means any humanly perceptible change in visibility (light extinction, visual range, contrast, coloration) from that which would have existed under natural conditions.

Visibility in any mandatory Class I federal area includes any integral vista associated with that area.

Glossary

Absorption coefficient effective cross-sectional area per unit volume resulting from photons being absorbed from a line of sight. Units are in inverse distance, usually inverse megameters or inverse kilometers.

Accumulation mode particles in the atmosphere have a range of sizes ranging from 1 nm to hundreds of micrometers. It is useful to characterize these particles based on their chemistry and role in the formation of other particles. The size of the accumulation mode is 0.1–1.0 μm.

Achromatic without color.

Acid precipitation typically, rain with high concentrations of acids produced by the interaction of water with oxygenated compounds of sulfur and nitrogen, which are the byproducts of fossil-fuel combustion.

Adverse impact visibility impairment that interferes with the management, protection, preservation, or enjoyment of the visitor's visual experience of the federal Class I area.

Aerosols gaseous suspension of ultramicroscopic particles of a liquid or a solid. Atmospheric aerosols govern variations in light extinction and therefore visibility reduction. Aerosol size distribution and chemistry are key parameters.

Air light light from the sun, clouds or terrestrial and aquatic surfaces that is incident on particles and gases in a sight path and subsequently scattered by the particles or gases in the direction of the observer.

Albedo fraction of total light incident on a reflecting surface that is reflected back omnidirectionally.

Angstrom 1/10,000,000,000 of a meter, 10^{-10} meters, abbreviated as A.

Anthropogenic alteration to the natural environment caused by human activity, that is, manmade.

Apparent spectral contrast percent difference in radiant energy associated with an object and its background when the object is observed at some distance r.

Apportionment act of assessing the degree to which specific aerosol components contribute to light extinction or aerosol mass.

Artifact any component of a signal or measurement that is extraneous to the variable represented by the signal or measurement.

Atmospheric attenuation reduction in the intensity of light by the earth's atmosphere as a result of scattering and absorption processes.

Atmospheric clarity optical property related to the visual quality of the landscape viewed from a distance (see optical depth and turbidity).

Average scene contrast index that captures the average contrast associated with an image.

Back scatter scattering in which the angle between the initial and final directions of motion of the scattered particles is greater than 90 degrees.

BackMC a radiation transfer model that utilizes a backward Monte Carlo technique.

BART best available retrofit technology.

Black carbon the most strongly light-absorbing component of ambient concentrations of particulate matter; it absorbs all wavelengths of light. It is formed from incomplete combustion of fossil and biofuels and biomass.

Bloch's law $R = I \times T$ where R is human response, I is intensity, and T is time. The same response to light can be evoked by decrease intensity but increasing the time of exposure or vice versa.

Visibility: The Seeing of Near and Distant Landscape Features. http://dx.doi.org/10.1016/B978-0-12-804450-6.00011-5

Boundary layer layer of stationary atmosphere that is bounded by a surface where the effects of viscosity are significant. In the earth's atmosphere, it is the layer near the ground that is affected by heating and cooling of the earth's surface, moisture, and momentum transfer to and from the earth's surface.

Brightness perception elicited by the luminance of a visual landscape feature, sky, or meteorological phenomenon.

Brown carbon class of organic carbon, known for its light brownish color, absorbs strongly in the ultraviolet wavelengths and less significantly going into the visible. Types of brown carbon include tar materials from smoldering fires or coal combustion, breakdown products from biomass burning, a mixture of organic compounds emitted from soil, and volatile organic compounds given off by vegetation.

CAA Clean Air Act.

CAIR Clean Air Interstate Rule.

Candle unit of light intensity based on the light emitted by a standard candle.

Chroma color purity.

Chromatic relating to physical processes involving color.

Chromaticity measure of the quality of a color independent of its luminance.

Class I areas comprises 156 national parks and wilderness areas that are provided visibility protection.

Class II areas areas provided a level of air quality protection less stringent than Class I areas.

Cloud condensation nuclei particles of liquids or solids upon which condensation of water vapor begins in the atmosphere.

Coarse particles usually, particles whose size range is between 2.5 and 10.0 μm ($PM_{2.5}$ to PM_{10}).

Color sensation that is evoked by the human visual system when light is sensed by the retina. It is said to have hue, saturation or strength, and value or shade.

Color contrast index that captures the average color difference within an image.

Colorado Plateau geographic region that encompasses many national parks with striking geological features such as Grand Canyon, Canyonlands, and Bryce Canyon national parks.

Colorfulness degree of difference between a color and gray.

Contiguous contrast fractional difference in radiance emanating from two landscape features that are adjacent to each other.

Contrast fractional difference in light arriving at an observer from two landscape features or from a landscape feature and some background such as the sky.

Contrast transmittance ratio between apparent and inherent spectral contrast. When the object is darker than its background, it has a value between 0 and -1. For objects brighter than their background, the value varies from 0 to infinity. When the contrast transmittance is equal to zero, the object cannot be seen.

CSAPR Cross-State Air Pollution Rule.

Current conditions contemporary, or modern, atmospheric conditions that are affected by human activity.

Deciview scene-independent metric that is proportional to perceived changes in visual air quality.

Deliquescence process that occurs when the vapor pressure of the saturated aqueous solution of a substance is less than the vapor pressure of water in the ambient air. Water vapor is collected until the substance is dissolved and in equilibrium with its environment.

Density mass per unit volume.

Diffraction various phenomena that occur when a wave encounters an obstacle or a slit. In classical physics, the diffraction phenomenon is described as the interference of waves.

Digital camera camera that uses an electronic detector to capture an image as opposed to photographic film.

Direct effects optical effects of aerosols on climate modification referring to absorption and scattering of solar radiation by airborne particles.

Distance mask mask that selects landscape features that correspond to fixed distances from the observation point or position from which the picture was taken.

DN digital number associated with each pixel.

"ΔE" parameter index of color and brightness differences for developed for colorimetry studies by Commission International de l'Eclairage.

Edge sharpness characteristic of landscape features. Landscape features with sharp edges contain scenic features with abrupt changes in brightness.

Electromagnetic waves synchronized oscillations of electric and magnetic fields that propagate through space at the speed of light. Electromagnetic waves include all forms of radiation, such as radio waves, microwaves, visible light, infrared radiation, x-rays, and gamma rays.

Elemental carbon see "Black carbon."

Equilibration balancing or counter balancing to create stability, often with a standard measure or constant.

Equivalent contrast any scene can be Fourier decomposed into light and dark bars of various frequencies and intensities modulated in accordance with a sine wave function. Equivalent contrast is the average contrast of those sine waves within a specified range of spatial frequencies.

Esthetic characterizing an appreciation of beauty.

Externally mixed particulate species that coexist as separate particles without comingling or combining.

Extinction coefficient sum of scatter and absorption. Total effective cross-sectional area per unit volume resulting from photons being absorbed and scattered from a line of sight. Units are in inverse distance, usually inverse megameters or inverse kilometers.

Eye–brain system part of the central nervous system that gives humans the ability to process visual detail to detect and interpret information from light to build a representation of the surrounding environment.

FFT Fast Fourier Transform is another operator used to extract radiance difference edges in an image.

Fine particles usually, particles whose size is less than 2.5 μm ($PM_{2.5}$).

Form perceived volume or depth of an image.

Forward scatter scattering in which the angle between the initial and final directions of motion of the scattered particles is less than 90 degrees.

Foveal vision the fovea area of the retina is a small central area of the retina composed of closely paced cones and is responsible for sharp central vision or foveal vision. A line of sight image is focused on the fovea of the retina.

Frequency rate at which successive crests of a wave repeat in a particular period of time. Speed = wavelength × frequency.

Gaussian plume model model that predicts the dispersion of emissions through the atmosphere according to a mathematical model that assumes a Gaussian profile.

Gaussian stimuli visual stimuli whose brightness profiles follow a Gaussian distribution.

Geometric mean mean of a lognormal distribution. Central tendency of a set of numbers based on their product as opposed to their sum as in arithmetic mean.

Geometric standard deviation standard deviation of a lognormal distribution.

Grating stimuli visual stimuli made up of sequential sine wave brightness profiles.

Haze horizontal obscuration by atmospheric suspensions of particles and gases, often referred to as aerosols.

Hue one of three parameters that describe a color in a quantifiable color space. The undiluted color of an object.

HVS human visual system.

Hydrophobic lacking affinity for water, or failing to adsorb or absorb water.

Hygroscopic ability or tendency to rapidly accelerate condensation of water vapor around a nucleus. Also pertains to a substance (e.g., aerosols) that have an affinity for water and whose physical characteristics are appreciably altered by the effects of water.

Image-forming information reflected light from a landscape feature or meteorological phenomenon.

IMPROVE Interagency Monitoring of Protected Visual Environments.

Index of refraction number indicating the speed of light in a given medium as a ratio of the speed of light in a vacuum to the speed of light in the medium.

Inherent spectral contrast percent difference in radiant energy associated with an object and its background at an observer distance equal to zero.

Integral vista vista integral to the enjoyment of the area.

Internally mixed situation in which individual particles contain one or more species. For example, water is internally mixed with its hygroscopic hosts.

Irradiance flux of radiant energy per unit area (normal to the direction of flow of radiant energy through a medium).

Joule unit of energy.

Just noticeable change (JNC) See JND.

Just noticeable difference (JND) measure of change in image appearance that affects image sharpness. Counting the number of JNDs (detectable changes) in scene appearance is regarded as an alternative method of quantifying visibility reduction (light extinction).

Koschmeider constant constant in the reciprocal relationship between visual range and the extinction coefficient (see visual range).

Lagrangian describes a mathematical function that summarized the dynamics of a physical system. Lagrangian air quality models attempt to dynamically model source emissions as they transport, transform, and disperse through the atmosphere.

Lambertian describes light reflected from a surface such that the apparent brightness of that surface appears to be independent of the angle of view.

Landscape feature single morphological part of a geographic landscape.

Layered haze pollutants emitted into the atmosphere under stagnant or stable meteorological conditions will not disperse vertically, resulting in a demarcation between a haze and its background. A layered haze can be trapped next to the ground or suspended in the atmosphere with a resulting haze layer "edge" above and below the haze. This type of haze is typically referred to as a plume and the resulting visibility impairment is referred to as plume blight.

Light electromagnetic radiation that can be seen by the human eye (0.4–0.7 μm)

Light absorbing carbon includes black and brown carbon.

Line in the context of a view or vista, a continuous demarcation between two points. A line has width, direction, and length.

Liquid water water present within a cloud, expressed as a percent of total cloud constituents, or liquid-phase water in an aerosol.

Lognormal distribution set of numbers that, when taking the logarithm of those numbers, its distribution is represented by a symmetric bell-shaped curve.

Lumen unit of luminous flux.

Luminance flux of radiation, weighted to the human eye response, emitted per unit solid angle in a given direction by a unit area of a source.

Mass absorption efficiency efficiency with which particles attenuate light by absorption on a per mass basis. The absorption coefficient associated with an aerosol distribution divided by its mass concentration. Units usually are m^2/gm.

Mass extinction efficiency efficiency with which particles attenuate light by extinction on a per mass basis. The extinction coefficient associated with an aerosol distribution divided by its mass concentration. Units usually are m^2/gm.

Mass scattering efficiency efficiency with which particles attenuate light by scattering on a per mass basis. The scattering coefficient associated with an aerosol distribution divided by its mass concentration. Units usually are m^2/gm.

Micron 1/1,000,000 of a meter, 10^{-6} meters, abbreviated as μm.

Mie scattering attenuation of light in the atmosphere by scattering due to particles of a size comparable to the wavelength of the incident light. This is the phenomenon largely responsible for the reduction of atmospheric visibility. Visible solar radiation falls into the range of 0.4 to 0.8 μm, roughly, with a maximum intensity around 0.52 μm.

Modulation contrast modulation contrast is defined as $(I_1 - I_2)/(I_1 + I_2)$ as opposed to universal contrast, which is defined as $(I_1 - I_2)/(I_2)$ or $(I_1 - I_2)/(I_1)$.

Modulation transfer function (MTF) mathematical function that describes contrast transmittance in spatial-frequency space. It is the ratio between scene equivalent contrast at the observer and equivalent contrast at the object. When the object of interest is small compared with its surroundings, the modulation transfer function and contrast transmittance reduce to the same value.

Monte Carlo computational algorithms that rely on repeated and random sampling to derive an average outcome.

Munsell color system a way of quantifying color. It is a color space that specifies color based on three color dimensions: hue, value, and chroma.

NAAQS National Ambient Air Quality Standard; designed to protect human health.

Nanometer 1/1,000,000,000 of a meter, 10^{-9} meters, abbreviated as nm.

Natural Conditions naturally occurring phenomena that reduce visibility as measured in terms of light extinction, visual range, contrast, or coloration.

Natural conditions prehistoric and pristine atmospheric states, that is, atmospheric conditions that are not affected by human activities.

NEI National Emission Inventory

Nephelometer instrument used to measure the light scattering component of light extinction.

Normal distribution set of variables whose distribution is represented by a symmetric bell-shaped curve.

Nucleation mode particles in the atmosphere have a range of sizes ranging from 1 nm to hundreds of micrometers. It is useful to characterize these particles based on their chemistry and role in formation of other particles. The size of the nucleation mode is 0.001–0.01 μm.

Object space space around an object in interest, such as a specific landscape feature.

Optical depth degree to which a cloud or haze prevents light from passing through it. It is a function of physical composition, size distribution, and particle concentration. Often used interchangeably with "turbidity."

Particulates microscopic solid or liquid matter suspended in the earth's atmosphere. Size ranges from 0.01 to 1000 μm.

Path function radiant energy gain within an incremental path segment of the atmosphere.

Path radiance or "airlight," is a radiometric property of air resulting from light scattering processes along the sight line, or path, between a viewer and the object (target).

Perception ability to see, hear, or become aware of something through the use of senses.

Phase shift the amount a wave has shifted from its original position.

Photoacoustic describes an instrumental technique used to measure absorption of atmospheric particulate matter.

Photographic camera camera that uses photographic film to capture an image.

Photometric describes a set of physical quantities for characterizing radiation or light weighted in proportion to the response of the human eye to different wavelengths.

Photon bundle of electromagnetic radiation that acts as if it were a particle; it is considered to be an elementary particle. All forms of electromagnetic radiation propagate through the atmosphere as photons or bundles of wavelike energy.

Pixel the smallest element of an image that is addressable in digital imaging.

Plume see "Layered haze."

Plume blight see "Layered haze."

PLUVUE plume visibility model; Gaussian plume model that is used to predict the appearance of emissions from stationary sources such as coal-fired power plants.

Primary particles particles suspended in the atmosphere as particles from the time of emission (e.g., dust and soot).

PSD prevention of significant deterioration.

Psychophysical investigation between the relationship of physical stimuli and sensations or perception they affect.

Quadratic detection model model used to predict the amount of change in equivalent contrast or perceived landscape structure required to evoke a single just-noticeable change in landscape appearance.

Radiance flux of radiation emitted per unit solid angle in a given direction by a unit area of a source.

Radiant flux radiant energy transmitted or received per unit time.

Radiation transfer physical phenomenon of the transfer of electromagnetic radiation through the atmosphere. It is affected by the emission, scattering, and absorption processes associated with atmospheric aerosols.

Radiometric describes a set of physical quantities for characterizing radiation or light at specific wavelengths.

Rayleigh scattering scattering of light by air molecules, also called blue-sky scatter.

Reasonably attributable attributable by visual observation or any other technique a state deems appropriate.

Reflection return of light without absorption by a body or surface.

Regional haze haze that covers a large geographic area and appears to be uniform in extent.

Regional haze visibility impairment that is caused by the emission of air pollutants from numerous sources located over a wide geographic area. Such sources include, but are not limited to, major and minor stationary sources, mobile sources, and area sources.

Regional Haze Rule requires that all states submit an implementation plan that provides for reasonable progress by 2064 toward achieving "natural background conditions" in national parks and wilderness areas on the average of the worst 20% visibility days while maintaining the air clarity of the best 20% days.

RGB the red, green, and blue values that make up a color image.

R_{oc} ratio of particle organic mass to carbon mass. POM = R_{oc}OC.

RPO regional planning organization.

Saccade quick, simultaneous movement of both eyes between two fixation points.

SBE scenic beauty estimation.

Scatter coefficient effective cross-sectional area per unit volume resulting from photons being scattered out of a line of sight. Units are in inverse distance, usually inverse megameters or inverse kilometers.

Scattering enhancement factor f(RH) ratio of the scattering coefficient associated with a distribution of particles that has absorbed water to that of a dry particle.

Secondary emissions emissions that occur as a result of the construction or operation of an existing stationary facility but that do not come from the existing stationary facility.

Secondary NAAQS Secondary National Ambient Air Quality Standard; provides public welfare protection against visibility impairment, damage to animals, crops, vegetation, and buildings.

Secondary particles particles formed in the atmosphere by a gas-to-particle conversion process.

Sight path line of sight between an observer and landscape feature.

Significant impairment visibility impairment that interferes with the management, protection, preservation, or enjoyment of a visitor's visual experience of a mandatory Class I area.

SIP state implementation plan.

Size parameter defined to be $2\pi r/\lambda$, where r is the radius of a particle and λ is the wavelength of light.

Sobel index the average of the radiance difference edges resulting from the application of the Sobel operator.

Sobel operator an image edge detection algorithm that extracts landscape edges in an image resulting from radiance differences.

Solid angle three-dimensional analog to an angle. The surface subtended by a cone when integrated over a sphere is equal to 4π steradians.

Spatial frequency reciprocal of the distance between sine wave crests (or troughs) measured in degrees of angular subtense of a sine wave grating. Spatial frequency is a general term for the frequencies associated with the image radiance in a scene along the path of radiance (path of sight). Landscape features contain multiple landscape scenic elements. Each element generates its own image radiance with its own frequency and intensity.

Standard visual range reciprocal of the extinction coefficient standardized to a Rayleigh atmosphere with a scattering coefficient of 0.01/km. The distance under daylight and uniform lighting conditions at which the apparent contrast between a specified target and its background becomes just equal to the threshold contrast of an observer, assumed to be 0.02.

Stationary source any building, structure, facility, or installation that emits or may emit an air pollutant.

Steradian unit of solid angle. A sphere has 4π steradians (sr).

Suprathreshold contrast measure of human eye sensitivity to contrast change from some baseline contrast. It is the smallest increment of contrast change perceptible by the human eye.

Talbot unit of luminous energy. One Talbot is one lumen-sec.

Tar balls amorphous carbon spheres that are particles of special morphology and composition that are fairly abundant in the plumes of biomass smoke. A special subset of brown carbon in that they preferentially absorb in the UV and shorter wavelengths.

Telephotometer photometer designed to measure the radiant energy arriving from a scene weighted in accordance with the response of the human eye–brain system to spectral radiance.

Textural content in the context of a view or vista, it refers to the scenic elements that are made up of high spatial frequencies or an area of an image with wide variation of discrete gray tones.

Threshold contrast measure of human eye sensitivity to contrast. It is the smallest increment of contrast perceptible by the human eye.

TOR thermal optical reflectance is a standard technique for measuring the amount of carbon mass collected on a filter substrate. It is based on thermal evolution of carbon as a function of temperature of the filter substrate.

Transmissometer instrument that measures atmospheric transmittance. From transmittance, the atmospheric extinction coefficient can be derived.

Transmittance fraction of initial light from a light source that is transmitted through the atmosphere. Light is attenuated by scattering and absorption from gases and particles.

Troland a way of correcting photometric measurements of luminance to equal retinal illuminance. Luminance scaled to the effective pupil size.

Turbidity condition that reduces atmospheric transparency to radiation, especially light. The degree of cloudiness, or haziness, caused by the presence of aerosols, gases, and dust.

Uniform haze haze that appears uniform in nature without vertical demarcations or edges.

Unimpaired visual range distance an observer can see without noticing visibility impairment.

Value lightness of a color.

VAQ visual air quality.

VISCREEN plume visibility screening model that calculates the potential visibility impact of emissions from stationary sources.

Visibility visual quality of a view, or scene, in daylight with respect to color rendition and contrast definition. The ability to perceive form, color, and texture.

Visibility impairment any humanly perceptible change in visibility (light extinction, visual range, contrast, coloration) from that which would have existed under natural conditions.

Visibility indexes indexes that have been formalized for aerosol, optical, and scenic attributes. Aerosol indexes include mass concentrations, particle compositions, physical characteristics, and size distributions. Optical indexes include coefficients for scattering, extinction, and absorption. Scenic indexes comprise visual range, contrast, radiance, color, and just noticeable changes.

Visibility reduction impairment or degradation of atmospheric clarity. Becomes significant when the color and contrast values of a scene to the horizon are altered or distorted by airborne impurities.

Vista distant view or site offering such a view.

Visual image processing digitizing, calibration, modeling, and display of the effects of atmospheric optical parameters on a scene. The process starts with a digital image of landscape features viewed in clean atmospheric conditions and models the effects of changes in atmospheric composition.

Visual impairment reduction in visual range and atmospheric discoloration according to the Clean Air Act of 1977. More generally, it is interpreted as any loss of color, shape, and form of a view or vista.

Visual resource like biological resources, scenic areas are considered to be a resource.

Volume scattering function differential scattering cross section per unit volume. Given an incident beam of light on a volume of aerosol, it describes the quantity of light scattered in any given direction.

Watt unit of power. One Watt is 1 joule/s. Used to express the rate of energy transfer.

Wavelength distance between successive crests of a wave, as in an electromagnetic wave or light. Speed = wavelength × frequency.

WinHaze Windows-based system to model the effects of haze on images of interest.

Subject Index

Printed in the United States
By Bookmasters